# Cage aquaculture

# Cage aquaculture

Malcolm C M Beveridge

Fishing News Books Ltd
Farnham · Surrey · England

British Library CIP Data

Beveridge, Malcolm C M
Cage aquaculture
1. Aquaculture — Equipment and supplies
I. Title
630′.9162 SH135
ISBN 0-85238-148-4

*Published by*
Fishing News Books Ltd
1 Long Garden Walk
Farnham, Surrey, England

Typeset by
Express Typesetters Ltd
Farnham, Surrey

Printed in Great Britain by
Henry Ling Ltd
The Dorset Press, Dorchester

# Contents

# Illustrations

# Tables

# Preface

Although of comparatively recent origin, cage aquaculture is playing an increasingly important role in the world production of food fishes, not only of the high market value marine species, such as salmon, which dominate interests in northern Europe and North America, but also the more ubiquitous carps and tilapias. Moreover, there is a growing awareness that the possibilities offered by cage aquaculture have only begun to be explored. For example, cages can provide low-cost alternatives to conventional land-based hatcheries or nurseries and may be employed in culturing even large, fast-swimming, pelagic species such as tunas. Moreover they may be used not only for rearing food fishes, but also to help combat eutrophication of lakes.

Since the late 1960s, many hundreds of technical papers and reports have been published which, although valuable or interesting, have often been of a highly specialised nature or have been largely unavailable to a wider audience. The pace of research has also accelerated and the originally rather narrow avenues of investigation have widened to encompass issues such as environmental impact and carrying capacity of sites.

The principal aim of this book is to attempt a synthesis of available information on cages and cage aquaculture. However, I have also tried to bring in and discuss other areas of research which are relevant to the subject, and to illuminate the problems facing cage operators and to suggest where solutions might lie. The book is not intended as a comprehensive do-it-yourself manual, since book learning is no substitute for practical experience, although it does deal to some extent with the practicalities of establishing and managing a cage farm. The book should be viewed more as a source or reference book which I hope may be of use to students, researchers, fisheries officers and fish farmers working with cages, as well as conservationists and resource managers.

I have included little information on cage or equipment suppliers, but recommend instead that the interested reader refers to trade papers, such as Fish Farmer or Fish Farming International, the European Aquaculture Trade Directory, published by the European Aquaculture Society, or the occasional products review published as a special edition of the American Aquaculture Magazine.

Malcolm C M Beveridge

# Acknowledgements

This book represents the outcome of more than five years of association with the Institute of Aquaculture, University of Stirling, where there is an unrivalled pool of expertise and enthusiasm for fish farming. In this environment I have been able to develop and research ideas and to discuss them at length with both staff and students, and to all I owe an enormous debt of gratitude. I would particularly like to thank the Director of the Institute, Professor Ron Roberts and my colleagues Dr James Muir, Dr Mike Phillips, Dr Lindsay Ross and Mr Allan Stewart for all their encouragement and advice.

Throughout much of this period I have been supported by the Overseas Development Administration of the British Government, and to them and the Fisheries Advisers Mr John Stoneman and Dr John Tarbit, I am most grateful. I would also like to acknowledge the assistance at various times of Rothesay Seafoods Ltd and the Highlands and Islands Development Board. The Food and Agriculture Organization of the United Nations kindly provided me with an André Mayer Research Fellowship at the College of Fisheries, University of the Philippines, which greatly broadened my outlook on cage aquaculture: in addition they authorised my use of various figures from *Advances in Aquaculture*, edited by T V R Pillay and W A Dill; and *Netting Materials for Fishing Gear*, by G Klust. British Council and ODA have also helped fund travel to various countries.

Many people have been involved in the long gestation of this book. I would particularly like to thank the following people who have kindly provided me with information and photographs: Dr Janet Brown, Dr Roy Clarke, Mr Richard Collins, Dr Nick Frerichs, Mr Don Griffiths, Dr Kim Jauncey, Dr Don Macintosh, Mr Ian Macrae, Mr Dave Sampson, Dr Chris Sommerville and Mr Billy Struthers (Institute of Aquaculture); Dr Derek Robertson and Mr Ian Semple (Howietoun Fish Farm, University of Stirling); Ms Susan Shaw (Dept. Business Studies, University of Stirling); Dr H Fran Henderson and Dr André Coche (FIRI, FAO, Rome); Dr Roger Pullin (ICLARM, Manila); Professor Tony Mines and Dr Gaudiosa Almazan (College of Fisheries, University of the Philippines); the staff at SEAFDEC (Binangonan Station, Laguna, Philippines); Mr Dick Coutts (ODA); Mr Andy Johnstone and Mr Jim Lindsay (HIDB, Scotland); Dr Richard Gowen and Dr Alastair Bullock (SMBA, Scotland); Dr Allan Hume (Torry Research Station, Scotland); Dr Albert Blair (DAFS, Scotland); Dr P H Milne (University of

Strathclyde); Dr T Jäger (Inst. fur Meereskunde, Kiel, FDR); Dr G Boeuf (IFREMER, France); Dr Magnus Enell (University of Lund, Sweden); Professor T Sano and Professor J Katoh (Tokyo University of Fisheries, Japan); Dr M Kuwa; Ms Renee Chou (Primary Production Dept., Singapore); Dr D Edwards; Mr Alastair Taylor and Mr Hugh Ford (Knox Nets, Scotland); Mr Tommy Rae (Rothesay Seafoods, Scotland); Mr Stuart Cannon (Kames Fish Farming Ltd, Scotland); Mr Paul Featherstone (Atlantic Sea Products, Scotland); Mr Barney Whelan (ESB, Ireland); Mr N Hellevang (Bergen Bartz, Norway); Dr John Hugenin (Woods Hole Engineering Associates, USA); Mr Ben Rañeses (St Peters Fish Farm, Philippines); Mr Job Bisuña (Jobski Fish Farms, Philippines); Dr Rafael Guerrero III (TRC, Philippines); Mr Noritada Endo (Nichiro Gyogyo Kaisha Ltd, Japan).

I would like to thank Maggie Beveridge for editorial assistance, Ms Hilary Dugua for research, Mr Graham Usher for help with line drawings and Mr Ron Stewart and Mr Brian Gourlay, STAVS, for help with photographic material. Mrs Moira Stewart, Mrs Laura Cumming, Mrs Beatrice Dale, Mrs Sheila McEwan and Miss Hazel Tetstall had the long and unrewarding task of committing my sheafs of handwritten notes to type, and to them all I am most grateful.

# 1 Cage aquaculture — origins and principles

## 1.1 Introduction

Aquaculture is the aquatic counterpart of agriculture and its origins extend back at least three thousand years (Bardach *et al*, 1972). However, unlike agriculture, which has been the most important way of obtaining food on land for several thousand years, aquaculture has until recently contributed little in real terms to world fish production. Instead of evolving towards cultivation, hunter-gatherer methods of procuring food from the aquatic environment developed along a different path; by the gradual improvement of tracking methods and increase in killing power. Farming of the aquatic environment was almost non-existent.

There are several reasons why agriculture and aquaculture did not develop in the same way. First, food in the lakes and seas has, until recently, been abundant, and there was thus little need to learn to farm. Moreover, much of the aquatic environment appeared hostile. It must have seemed impossible that a structure which could hold fish securely and withstand the prevailing forces of the tides and currents, waves and storms, could be built in the sea. There were other technical problems too, to overcome. Whilst the breeding and welfare of animals and the harvesting and planting of seeds was comparatively easy to achieve on land, it has proved difficult to breed many aquatic organisms and to hatch their eggs and successfully rear their offspring. These problems were probably in part due to the fact that man was dealing with organisms which were very different from himself and in an environment which he feared and until recently did not understand.

World demand for fish, both as a source of food for human consumption and for reduction to fishmeal has grown at a steady pace since the end of World War 2, and has been met by the expansion of capture fisheries. Growth in the industry was most rapid during the 1960s when yields increased by an average of 5.8% per annum (Lawson, 1984) (*Fig 1.1*). During the 1970s and 1980s, growth has been slow and erratic, due largely to the decline of fisheries exploited for fishmeal and oil production, and although this same period has seen some increase in catches of fish for human consumption, there has actually been a decline in the overall rate of growth (Robinson, 1982).

The principal cause of the decline in capture fisheries production is the dwindling number of conventional stocks which can sustain further increases

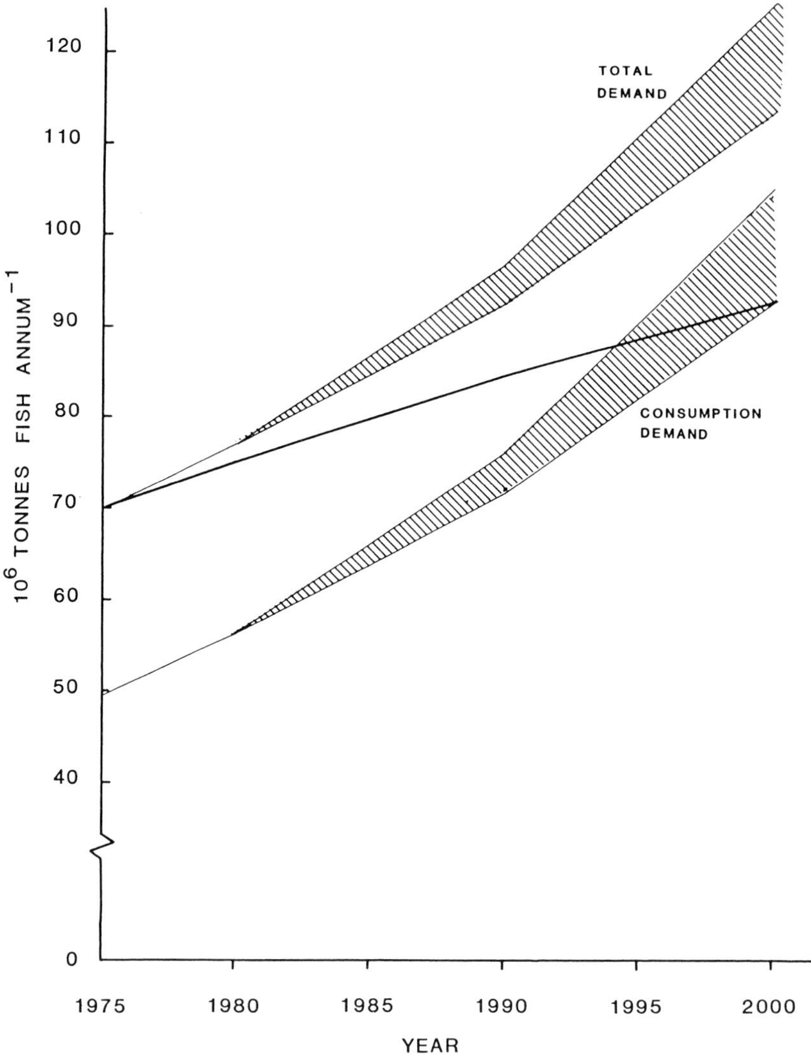


*Fig 1.1* Projected growth in world fisheries (black line), world total fish demand (including fishmeal) and consumption demand (data from Robinson, 1982).

in catch, and the situation has been exacerbated by steep increases in fuel oil prices, and the development of Economic Exclusion Zones (EEZs). The latter factors have hit major fishing nations like Japan, which depended to a considerable extent on distant water fishing, and has curbed their growth.

Optimistic predictions for the future suggest that a sustained growth of around 1% per annum might be possible, providing appropriate management of stocks, development of new fisheries, and marketing of novel fish products can be achieved (Robinson, *ibid*). Demand for fish for human consumption however, is projected to increase by around 2-2.4% per annum, and by the end of the century, capture fisheries production will be insufficient to meet this demand. If fishmeal fisheries production is assumed to remain constant during the next 15 years, then by the year 2,000, 72 million tonnes of fish from capture fisheries will be available for human consumption, leaving a discrepancy of at least 40 million tonnes (*Fig 1.1*).

Thus there has been a growing awareness that world capture fisheries stocks are not inexhaustible, and attention is increasingly being turned to the possibilities offered by aquaculture.

## 1.2 Principles of aquaculture

Both fisheries and aquaculture have a similar aim: to maximise the yield of useful organisms from the aquatic environment. The classical theories of Russell (1931) and Beverton and Holt (1957) show that the size of exploitable stocks is determined by four factors — recruitment rate, growth rate, natural mortality rate and fishing mortality rate (*Fig 1.2*). Capture fisheries try to maximise yields by increasing the fishing mortality rate, partly at the expense of natural mortality, although if too many fish are killed, then recruitment and growth are unable to compensate and the stock dwindles. Aquaculture, on the other hand, tries to achieve increased yields by manipulation of growth, reproduction and recruitment, and natural mortality rates.

Aquaculture probably evolved by initially achieving control over natural mortality, through the capture and temporary holding of organisms in

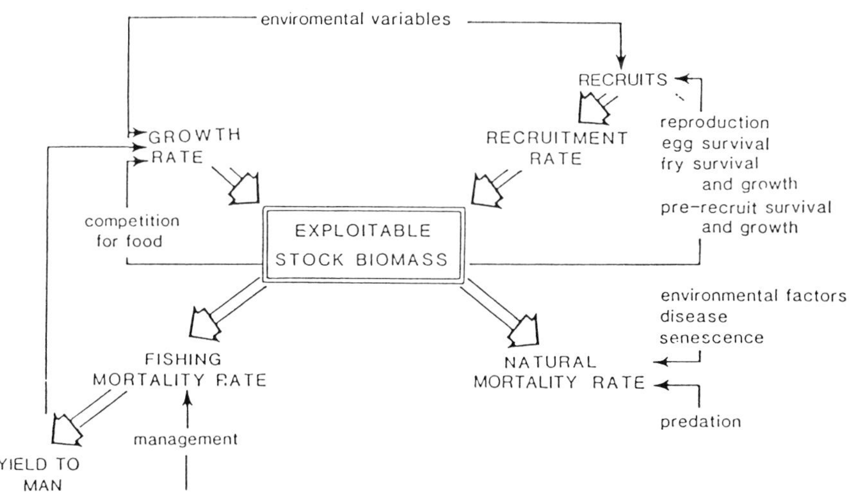


*Fig 1.2* Factors governing the exploitable stock biomass (from Beveridge, 1984b).

man-made facilities whilst they increased in biomass. The simplest facilities to construct would have been earth ponds. In some parts of the world these would have been little more than mud walls constructed to temporarily hold water and fish following the seasonal flooding of a river. Manipulation of growth through feeding with household scraps would have been an easy next step. However, control of spawning and recruitment is a comparatively recent achievement, as it is difficult to persuade many species to breed in captivity. There are also many technical problems involved in the hatching of eggs and the maintenance and feeding of larval and juvenile stages (Bardach *et al*, 1972).

Aquaculture has gradually evolved to gain control over all three of these population-determining processes. Great advances have been made in the fields of nutrition, genetics, engineering, physiology and biochemistry, which have all contributed towards improved yields. However, since inputs of both energy and effort are involved in achieving total control, several types of aquaculture have flourished by opting for lower yields at a considerable saving in human resources and expense. For example, ranching, or the release of juvenile fish into the wild, involves little more than increasing recruitment to the natural exploitable stock, and does nothing to reduce natural mortality (predation, disease), or influence growth. However, there are considerable financial savings in terms of feeding and construction of holding facilities. Similarly, attention is once more being turned towards fish culture that reduces expenditure on feeds. Feed costs can account for as much as 60% of the production costs at a farm (ADCP, 1983). However, they can be markedly reduced if advantage is taken of naturally available foods. In ponds, fertilisation with inorganic or — better still — organic fertilisers, will stimulate the growth of organisms at the base of the food web, whilst in lakes and rivers, cages or pens can be used to hold fish which will crop available food items such as algae, zooplankton or suspended organic material.

In summary, aquaculture, or the farming of aquatic organisms is achieved through the manipulation of an organism's life cycle and control of the environmental factors which influence it. Three main factors are involved: (i) control of reproduction, (ii) control of growth and (iii) elimination of natural mortality agents. Control of reproduction is an essential step, otherwise the farmer must rely on naturally spawning stocks. The supply of fry from the wild may be restricted to a particular season and a particular area, and there may also be shortages due to over-exploitation of wild stocks. For technical reasons, this step has not yet been achieved in the culture of many, particularly marine, species (see Chapter 8). Growth can be increased through selection of broodstock (control of breeding is therefore a prerequisite), and through feeding. As will be demonstrated, the culture of carnivorous species is usually tied to the supply of high-protein, fishmeal-based diets. However, there is considerable scope for reduction in feed costs providing the appropriate omnivorous/detrivorous/planktivorous species and systems are used.

Finally, rearing systems are essential to all types of aquaculture except ranching. Such systems are designed to hold organisms captive whilst they increase in biomass, by minimising losses through predation and disease, and exclusion of competitors (Reay, 1979). As a secondary consideration, rearing systems should also be designed in such a way as to facilitate harvesting.

## 1.3 Rearing facilities

Cage culture is a method of farming aquatic organisms in a particular type of rearing facility. Rearing facilities for fish can either be land-based or water-based, the former type including ponds, raceways, tanks and silos, and the latter comprising enclosures, pens and cages. In standard English dictionaries, the terms 'enclosure', 'pen' and 'cage' appear to be virually synonomous, and indeed can often be used interchangeably. However, in the field of aquaculture this has given rise to a certain amount of confusion, the term 'enclosure' often being used to describe something which could either be a cage or a pen, and the word 'pen' being used by North American researchers to denote a large sea cage. For the sake of clarity, therefore, a few moments would be well spent on defining exactly what will be meant when these labels are used in this book.

The term 'enclosure' I have taken to mean an enclosed natural bay, where the shoreline forms all but one side, which is typically closed off by a solid, net or mesh barrier (*Fig 1.3a*). In pen culture, the rearing facility is almost entirely man-made, the sides being constructed from wooden poles, mesh or netting (*Fig 1.3b*). The bottom of the structure, however, is formed by the sea bed. Cages, on the other hand, are enclosed on the bottom as well as the sides by wooden, mesh or net screens (*Fig 1.3c*).

There are other differences between these water-based rearing facilities. Pens and enclosures tend to be bigger, ranging in size from around 0.1ha to some which are well over 1,000ha. Cages, however, are smaller, and typically have a surface area somewhere between $1m^2$ and $1{,}000m^2$. Moreover, because of their small size, cages are better suited to intensive culture methods than pens.

## 1.4 The origins of cage culture

Cages were probably first used by fishermen as a convenient holding facility for fish, until sufficient quantities were caught to make a journey to market worthwhile. The earliest types of holding cage may have been little more than modified fish traps, and such traditional types of holding facility have been in use in many parts of the world for generations. Typical examples can be found today in northern Lake Chilwa, Malawi where fishermen may spend several days fishing in the dense *Typha* reed swamps prior to transporting their catches by canoe to distant fish landing sites. Strictly speaking, the fish boxes used by fishermen for holding lobsters could be regarded as cages.

True cage culture, in which fish or other organisms were actually held for long periods of time whilst they increased in weight, is of comparatively recent origin, and seems to have developed independently in a number of countries, all in Southeast Asia. In the Great Lake region of Kampuchea,

a

b

c

*Fig 1.3* Water-based aquaculture facilities (a) enclosure (Lake Buhi, Philippines); (b) fishpen (Laguna de Bay, Philippines); (c) cages (Ireland — courtesy M J Phillips).

floating cages have been in use since the end of the last century (Lafont and Savoeun, 1951; Hickling, 1962; Ling, 1977; Pantulu, 1979). Snakeheads (*Channa* spp.), catfishes (*Pangasius* spp; *Clarias* spp.) and marble-headed gobies (*Oxyeleotris marmorata*) were held in wood or bamboo cages, fed on a mixture of kitchen scraps augmented by trash fish, and transported by river to the markets of Phnom Penh. The cages were either towed behind the boats, or occasionally incorporated into the vessel to form a sort of well boat (*Fig 1.4a, b*). During the present century, this type of cage culture spread to most parts of the lower Mekong delta and into Vietnam (Pantulu, 1979).

In Mungdung Lake, Jambi, Indonesia, floating bamboo cages have also been in use since the early 1920s (Reksalegora, 1979) to rear *Leptobarbus hoeveni* fry captured from the lake. A different form of cage culture appeared in Bandung, Indonesia around 1940. Here, small bamboo and 'bulian' wood cages were anchored to the bottom of organically polluted rivers and canals, and stocked with common carp, *Cyprinus carpio*, which fed on wastes and invertebrates carried in the current (Vass and Sachlan, 1957).

Traditional cage culture, distinguished by its reliance on natural construction materials and on natural or waste feeds, is still practised in many parts of Indonesia and Indo-China. However, although moderately successful, these methods of rearing fish had a largely localised influence and did not directly give rise to the current cage fish farming industry. Modern cages utilise synthetic mesh or netting materials and have collars largely fabricated from synthetic polymers and metals, although wood is still widely

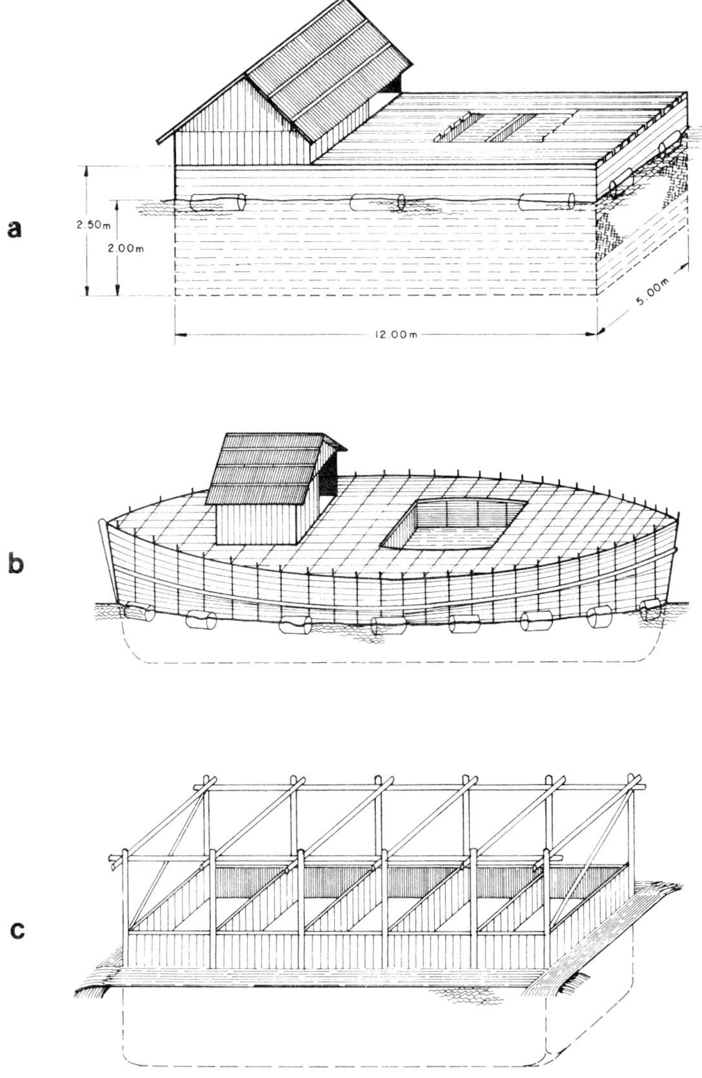


*Fig 1.4* Traditional fish cage designs, Indochina. (a) Southern Vietnam; (b) boat-shaped cage from Kampuchea; (c) battery of small cages, Kampuchea. (from Pantulu, 1979).

used in many designs. It is difficult to pinpoint the origins of modern cage fish farming, although it probably first began in Japan in the early 1950s. According to Milne (1974), Professor Harada, Director of the Fisheries Laboratory at Kinki University first started experimenting with cage fish culture in 1954, and commercial culture of yellowtail *Seriola quinqueradiata* followed three years later. In Norway, cages were being used to culture

Atlantic salmon (*Salmo salar*) in the early 1960s and in Scotland the White Fish Authority commenced salmon cage rearing trials around 1965. Surprisingly, tilapia (*Oreochromis* spp.) culture in cages is of even more recent origin, and owes its beginnings to work carried out at Auburn University in the late 1960s (Schmittou, 1969).

The growth of modern cage farming is thus very much a phenomenon of the 1970s and 1980s. In the following chapter, the relative importance of aquaculture in cages and the reasons underlying its current status are discussed.

# 2 Cage culture — an overview

## 2.1 Introduction

There is an enormous variety in the size and design of cages. They are extremely versatile and lend themselves to being used in many different ways. This section is intended as an overview of cages and cage culture and many of the themes touched on here will be expanded later in the book.

## 2.2 Diversity of cage types

Cages have progressed a great deal from their humble origins and today there is an enormous diversity of types and designs. For convenience, it is possible to impose upon them a classification system of the type shown in *Fig 2.1*. There are four basic types: fixed, floating, submersible and submerged. Fixed cages consist of a net bag supported by posts driven into the bottom of a lake or river (*Fig 2.2a*). They are commonly used in some tropical countries, such as the Philippines, where they have proved to be inexpensive and simple to fabricate compared with other designs, although they are more limited in size and shape, and restricted to sheltered shallow sites with suitable substrates.

The bag of a floating cage is supported by a buoyant collar or, in some cases, frame. This type is by far the most widely used and can be designed in an enormous variety of shapes and sizes to suit the purposes of the farmer. They are also less restricted than most other types of cage in terms of suitability of sites. Some floating types rotate, as a means of controlling fouling (see Chapter 7). The design in *Fig 2.2b* rotates about a central axis mounted on the collar, although other designs are rotated by means of moving the supplementary flotation elements or by adjusting the buoyancy of the frame members (*Fig 2.2c*). The much more widely used non-rotating floating types can be constructed with wide or narrow collars. The former are common on larger cages and serve as work platforms, thus facilitating many of the routine farm tasks (*Fig 2.2d*). Most wide collars are designed to be rigid, although some of those which support net bags are flexible so that they may be used at more exposed sites (*Fig 2.2e*) (see Chapters 3 and 7). Simple and inexpensive flexible collar narrow cages can be fabricated using rope and buoys (*Fig 2.2f*), but in practice they have been found to be difficult to manage. Rigid narrow collars, constructed from glass fibre or steel section

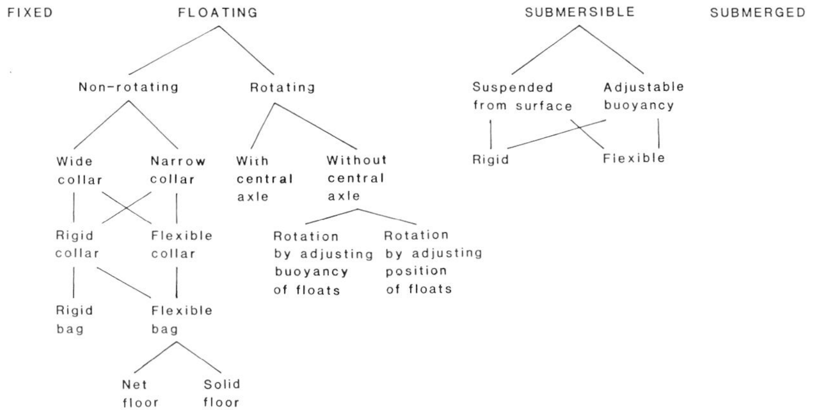


*Fig 2.1* A classification system for cages (developed from Kerr *et al*, 1980).

and buoys (*Fig 2.2g*) are popular in western Europe although all routine operations must be performed from a boat or pontoon. Rigid mesh designs must, of course, utilise a rigid collar. Some of the floating net bag designs, such as those developed for flatfish culture, have a solid bottom (Hull and Edwards, 1979).

Neither net nor rigid mesh bag submersible cages have a collar, but instead rely on the frame or rigging to maintain shape. The advantage of this type of cage over other designs is that its position in the water column can be adjusted to take advantage of prevailing environmental conditions. Some designs rely simply on the bag being suspended from buoys or a floating frame on the water surface (*Fig 2.2h*) whilst others have variable buoyancy (*Fig 7.24*). In some designs, however, it has proved difficult to maintain the shape of the bag when the cage is submerged. Moreover, not all species have readily adapted to culture in submerged holding facilities.

Whilst a number of submerged cage designs have been proposed (*eg* Hugenin and Rothwell, 1979), few have actually been built or tested. Simple submerged cages, however, are widely used in flowing waters in parts of Indonesia and the USSR (Vass and Sachlan, 1957; Martyshev, 1983). Most designs are little more than wooden boxes with gaps between the slats to facilitate water flow, and are anchored to the substrate by stones or posts (*Fig 2.2i*).

## 2.3 Cages and cage culture

Because they are a relatively inexpensive and convenient way to hold captive aquatic organisms, cages have been used for a variety of purposes, some of which are totally unrelated to aquaculture. For many decades, cages have been widely used to hold and transport bait fishes for tuna pole and line fishing (Ben Yami, 1978) although today their use has been largely superseded by live-bait holds in boats. More recently, cages have been

*Fig 2.2* Different types of cages (a) fixed cage (SEAFDEC, Laguna de Bay, Philippines); (b) rotating cages with central axle, (Kiel, Germany); (c) submersible, rotating cage, without central axle (Dunstaffnage, Scotland — courtesy A Blair); (d) floating, wide-collar milkfish broodstock cage (SEAFDEC, Philippines — courtesy R S V Pullin); (e) flexible wide collar (Ireland — courtesy B Whelan, E S B); (f) flexible narrow collar (Lake Titicaca, Bolivia); (g) rigid, narrow frame (Scotland — courtesy, I Macrae); (h) submersible, suspended from surface (Germany); (i) submerged (West Java, Indonesia — courtesy C Sommerville).

g

h

i

developed to hold fish for water quality monitoring of power station effluents (Holt, 1977; Chamberlain, 1978), and to treat the symptoms of cultural eutrophication (Yang, 1982). Cages have also proved invaluable in experimental work, where it is important that there are no significant differences in environmental conditions between groups of organisms being studied (see Collins, 1975, for discussion).

Like other forms of aquaculture, cage culture can be used to rear food organisms or to rear animals for restocking, and may be classified on the basis of feed inputs as extensive, semi-intensive and intensive. In extensive culture the fishes must rely solely on available natural foods such as plankton, detritus and organisms carried in the drift. Semi-intensive culture involves the use of low protein (<10%) feedstuffs, usually compounded from locally available plants or agricultural by-products to supplement the intake of natural food whilst in intensive culture operations, fish rely almost exclusively on an external supply of high protein (>20%) food, usually based on fishmeal (see also Coche, 1982).

Extensive cage culture is restricted to fresh waters and may be practised in two types of environment: highly productive lakes and reservoirs, and water bodies which receive considerable quantities of sewage or domestic wastes. Primary production, which fuels all successive energy transactions in aquatic food webs, other than in waste-fed systems and systems with high allochthonous (*ie* externally-produced materials, such as leaves) inputs, is dependent upon the availability of essential nutrients such as phosphorus and nitrogen compounds, and light and temperature (Le Cren and Lowe-McConnell, 1980; OECD, 1982). These systems with high nutrient loadings, such as those located in fertile watersheds are likely to be highly productive (see also Chapter 5). However, productivity is also strongly correlated with latitude (Brylinsky, 1980) and between temperate (23-67°) and tropical zones (23°N-23°S) there is considerable increase in the range of annual primary production values (*Fig 2.3*). Hence, tropical water bodies with high nutrient loadings offer the best opportunities for extensive fish culture.

Commercial extensive cage culture is at present only widely practised in the Philippines (Beveridge, 1984b), although cages of bighead carp are successfully used at several reservoirs in Singapore to control excessive plankton blooms in municipal water supplies (Yang, 1982). In western Europe, extensive rearing of juvenile planktivorous stages of salmonids, coregonids and pike (*Esox lucius*) is carried out on an experimental/semi-commercial basis (Bronisz, 1979; Uryn, 1979; Jäger and Kiwus, 1980; Holm and Møller, 1984). Attempts to grow bighead carp in cages without supplementary feeding in the United States, however, have been disastrous (Engle, 1982).

Both sewage-fed ponds and streams and rivers subject to high loadings of domestic waste have proved suitable for extensive cage culture (Vass and Sachlan, 1957; Gaigher and Krause, 1983), although there is some concern about the public acceptability of fishes grown in such systems. Edwards *et al* (1984) in Thailand have circumvented this problem by managing sewage-fed ponds to maximise production of fish which are then harvested and incorporated into diets for other fish.

Planktivores, detrivores and omnivores, such as many of the carps and tilapias, may be suitable candidates for extensive culture, although little research has been carried out to date to determine the best species for different conditions. Sometimes the choice can be crucial. For example, in flowing water (lotic) systems, there are usually few planktonic species present except in slow-flowing, lowland reaches. Hence extensive cage culture of plankton-feeding species in rivers and streams is likely to be impractical. Trials conducted by Othman *et al* (1985) in the Tengi River, Malaysia, with bighead carp illustrate this. Cages were stocked with 25.3g fish, at a stocking rate of 15 fish $m^{-3}$. During the 2 month trial, 95% of the carp died, and the average weight of the survivors was 19.5g.

In the Philippines, production of tilapia by means of extensive cage culture is claimed to be very high — up to 1.90kg $m^{-3}$ $month^{-1}$ (Beveridge, 1984b) although it is unlikely that production at this level would be sustainable in the long term. Nevertheless, extensive cage culture may be of value in certain

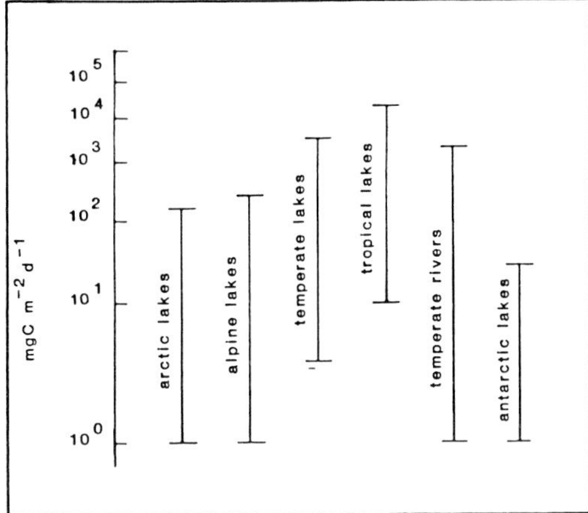


*Fig 2.3* Primary productivity of freshwaters (from Beveridge, 1984b).

areas of the world and for certain sectors of rural communities. It may prove to be a viable alternative to the management of a resource as a fishery (Beveridge, *ibid*) and it may be prudent to carry out some extensive culture of fishes in cages in conjunction with intensive lake-based fish rearing, thus reducing the environmental impact of wastes whilst increasing the profitability of the venture. It may also be advisable to switch between semi-intensive and extensive cage culture at different times of the year, so as to take advantage of periods of high phytoplankton densities. Research efforts at present are aimed not only at determining which species perform best under which conditions, but also at the effects of increasing predation pressure on one particular community or size group of aquatic organisms in order to better evaluate the potential and economic viability of this method of fish culture.

In tropical freshwaters, semi-intensive rearing of fishes is the most common method of cage culture and species which feed low in the food chain such as *Oreochromis niloticus*, *O.mossambicus*, *O.aureus*, and bighead, silver (*Hypophthalmicthys molitrix*), and common carp are fed an almost infinite variety of materials including rice bran, brewery and domestic wastes (Jangkaru and Djajadiredja, 1979; Pantulu, 1979; Dela Cruz, 1980; Coche, 1982) Semi-intensive cage fish rearing is also practised to a limited extent in eastern Europe (Muller and Varadi, 1980; Martyshev, 1983). However, apart from some experimental work with herbivorous species such as siganids and mullets (Pitt *et al*, 1977; Tahil, 1978) semi-intensive culture is not practised in marine environments (*Fig 2.4*). Future expansion in this method of cage fish rearing is likely to be restricted to tropical fresh waters since few commercially-important marine species, other than those mentioned above, and few temperate freshwater species can utilise the natural foods available to caged species, or the typical low protein plant-based supplemental feeds.

The supplementary feeds used are largely determined by availability and tend to be delivered on an *ad hoc* basis rather than according to any predetermined set of rules. Indeed, there is little known about the role of supplementary feeding in semi-intensive culture, and so it is impossible to provide any firm guidelines. Methods of administering feeds are discussed in Section 6.3. Most types of freshwater sites have proved to be suitable for semi-intensive cage culture, although precautions may have to be taken to avoid excessive feed losses (see also Section 6.3).

Intensive cage culture is largely restricted to the rearing of high value carnivorous species. In freshwater, salmonids, channel catfish (*Ictalurus punctatus*) and increasingly, common carp are reared intensively, whilst in the marine environment Atlantic salmon, yellowtail (*Seriola quinqueradiata*) and groupers ( *Epinephelus* spp.) are the main intensively farmed species. Some intensive rearing of caged tilapia — normally a low value species — is practised in parts of the United States, Mexico and Southeast Asia (Taiwan, Hong Kong, Japan), where they fetch a high market price. There may also be restrictions in the length of the growing season available to farmers in these countries, thus encouraging the use of intensive feeds.

Trash fish is still the main feed type used for yellowtail and grouper culture. Specially-formulated pelleted dry diets have been developed and are widely-available for salmonid and channel catfish culture. Intensive culture is not recommended in lotic sites, where feed losses can be excessive.

In cage culture, monoculture is the rule and polyculture the exception. There are several reasons for this. First of all, there are fewer feeding niches to exploit in cages than in pens or ponds, even when the cage bottom is buried in the substrate. Thus, natural foods are unavailable to caged benthic

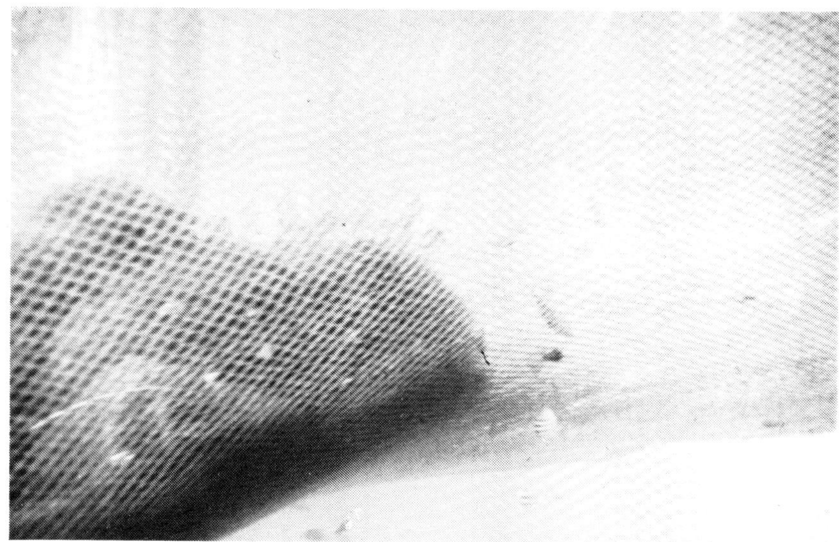

*Fig 2.4* Siganids feeding on filamentous algae (courtesy R S V Pullin).

or macrophyte-feeding species. Whilst it is possible to establish a polyculture of several plankton-feeding species in extensively or semi-intensively managed cages, to date, this type of system has not been properly evaluated. Some polyculture of carps in cages is practised in China, with yields of up to 7.5kg $m^{-3}$ $y^{-1}$ from extensive systems and up to 13.5kg $m^{-3}$ from semi-intensively managed cages (FAO, 1983). Experimental polyculture of Chinese carps in cages has also been attempted in Hungary (Muller and Varadi, 1980).

In intensive cage culture there at first sight appears to be even less reason for polyculture. However, with increasing pressure to improve profitability and to reduce waste loadings there is some incentive to stock herbivores to control fouling and benthos/detritus feeders to utilise uneaten food (see Chapters 5 and 7). Recently, commercial trials began in Scotland to evaluate the feasibility of growing two high-value species together, turbot and Atlantic salmon (Anon, 1985).

Cages have proved suitable for both small-scale artisanal and large-scale commercial aquaculture production, the materials used to fabricate the cages and the size for the rearing units being chosen accordingly. Several companies in Scotland today produce in excess of 1,000 tonnes of Atlantic salmon in cages — although not all at the one site — and have an annual turnover of close to £4 million.

## 2.4 Cage culture and aquaculture

A scan through the taxonomic index in Coche's bibliography on cage culture yields a figure of more than 130 fish species and nearly a dozen species of prawn, lobster and crab that have been grown in cages, on an experimental basis at least (Coche, 1983). Surprisingly, some of the most unlikely candidates for domestication and culture in enclosures are included in this list. Fast swimming pelagic species, such as the tunas (*Thunnus thynnus*, *Euthynnus pelamis*) have been found to adapt readily to the confines of a floating cage, albeit that the cages used are enormous (Piccinetti *et al*, 1980; Butler, 1982). Bottom-dwelling flat fishes (*eg Scophthalmus maximus*), too, have been successfully cultured in cages fitted with a solid bottom (Hull and Edwards, 1979). Some fishes such as the estuarine groupers (*Epinephelus salmoides*) are crepuscular predators and prefer to hide among rocks and corals and ambush their prey. However, even this behaviour seems to be of little disadvantage when they are crowded together in cages (Chua and Teng, 1980), although the provision of artificial hides in the form of used car tyres is claimed to increase production (Teng and Chua, 1979). Prawns are another group of animals which are difficult to culture at high densities since they fight and exhibit cannibalistic behaviour. However, if the bottom of the cage is buried in mud, then prawns can be successfully reared even in small units (see Chapter 8).

Indeed, few of the commercially important cultured species are absent from Coche's list, the most notable exception being the eels ( *Anguilla* spp.) whose body shape and environmental requirements seem not to predispose them to culture in net or mesh enclosures. In Scotland, attempts to use cages in freshwater lochs as temporary holding facilities for eels caught in fyke nets have resulted in large mortalities, although a few eels are occasionally observed in cages of rainbow trout where they have presumably been attracted by the presence of pelleted feeds.

In a recent review of world finfish, crustacean and mollusc culture, FAO estimated that the annual production in 1983 was 8.1 x $10^6$ tonnes (FAO, 1985). Around 50% of this figure is accounted for by molluscs and 50% by finfish, with less than 1% (*ie* <80,000 tonnes) of prawns and other crustaceans. Production in freshwater probably accounts for 80-90% of world finfish culture, and around 95% of this is from ponds. Thus cage culture will account for at most 3-4%, or 100-140,000 tonnes of freshwater finfish per annum. In certain sectors of the industry, however, cage culture assumes an extremely important role. In Scotland, for example, more than 40% of rainbow trout production is from freshwater cages (*Table 2.1*).

Culture of finfishes in marine and brackish waters is currently somewhere in the region of 500,000 tonnes per annum. Around 40% of this is derived from cages, the majority of the balance being produced in tropical coastal ponds. Japan's yellowtail and western Europe's Atlantic salmon (*Salmo salar*) industries are almost exclusively based on the use of cages.

The culture of prawns and other crustaceans in cages is still in its infancy, commercial production being limited to a few tonnes of penaeid species in Singapore.

## 2.5 Advantages and disadvantages of cage culture

As stated above, all culture systems are designed to hold organisms captive whilst they increase in biomass, and in order to evaluate the performance of cages, they must be judged against other available systems such as pens, ponds, tanks and raceways in terms of:-

(i) level of technology required for construction,
(ii) ease of management,
(iii) adaptability,
(iv) quality of the fish reared,
(v) resource use,
(vi) social implications,
(vii) economic performance.

Whilst some of the designs used for rearing fish in offshore locations are fairly sophisticated, the majority of cages are fairly simple to construct and can be put together in a day or so using local, unskilled labour. They are also easily managed. Observation of stock is facilitated and, unlike pens, fish can be harvested relatively easily using little more than a scoop net. Land-based culture systems are relatively difficult to change once they are installed. However, cage farms can be expanded simply by adding a few more cages as experience grows and circumstances allow and cages moved around with little restriction.

Fish grown in cages can succumb to fin and skin damage through abrasion

*Table 2.1* Production of rainbow trout in Scotland, 1979-1984. Data courtesy of Department of Agriculture and Fisheries for Scotland (from Phillips *et al*, 1985).

| *Year* | *Pond culture (t)* | *Tank culture (t)* | *Freshwater cages (t)* | *Marine cages (t)* | *Total production (t)* |
|---|---|---|---|---|---|
| 1979 | 565 | 377 | 240 | 97 | 1,279 |
| 1980 | 601 | 560 | 470 | 86 | 1,717 |
| 1981 | 820 | 608 | 721 | 112 | 2,261 |
| 1982 | 654 | 404 | 728 | 133 | 1,919 |
| 1983 | 828 | 265 | 843 | 73 | 2,009 |
| 1984 | 866 | 197 | 932 | 87 | 2,082 |

(Moring, 1982) although these disfigurations can be kept to a minimum if cages are sited and moored properly, if recommended rearing densities are adhered to and if fish are carefully handled (Boydstun and Hopelain, 1977). Numerous studies have confirmed that cage-reared fish are often superior to fish reared in other systems and even wild fish, in terms of condition factor, appearance and taste (Holt *et al*, 1976; NORDA, 1984).

It is often argued that the over-riding advantage of cages over other systems is that they make use of existing water bodies. This can indeed be a major benefit in giving non land-owning sectors of the community access to fish farming and may be particularly important in areas where fisheries are in decline, thus greatly increasing the income of fishermen's families (see Gonzales, 1984, for example). Moreover, fishermen possess most of the skills necessary to construct and operate fish cages. However, it must also be recognised that cage fish farms have an impact on the aquatic environment due both to the presence of the cages and to the method of culture (see Beveridge, 1984b, for review). Cages take up space which can disrupt access and make navigation difficult, reduce the landscape value of a site (Beveridge and Muir, 1982) and also alter current flows and increase local sedimentation rates (IDRC/SEAFDEC, 1979). Irrespective of the method of rearing, cage culture can introduce or disrupt disease and parasite cycles, change the aquatic flora and fauna and alter the behaviour and distribution of local fishes (Loyacano and Smith, 1976; Phillips *et al*, 1985b; McGuigan and Sommerville, 1985). Associated with intensive cage culture is the release of uneaten food and faeces into the environment, stimulating primary production and adversly affecting water quality, whilst in freshwater lakes where extensive cage rearing of fishes is practised, algal populations can be over-grazed, effectively reducing primary productivity (Beveridge *et al*, 1982; Penczak *et al*, 1982; Beveridge, 1984b; Phillips *et al*, 1985b).

The impacts of cage fish farming on the aquatic environment can not only bring about conflict with other interests, such as fishing or recreation, but can also exert a negative feedback effect on the cage operations. It is thus important to understand and be able to quantify the impacts of cage fish rearing, and to develop guidelines for the rational exploitation of aquatic resources. Although research is still at an early stage, a number of predictive models have been proposed, and are currently being evaluated (see Chapter 5).

There have been numerous detailed economic evaluations of cage culture. Capital costs vary with cage size and materials used. Hugenin and Ansuini (1978) demonstrated an almost 3-fold difference in costs per unit volume between two cages constructed from identical materials but which differed in size. Similarly, a cage constructed from rigid copper-nickel (Cu-Ni) mesh costs around three times more than a cage of similar size which utilises a synthetic fibre net bag (Hugenin *et al*, 1981), whilst those cages designed to withstand exposed conditions can cost twice as much per unit volume as more conventional designs.

Operational costs are also highly variable, and are largely determined by species, site, method of culture, management and scale of operations. Seed costs may vary between 10 and 40% of operating costs, (Hugenin and Ansuini, 1978; Escover and Claveria, 1985) depending on species and arrangements for acquiring fry. For example, in Europe a marine-based rainbow trout farm will have much lower seed costs than an Atlantic salmon farm, since smolts cost around four times as much as 50 gram rainbow trout. In intensive cage farming operations, feed costs can account for more than 50% of operating costs (Collins and Delmendo, 1979; Brown, 1983) compared with only 3-4% at semi-intensive cage tilapia operations (*Table 2.2*). Marine cage farms, which are much more likely to experience problems with fouling, are generally more labour intensive than freshwater operations, as are farms which have problems with poachers or predators (Escover and Claveria, 1985). Depreciation costs can be as high as 45% of the operating costs (Aragon *et al*, 1985), indicating a short service life for most cages, although it is argued that copper nickel mesh cages and some of those designed for exposed locations have considerably longer useful lives. There are also differences in operating costs between small and large operations, as illustrated by data for semi-intensive tilapia farms in the the Philippines (*Table 2.2*).

Despite such wide-ranging differences, is it possible to reach any conclusion about the economics of cage fish farming compared with pen, pond, raceway or tank culture? Both the capital investment and the operational costs of intensive cage salmon culture are significantly lower than the alternative methods of rearing which must rely on pumped seawater and land-based tanks or raceways (Shaw and Muir, 1986). Economic comparisons of intensive channel catfish production in the USA using cages, raceways and pens confirms that initial capital costs are lower for cages although operating costs are slightly higher than for raceways (*Table 2.3*). Nevertheless, the profitability of the cage-based farm was marginally higher. More recent studies confirm that channel catfish production in cages is at least as profitable as other methods of rearing (Lowell, *et al*, 1982).

Data for less intensively managed systems are difficult to find. Using information from Guerrero (1981) for tilapia farming in the Philippines, pond farming appears to be the more attractive option (*Table 2.4*). However, these costs do not include land purchase or rental charges and the latter can account for 30% of operational expenditure (Rosario, 1985). Capital costs are based on $1m^3$ cages which are considerably more expensive to construct than a single large cage. Moreover, feed costs quoted for farms in Nueva Ecija are around 10 times greater than those incurred in Laguna Province (see *Table 2.2*) on a percentage operational costs basis.

In summary, the use of cages is by far the most economically feasible method of intensively rearing fishes such as salmonids, yellowtail and grouper in marine waters, and can be a comparatively profitable means of producing other species, although this depends very much upon local circumstances.

The major disadvantages of water-based culture systems are summed up by the word 'vulnerable'. Cages are often sited in public or multi-user water bodies and farmers may be powerless to stop or control any pollution that occurs. Pollution problems are a major problem for cage fish farmers in Japan (Tabira, 1980; Nose, 1985) causing thousands of dollars of damage each year.

*Table 2.2* Operating costs, expressed as percentages, of small ($\bar{x}$ = $420m^2$), medium ($\bar{x}$ = $848m^2$) and large ($\bar{x}$ = $2{,}499m^2$) cage fish farms located around San Pablo City, Laguna, Philippines, 1982. Data from Aragon *et al*(1985).

| | *Small* | *Medium* | *Large* |
|---|---|---|---|
| Seed | 34 | 42 | 38 |
| Feed | 3 | 4 | 3 |
| Labour[1] | 19 | 11 | 8 |
| Depreciation | 30 | 31 | 45 |
| Interest | 10 | 9 | 5 |
| Miscellaneous | 5 | 3 | 1 |
| | 100 | 100 | 100 |

1 Labour costs include estimates for unpaid labour, which can account for 50% of total labour costs.

*Table 2.3* Production costs ($) of 50,000 channel catfish in cages, pens and raceways. Data based on a 160 day growth period and a mean weight at harvest of 800g. No cost for water is included. Data from Collins and Delmendo (1979).

| | *Cages* | | | *Pens* | | | *Raceways* | | |
|---|---|---|---|---|---|---|---|---|---|
| | *Initial costs* | *Annual costs* | % | *Initial costs* | *Annual costs* | % | *Initial costs* | *Annual costs* | % |
| Capital costs | 11,500 | 2,764 | 10.8 | 14,500 | 3,232 | 11.8 | 47,000 | 3,908 | 15.2 |
| Operational costs | | 22,765 | 89.2 | | 24,136 | 88.2 | | 21,866 | 84.8 |
| Gross income | | 45,000 | | | 45,000 | | | 45,000 | |
| Net income | | 19,471 | | | 17,632 | | | 19,226 | |
| Ratio to total annual operational costs | | | 76.0 | | | 64.0 | | | 75.0 |
| Ratio to gross income | | | 43.0 | | | 38.1 | | | 43.0 |

*Table 2.4* Comparative costs (peso) of producing 3 tonnes of tilapia per annum in cages and ponds in Nueva Ecija, Philippines, 1980. Data based on a 1 ha pond and 42 $1 m^3$ cages (from Guerrero, 1981).

| | *Cages* | *Pond* |
|---|---|---|
| Capital costs | 4,200 | 27,000 |
| Fixed costs | 1,400 | 2,880 |
| Variable costs | | |
| Fingerlings | 6,300 | 4,000 |
| Fertiliser | — | 150 |
| Feed | 7,728 | 600 |
| Labour | 1,950 | 845 |
| Total costs | 17,378 | 8,475 |
| Gross income | 24,000 | 24,000 |
| Net income | 6,622 | 15,525 |
| Return | 38% | 183% |

Cages and pens are more susceptible to storm damage than ponds, tanks or raceways and this is cited as a major reason why many lake-based tilapia hatchery owners in the Philippines have considered moving operations back to the land. Cages are also highly vulnerable to poaching and vandalism which may be such a major problem as to preclude their use in certain parts of the world.

# 3 Cage construction

## 3.1 Introduction

In theory, the shape and size of fish cages should be dictated by a combination of factors, including:-

— species,
— prevailing environmental conditions,
— method of culture (extensive/semi-intensive/intensive),
— properties, costs and availability of materials,
— local skills.

The design must also take cognizance of the type of fish culture, *eg* large commercial farm or artisanal operation. In practice, ignorance of the effects of cage size or shape on fish production and lack of research into the interaction between impinging environmental forces and cage structures has resulted in cage designs having largely evolved empirically. Hence, until recently it could be said with little fear of contradiction that all cage designs, traditional and otherwise, reflect more the limitations of the materials used, local skills and the prevailing socio-economic situation rather than the requirements of the fish or the environmental conditions in which the cages are to be sited. This is not really surprising: the forces acting on a cage and mooring system are exceptionally complex and difficult to quantify, and analysis of the responses of a cage structure to these forces requires the development and use of sophisticated test rigs and computer models. A thorough evaluation of holding facilities for fishes requires not only studies into water quality at different stocking densities but also investigation of the fishes' behavioural and physiological responses. Research and development work of this nature is expensive and as Kerr (1981) has indicated, is unlikely to be carried out whilst the fish farming industry remains small and fragmented and unable to make a concerted effort to provide funding. Moreover, there has been little incentive to carry out this type of investigative work since most designs work reasonably well, at least in the comparative shelter afforded by most coastal and inland sites.

However, the situation is changing. The industry, particularly in Western Europe, North America and Japan, has expanded rapidly in the past five years, and has attracted the interests of a growing number of large multi-nationals, often looking for another outlet for a product or technology, and with money and facilities to carry out research and development. There is, too, growing pressure to culture fish in more exposed, offshore locations and increasing concern over security of stock and staff safety on farms. Thus in future cages will have to be more stringently designed, tested and cost-effectively manufactured and they will have to satisfy not only the fish farmers but also the insurance companies and government safety standards. Plans to introduce a system of licencing for cages, similar to that already governing shipping, have been mooted by the Norwegian government and may well set a precedent for other countries.

This chapter is intended as a brief review of cage and mooring system design and engineering and will concentrate on the most widely used floating types.

## 3.2 Size, shape and materials

All cage designs should start with the bag which, as stated above, should be primarily designed to meet the requirements of the fish. However, as was concluded in Chapter 2, present knowledge is sufficient to provide only the most general of guidelines. Circular holding facilities are purported to be best for schooling fishes such as the salmon and milkfish which tend to swim in circles when enclosed (Sutterlin *et al*, 1979; Yu *et al*, 1979), and are thus believed by many to provide a less stressful environment. However, octagonal or square designs probably work equally as well. Other, less active species such as the tilapias and common carp are thought to be less particular with regard to the shape of the holding facility.

Despite various arguments for constructing circular cage bags (they are stronger and make the most efficient use of available netting materials, resulting in the lowest costs per unit volume, see *Fig 3.1*), there are a number of disadvantages which should be taken into consideration. Because the surface area: volume ratio is comparatively small, water exchange is believed to be relatively poor (Howard and Kingwell, 1975) and, as will be discussed below, the technology required to construct a supporting collar for a circular enclosure is considerably more sophisticated than for other shapes.

The influences of surface area and depth on the health and production of fish have been little studied to date (Phillips, 1985). Cages of less than 1.5m depth have been shown to retard growth and change body shape of common carp and tilapia (Maruyama and Ishida, 1976, 1977), whilst depths of greater than 10-12m are probably poorly utilised by most fish species unless during period of adverse surface temperature or water quality conditions. Cages for holding flatfish, such as turbot are an exception. Cage bag depths of between 0.9 and 1.6m are recommended and give the fish sufficient shelter from surface effects, whilst providing an adequate water volume for exercise and feeding (Kerr *et al*, 1980).

Bigger cages have been shown to be better suited to the larger and faster swimming species such as the tunas (Matsusato, 1980), although for most species the beneficial effects of large capacity cages on growth and production have yet to be proven. From the fish farmer's point of view, the principal advantage of increasing bag size is that costs per unit volume becomes increasingly cheaper (*Fig 3.1*) and this is the main reason for the trend in sectors of the industry towards the use of progressively larger rearing units. In Scotland, for example, the traditional 300-400m$^3$ Atlantic salmon rearing cages are being replaced at many farms by cages of at least twice this volume (see Chapter 8).

However, large cages require more sophisticated materials and technology. They are also more difficult to manage, requiring two or three staff to perform even the most simple of routine operations. Inspection of nets and removal of mortalities may have to be carried out by divers, whilst monitoring fish health becomes very difficult. Grading, too, is a problem and for this reason it has been suggested that the new generation of giant cages may be best suited to the culture of species such as yellowtail, which do not require periodic sorting and harvesting. Water exchange is likely to be poorer, since surface area:volume ratio decreases with increasing size. Other doubts concerning the use of increasingly large rearing units have been voiced by the insurers who are concerned about the risks involved in holding tens of tonnes of fish in a single cage.

In other words, many of the advantages that cages are purported to have over other rearing systems (see Chapter 2) disappear when large units are employed.

Ideally, the materials used to form the cage bag should be:

(i) strong
(ii) light
(iii) rot, corrosion and weather resistant
(iv) fouling resistant
(v) easily worked and repairable
(vi) drag free
(vii) smooth textured and thus non-abrasive to the fish
(viii) inexpensive

Some would also argue that the materials should be rigid and thus resistant to deformation in strong currents, whilst others would conversely argue that a flexible material is better since it absorbs current energy (see later) and facilitates harvesting.

To a greater or lesser extent, similar properties should be present in the materials used to fabricate the cage frame and collar.

No single material possesses all of the qualities listed above, and so all are to some extent a compromise. Undoubtedly, however, some materials are better suited to certain species, sites and purposes.

## 3.3 Traditional designs

As traditional cages developed from fish traps and fish holding facilities, so they borrowed heavily from them in design and were fabricated using existing skills from whatever materials were to hand and were familiar to the builders. Thus natural materials such as grasses or wood were used not only

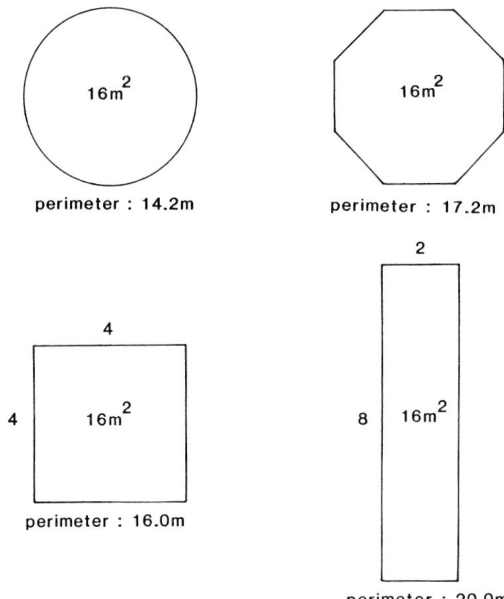


*Fig 3.1* The perimeter lengths of various cage shapes of identical surface area.

to form the collar or frame, but also to enclose the fish. Reeds (eg *Phragmites* and *Cypsus* spp), jungle vines and creepers such as rattan ( *Calamus* spp.) and ridan (*Salacca glabescens*), bamboos, hardwoods and softwoods have all been used in this way (Huet, 1956; Hickling, 1962; Watson and Raja, 1979; Pantulu, 1979; Murugesan and Parameswaran, 1983; Fedoruk and Srisuwantach, 1984). Reeds and vines, however, have all proved to be unsatisfactory: they are largely composed of cellulose and decompose in a matter of months in tropical waters, principally through the action of cellulytic bacteria, although life expectancy may be extended a little by application of tar-based compounds (Klust, 1982). Used on their own, they are also easily damaged by the caged fish and predators.

Bamboo, on the other hand, is a much more suitable material being strong, cheap, widely-available, and easily worked with simple tools (see also later). Typically, slats are cut and planed using a knife or machette and either fixed directly to the cage frame, or woven into matting, sometimes in combination with synthetic or natural fibres, and subsequently fixed to the frame. However in some parts of Thailand, traditional cages are virtually woven entirely from bamboo slats to form a basket-shaped structure (*Fig 3.2*).

Providing the bamboos have been properly selected and treated prior to use the resultant mesh will last for up to 2 years. Moreover, in trials comparing growth, survival and production of several species of fish in similar sized cages, all-bamboo constructions were found to be as good as, and in some cases superior to, those fitted with nylon mesh, and moreover cost considerably less (Lipton, 1984; Natarjan *et al*, 1984). However, it is sometimes difficult to find straight bamboos, resulting in non-uniform mesh sizes which may permit fry to escape. Bamboo slatted cages are also labour-intensive to construct, heavy and cumbersome to manage — particularly if more than a few cubic metres in volume — and unsuitable for use in the more corrosive brackish and marine environments.

Hardwoods are generally denser than bamboo or softwoods and are thus stronger and more resistant to impact damage and to infestation by boring organisms, such as the mayfly *Povilla adusta* which is prevalent in many African waters (Coche, 1979) (see *Fig 4.7*), and to fungal attack. A variety of species have been used throughout the tropics. Planks are simply lashed or nailed together (zinc nails are recommended) to form box-like constructions which are either floated using supplementary flotation materials, or anchored to the bottom with stones and posts (*Fig 3.3*). Economic analyses suggest that although hardwood cages typically cost around 10-15 times more than those fabricated from bamboo, they have a useful working life of 9-12 years in freshwaters (Sodikin, 1977; Fedoruk and Srisuwantach, 1984).

Traditional floating cages utilise a variety of materials such as bundles of bamboos, or hardwood logs and oil drums which are lashed to the sides of the structure for supplementary flotation (*Fig 3.3*). Simple anchoring systems — ropes and block weights, or posts driven into the substrate — are most commonly used.

## 3.4 Modern designs

### 3.4.1 CAGE BAG

The International Organisation for Standardisation (ISO) defines netting as 'a meshed structure of indefinite shape and size', but the term will be used here in a more restricted sense to differentiate flexible netting composed of natural and synthetic fibres such as cotton and nylon from rigid and semi-rigid materials such as extruded plastics and metals. These latter materials will be referred to in this book as rigid meshes.

Netting and rigid mesh materials can be fabricated from a range of man-made and natural materials and are available in a variety of different forms. Rigid mesh materials have been developed for many different purposes with the result that there is no single accepted system for their characterisation. They are usually described in terms of density (weight $m^{-2}$ of panel) and the diameter of the mesh bars, the specified mesh size generally being a measure of the distance between parallel bars (*Fig 3.4*). For netting, ISO have laid down stringent regulations governing terminology. Mesh size is illustrated in *Fig 3.4*. The yarn used for net fabrication is usually described in terms of tex, which is a measure of linear density (1 tex = 1g $100m^{-1}$).

#### *3.4.1.1 Flexible mesh materials (netting)*

Natural fibres are rarely used today for cage netting. They suffer from a number of disadvantages, not least of which is their susceptibility to rotting and consequent loss of strength (Klust, 1982). They also cannot be used for

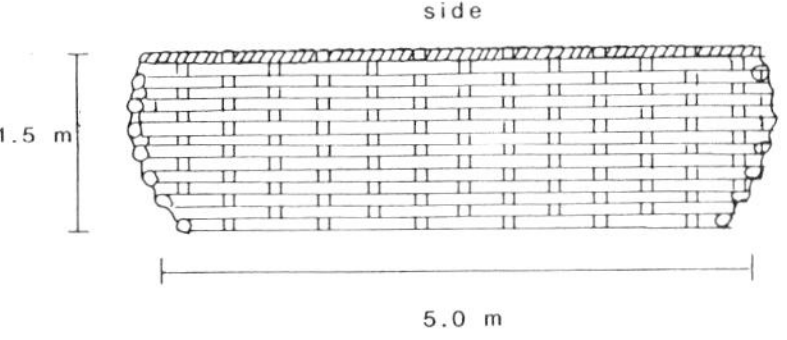


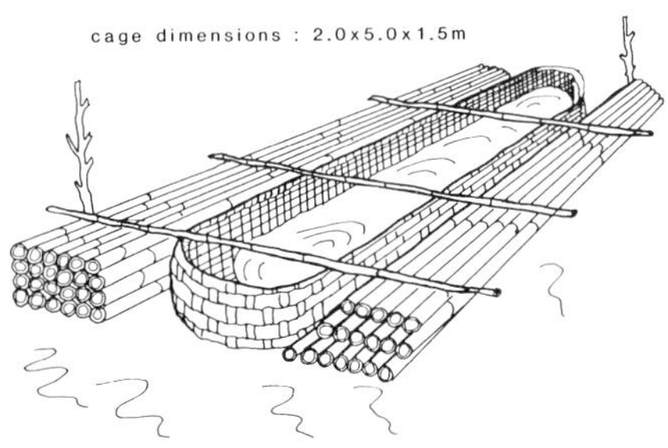


*Fig 3.2* Traditional basket-shaped bamboo sand goby cage from central Thailand (redrawn from Fedoruk and Srisuwantach, 1984).

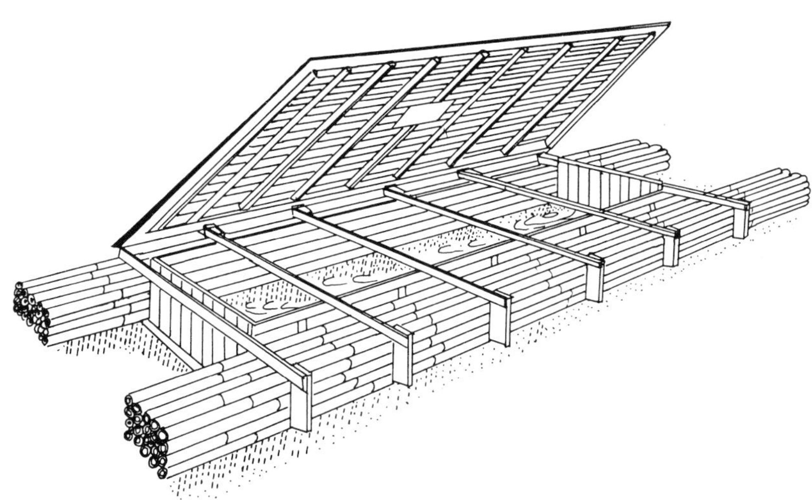

*Fig 3.3* Slatted hardwood sand goby cage (redrawn from Fedoruk and Srisuwantach, 1984).

fabricating knotless materials which are strongly recommmended for use in cage fish culture (see later).

Modern netting is composed of synthetic fibres which are manufactured from coal or oil-based raw materials. The most common types of synthetic fibres used today in the manufacture of fishing nets are polyamide (PA), polyester (PES), polyethylene (PE) and polypropylene (PP), which unfortunately, for commercial reasons, trade under a plethora of names. These vary not only from country to country but also within a country from manufacturer to manufacturer. Klust (1982) lists 274 brand names for PA netting yarns (both PA6 and PA6.6), 100 for PES, 78 for PE and 136 for PP! The use of one of the generic terms for PA fibres, nylon, has become so widespread that it is applied to all polyamide materials and henceforth will be used in this book.

Not only do the various netting materials have different properties, the textile materials used in their fabrication are available in a number of different forms which partly determine the ultimate characteristics of the netting. Nylon and PES netting are usually manufactured from continuous filament (also known as multifilament) yarn, composed of extremely long, fine fibres, generally less than 50 μm in diameter. All filaments run the entire length of the yarn and there are therefore exactly the same number of filaments present throughout the yarn. Although some PP netting yarn is manufactured from continuous filaments, some is also manufactured from split fibre yarn which is produced either by twisting plastic film under tension or by mechanically fibrillating the film immediately after extrusion. The resultant fibres are considerably coarser than continuous filaments and vary a great deal in their thickness. Split fibre PE netting yarns are also occasionally used for netting although most PE netting is still manufactured from folded monofilaments. Monofilaments are similar to the continuous filaments

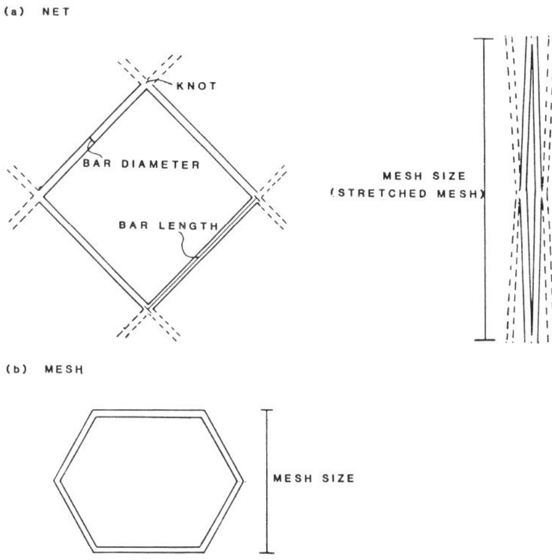


*Fig 3.4* Net and mesh terminology.

described above, but are considerably thicker (100-500 μm in diameter) and are sufficiently strong to function on their own as a yarn. However, the PE netting used for cages is usually manufactured from a twisted monofilament yarn.

The yarns described above are what are known as single yarns and form the basic components of netting yarns. There are two principal types of netting yarn used to make knotted netting; braided and cabled. Most knotted nets used in cage fish farming are fabricated from cabled yarns, which are produced in three stages: (1) the fibres are combined as described above to form single yarns; (2) several — usually three — single yarns are twisted together to form a folded yarn or netting twine; (3) several — usually three — of these folded yarns or netting twines are twisted together by a second twisting operation to form a cabled netting twine or cabled yarn (*Fig 3.5* ). Note that the twisting is carried out in alternative directions at each stage of the manufacture.

Single or double weaver's knots are used to construct most knotted netting (*Fig 3.6*), although in Japan, reef knots and square knots are also used. In factory produced nets, this process is entirely mechanised.

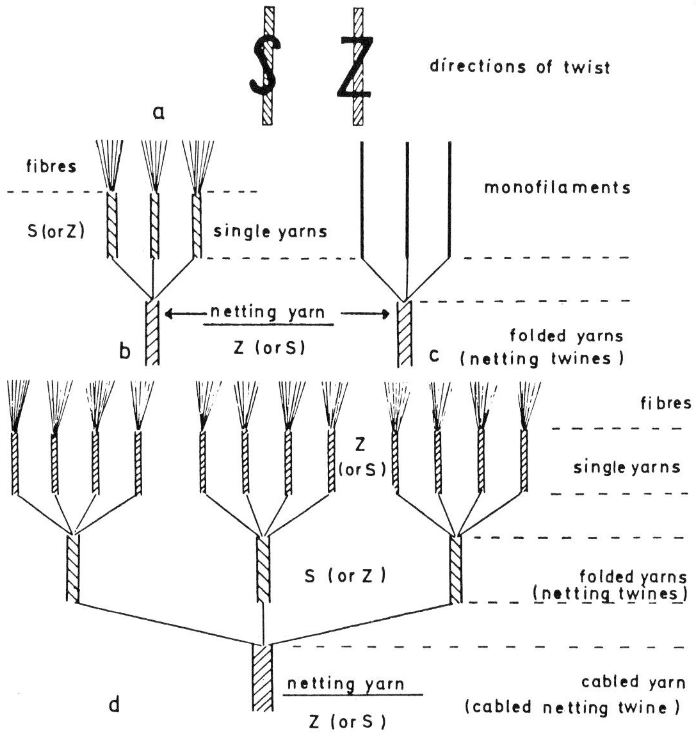


*Fig 3.5* Construction of twisted and folded netting yarns (from Klust, 1982).

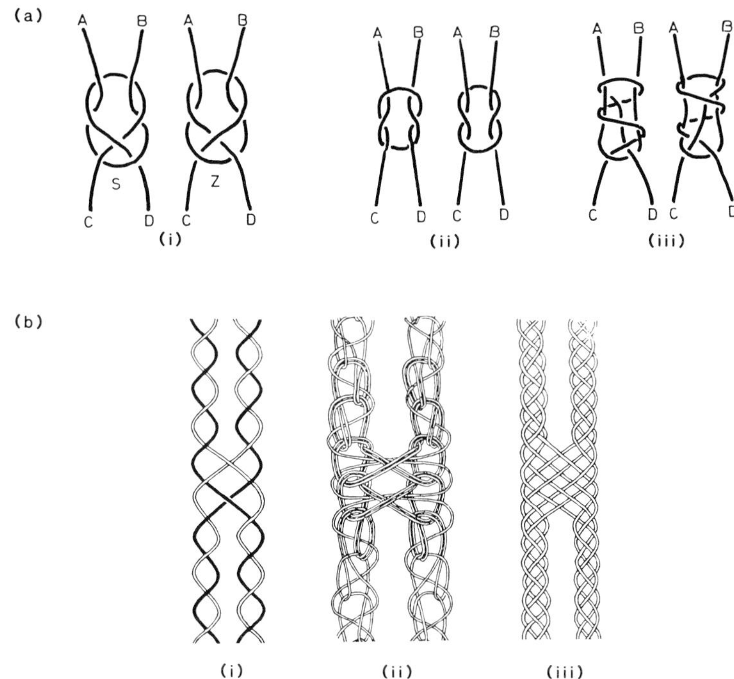


*Fig 3.6* (a) Types of knot used in netmaking (i) single weaver's or English knot; (ii) reef or Japanese knot; (iii) double weaver's knot. (b) Examples of various methods for the construction of bars and joints in knotless netting (i) Japanese twisting; (ii) Raschel; (iii) braiding (from Klust, 1982).

The yarns used for the fabrication of knotless netting are either single or folded PES or nylon yarns. Two types of knotless netting are commonly used for cages; the Japanese twisted and the Raschel. Examples are shown in *Fig 3.6* to illustrate how they are constructed. According to Klust (*ibid*) there are many more variations possible in the construction of bars and joints in the Raschel than in the Japanese twisted netting and it is also easier to repair. The Raschel type is widely available throughout the world, whereas the Japanese twisted knotless netting is mainly manufactured by Japanese companies.

The above section is intended to serve as an introduction to modern synthetic fibre netting, and readers who are interested to discover more are recommended to consult Garner (1962) and Klust (1982). For most fish farmers, however, it is probably sufficient to know the properties of different netting materials, so that the appropriate choice can be made. The factors which are most important were briefly outlined above. Ideally, cage nets should be slightly denser than water, thus facilitating hanging, but not so dense as to make handling excessively difficult or labour-intensive, or to significantly affect the flotation or mooring systems. The netting should be strong enough to resist ripping by floating objects and predators and be capable of supporting a portion of the biomass of enclosed fish when the nets are lifted for harvesting. High extensibility, so that the netting can endure stretching without breaking, and resistance to abrasion are also desirable. Flexural stiffness, or resistance of the netting to lateral and bending deformation is important since a stiff net will resist collapsing in currents. On the other hand, the netting should be soft and smooth textured so as to minimise abrasion damage to the skin, fins and eyes of the caged fish. However, softness and flexural stiffness are to some extent contradictory in synthetic fibre netting. Other desirable properties are that the netting should have as low a resistance as possible to water currents so as to minimise deformation and dynamic loadings, and that it should be resistant to fouling, thus minimising increases in loading to the cage mooring system whilst maintaining the best possible water flow through the cage. Finally, durability, availability, cost and maintenance requirements are extremely important considerations.

The characteristics of netting are determined not only by the properties of the yarn which are a product of the material used, the type of fibre (*eg* monofilament or multifilament), diameter of yarn and degree of twisting, but also by the method of net fabrication (*eg* knotted or knotless, Raschel or Japanese twisted) and mesh size. The addition or application of chemical stabilisers, dyestuffs, antifoulants and stiffening compounds can also modify these properties.

The general characteristics of various netting fibres are summarised in *Table 3.1*. Direct comparison between knotted and knotless netting is more difficult because many of the mechanical tests commonly carried out on knotted netting cannot be directly applied to knotless materials. Nevertheless, it is possible to make a number of qualitative comparisons. The principal problem with knotted netting is the knots, which are costly to make, both in terms of time and materials, result in a significant increase in weight (up to 100% increase in weight per unit area for some types) and bulk, and are the weakest part of the net. Furthermore, they cause a major increase in drag, foul and abrade more readily than knotless materials and have been widely reported as damaging to caged fish. By contrast, knotless netting is often cheaper and because it is knotless, panels can be machine stitched to form extremely strong enclosures (see later). It is thus by far the more suitable material for fabricating cage bags. Unfortunately, knotless materials are not readily available in all parts of the world.

Knotless netting is generally only fabricated in nylon, PE or PES. The latter is extremely expensive and considerably heavier than nylon (*Table 3.1*), although some experts believe that it may be a slightly better material. However, the economic advantages offered by nylon far outweigh any disadvantages that it might have in terms of stiffness or resistance to weathering. PE is a suitable alternative to nylon, being much cheaper and more readily available in many parts of the world. In most other respects, however, it is inferior.

*Table 3.1* Characteristics of synthetic fibres. Note that the two types of nylon are referred to, although both have near identical properties. (Modified from Milne, 1972 and Klust, 1982.)

| | *P.A.6.6* | *P.A.6* | *P.E.S.*[1] | *P.E.*[2] | *PP* |
|---|---|---|---|---|---|
| Fibre density ($g.cm^{-3}$) | 1.14 | 1.14 | 1.38 | 0.96 | 0.91 |
| Breaking strength | very high | very high | high | high | high[3] |
| Breaking strength, wet, in % of dry breaking strength | 85–95[3] | 85–95[3] | 100 | 110 | 100 |
| Weight in water in % of air–dry weight | 12 | 12 | 28 | 0 (buoyant) | 0 (buoyant) |
| Extensibility, wet | high | high | low | intermediate between PA and PES | low |
| Stiffness | flexible | flexible | moderately stiff | stiff | stiff |
| Softness | soft | soft | moderate | moderate | hard |
| Resistance against weathering without treatment or colouring | medium | medium | high | medium | low–medium |
| Resistance to fouling | moderate | moderate | — | low | moderate |

1 PES type Terylene — Dacron — Diolen — Tergal — Tetoron — Trevira
2 PE type high density, polymerised at low pressure
3 Continuous filament form

### *3.4.1.2 Rigid mesh materials*

Modern rigid mesh cages consist of mesh panels attatched to a frame and supported by a floating collar. Two basic types of material have been used — rigid plastic and metals.

Rigid plastic mesh materials such as netlon are formed by an extrusion process. They are mostly square mesh, are readily available in a range of mesh sizes and are semi-rigid rather than rigid and thus may be conveniently stored in rolls. They have a density intermediate to that of flexible synthetic fibre netting and rigid metal meshes — somewhere in the order of 0.35 — 0.50kg $m^{-2}$. However, netlon has not proved to be a popular material for cage construction and thus its durability and performance characteristics have not been fully evaluated. Although it does not corrode and is highly rot resistant it is, like all other synthetic polymers, prone to weathering. It is more resistant to fouling than untreated flexible netting but less so than the galvanised steel or copper alloy meshes (see *Table 3.5*).

Metal meshes, on the other hand, have been extensively tested for use in the construction of rigid cages. Metals suffer from one major disadvantage in that they are subject to chemical attack or corrosion when immersed in water. The ease with which a metal corrodes depends upon its electronegativity (its capacity for attracting electrons) and its standard electrode potential. When a metal is immersed in water a potential difference or electrode potential is established between the metal and the liquid and metal ions pass into solution leaving the metal with an excess of free electrons:-

$$M \rightleftarrows M^{+} + e^{-}$$

The electrode potential is dependent upon the electronegativity of the metal and the temperature and ionic concentration of the solution it is immersed in. Electrode potentials are measured by immersing the metal in a solution of its own ions and using a standard hydrogen or calomel electrode as reference. A list of electrode potentials for some common metals is given in *Table 3.2*.

The metals are arranged in order of increasing negativity and reactivity, with gold at the top and potassium at the bottom. If two different metals are connected together in a conducting medium, current flows from the metal with the more negative potential (*ie* anode) to the metal with less negative potential (the cathode) resulting in the corrosion of the more anodic material. For example, by coupling zinc (−0.76) to copper (+0.34) in an electrolytic medium, the zinc will form the anode and thus suffer corrosion, whilst the copper will become the cathode and thus be protected. The electromotive force (EMF) between the metals will be 0.34 − (−0.76) = 1.1v. If iron (−0.44) replaces zinc as the anode, then the EMF will be smaller (0.78v) and the resultant rate of corrosion will be reduced.

Because the rate of corrosion is dependent upon the nature of the electrolytic medium — and in particular its ionic composition, temperature and oxygen concentration — and the structure and chemical uniformity of the metals, a range of Eo values exists for each metal and its alloys. The galvanic series, as it is known, is based upon empirical findings. *Figure 3.7* shows the galvanic series for metals and their alloys in flowing seawater conditions, measured against a calomel electrode. Although the Eo values given in the figure may change outside the given temperature, current and salinity regime, the relative galvanic positions will remain the same.

Because galvanic cells can be established by connecting together two dissimilar metals in a conducting medium, such as sea or brackish water, care should be taken when choosing metal components for a cage system (see below). However, galvanic cells can also be caused by:-

(i) differences in the microstructure of a metal or alloy component.

(ii) differences in the ionic composition of the conducting medium. In seawater, areas in contact with less saline water form the anode, whilst those in contact with higher salinities act as cathodes.

(iii) differences in the dissolved oxygen (DO) concentration of the conducting medium. Low DO areas act as anodes whilst high DO areas become cathodes.

(iv) the existence of stress cells within the metal or alloy. Differences in grain size (fine grain — anode; coarse grain — cathode), and the presence of imperfections (defects — anode; perfect — cathode), strains (cold-worked — anode; annealed — cathode) and stresses (loaded areas — anode; non-loaded areas — cathode) can all lead to corrosion (see ICE, 1979; Hey and Pagan, 1983, for reviews).

*Table 3.2* Standard electrode potentials (Eo) of some common metals measured relative to the hydrogen electrode at 25°C.

| *Metal* | *Electrode reaction* | *Eo* | *Metal* | *Electrode reaction* | *Eo* |
|---|---|---|---|---|---|
| Copper | $Cu = 2Cu^{2+} + 2e^-$ | +0.34 | Iron | $Fe = Fe^{2+} + 2e^-$ | −0.44 |
| Hydrogen | $H = H^+ + e^-$ | 0 | Chromium | $Cr = Cr^{3+} + 3e^-$ | −0.71 |
| Lead | $Pb = Pb^{2+} + 2e^-$ | −0.13 | Zinc | $Zn = Zn^{2+} + 2e^-$ | −0.76 |
| Tin | $Sn = Sn^{2+} + 2e^-$ | −0.14 | Aluminium | $Al = Al^{3+} + 3e^-$ | −1.66 |
| Nickel | $Ni = Ni^{2+} + 2e^-$ | −0.25 | Magnesium | $Mg = Mg^{2+} + 2e^-$ | −2.37 |
| Cadmium | $Cd = Cd^{2+} + 2e^-$ | −0.40 | Potassium | $K = K^+ + e^-$ | −2.93 |

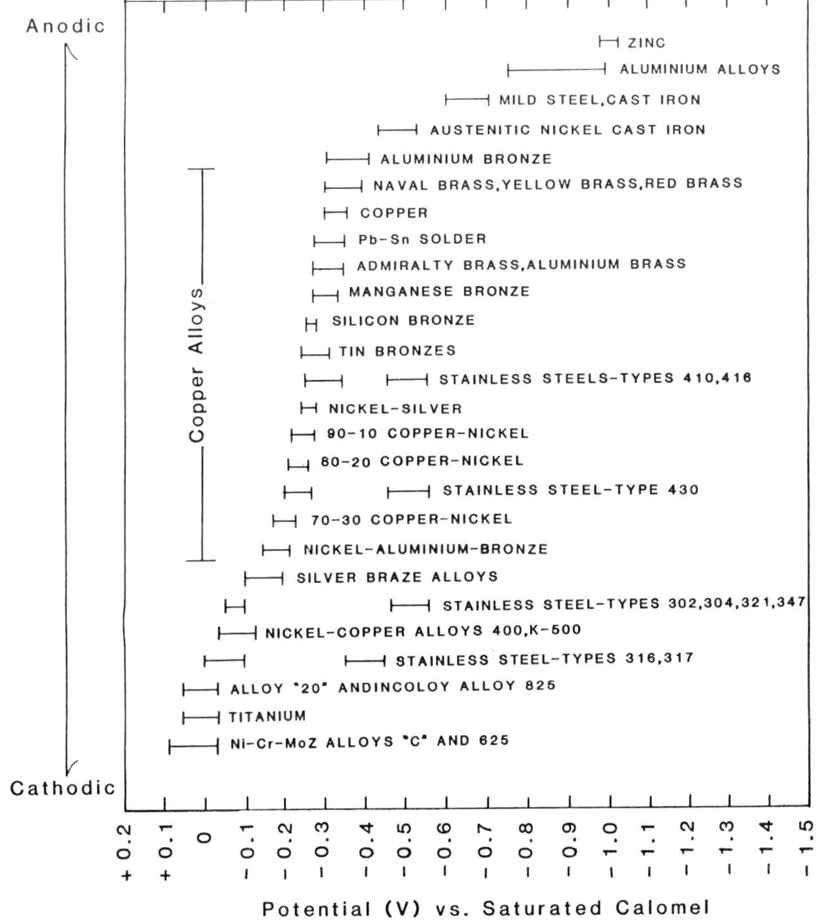


*Fig 3.7* Galvanic relationships in flowing seawater (2.5 – 4.0m $s^{-1}$; 10-27°C). Both active and passive potential ranges are given for stainless steels. (Redrawn from Woods Hole Engineering Associates, 1984).

Some metals are particularly unreactive and are thus relatively resistant to corrosion (*eg* copper, tin), whilst others, although subject to corrosion by virtue of their high electrode potentials, are protected by the formation of a surface oxide coating when immersed in water (*eg* aluminium, zinc). The corrosion resistance of a metal may also be improved by alloying it with another, more corrosion resistant material (*eg* the addition of 12-20% chromium in stainless steels, or copper in ferritic steel). Cathodic protection by sacrificial corrosion, in which a base metal is connected to the part to be protected and thus corrodes in its place, is a technique commonly employed to protect marine installations such as pipelines and ships. Magnesium alloy anodes, containing 6% aluminium, 3% zinc and 0.2% manganese are widely used for this purpose (Chong, 1977). Another technique to suppress corrosion is the use of protective coatings. Metallic coatings may be applied by dipping, electroplating, spraying, cladding or cementation, whereas inorganic oxide, phosphate or chromate coatings are formed by chemical treatment. A variety of paints, varnishes, plastics, lacquers and rubber compounds may also be used to isolate the metal from the corrosive environment, although few of the above protective coating methods are appropriate for metal meshes.

In summary, the most suitable materials for construction of rigid mesh cages, from a corrosion-resistance point of view, are those with low electrode potentials. Alternatively, a coating or sacrificial corrosion technique must be employed to protect the mesh. Unfortunately, many of the most eligible materials are unsuitable, because of their cost, weight, strength or brittleness, and as a result only three types of metal mesh have been widely employed in the fabrication of cages:-

(i) 90:10 copper nickel wire and expanded metal mesh (alloy CA-706). Although a range of copper nickel (Cu-Ni) alloys has been tested (Ansuini and Hugenin, 1978) 90:10 Cu-Ni has proved the most appropriate for use in aquaculture, combining structural strength, lightness and corrosion resistance with antifouling properties. Although there have been no controlled comparative tests of durability of wire and expanded Cu-Ni mesh materials, the latter is probably superior since there are no crevices in which corrosion can occur. Expanded Cu-Ni mesh is also widely available in a variety of mesh sizes (3mm to 5cm) and gauges. However, almost all of the research and development work has been carried out on hexagonal mesh material (gauge 0.89mm, strand width 1.27mm), with a nominal mesh size of 1cm. This material weighs 1.86kg $m^{-2}$ and has an ultimate yield strength of 3.3–4.3 x $10^8$ N $m^{-2}$ (Woods Hole Engineering Associates, 1984). Most experimental and commercial designs to date have used material with the above specifications.

The long term stabilised corrosion rate for 90:10 copper nickel is around 2.5 x $10^{-3}$mm $y^{-1}$ or less in flowing seawater conditions in a temperate climate (Effird, 1977).

Quantitative data on rates of fouling has been published by Hugenin and Ansuini (1984) and suggests that under typical temperate flowing seawater conditions fouling will rarely cause more than a 10% blockage of the meshes.

(ii) Galvanised steel weldmesh or chain link netting. The raw material for these types of meshes is cold-drawn mild steel wire which should be galvanised with zinc by the hot-dipping process. The wire can then be woven to form diamond mesh chain link, or welded together to form square mesh (*NB* galvanising of the latter is usually carried out after welding). For equivalent mesh sizes, chain link is generally made of finer gauge wire, and is consequently lighter but less strong (*Table 3.3*).

The service life of galvanised steel mesh in a particular environment is dependent upon (i) the type of mesh, and (ii) the quality of galvanising. According to Milne (1970, 1972), chain link has a far shorter service life compared with weldmesh due to abrasion at the links accelerating corrosion. Even using materials galvanised to a high standard, he found that chain link retained only 50% of its initial strength after 1 year's sea immersion, whereas weldmesh made of the same material after immersion for the same period retained 80% of its initial strength. Kuwa (1983) has shown that the service life of the galvanisation is dependent upon its thickness as well as the method of application (hot-dipping is superior to electroplating or spraying). In Japanese waters it was found that at 400g $m^{-2}$ the service life was about 15 months, with 260g $m^{-2}$, about 10 months, and with 250g $m^{-2}$ the galvanising lasted 9 months. After consumption of the galvanised coating, corrosion of the steel wire was rapid — 0.4-0.6mm $y^{-1}$.

In practice, galvanised chain link is replaced each 12-15 months (Milne, 1979). However, trials conducted by Kuwa (1983) have shown that cathodic protection using aluminium alloy anodes can effectively double the service life of the material in seawater. Details of the anodes are shown in *Fig 3.8* and specifications given in *Table 3.4*.

The fouling rates for galvanised steel mesh materials are shown in *Table 3.5*.

(iii) plastic-coated galvanised chain link or weldmesh. The chain link

*Table 3.3* The physical properties of various rigid mesh materials used in cage fabrication.

| *Material* | *Mesh size (mm)* | *Gauge (mm)* | *Density ($kg\ m^{-2}$)* | *Strength* | *Source* |
|---|---|---|---|---|---|
| Netlon | 50 | 3.3 | 0.34 | — | Milne, 1970 |
| 90:10 Cu–Ni expanded metal | 10 | 1.3 | 1.86 | 3.3–4.3 × $10^8$ $Nm^{-2}$ [a] | Woods Hole Eng. Associates, 1984 |
| Galvanised steel weld mesh | 25 | 2.5 | 3.40 | 205kg[b] | Milne, 1970 |
| Galvanised steel chainlink | 25 | 2.0 | 2.03 | 127kg[b] | Milne, 1970 |
| | 32 | 2.6 | | | Kuwa, 1983 |
| | 45 | 3.2 | | | Kuwa, 1983 |
| | 55 | 4.0 | | | Kuwa, 1983 |
| Plastabond (PVC coated chainlink) | 76 | 2.5 | 3.25 | 127kg[b] | Milne, 1970 |

a — Refers to yield strength of a panel
b — Refers to tensile wet break strength

*Table 3.4* Details of anodic protection for two fish cages (modified from Kuwa, 1983).

| *Scope of protection* | *Submerged metal mesh bag and frame* | |
|---|---|---|
| *Species, 1st season* | *Yellowtail* | *Japanese striped knifejaw* |
| *Species, 2nd season* | *Japanese striped knifejaw* | |
| Protective area | 119m$^2$ | 135m$^2$ |
| Current density | 130mA m$^{-2}$ | 110mA m$^{-2}$ |
| Current required | 15.5A | 14.9A |
| Anode material | aluminium alloy | aluminium alloy |
| Anode specifications | 8 × 9.30kg pieces | 8 × 5.67kg pieces |
| Anode installation | suspension type | suspension type |
| Date of installation | 4 months after cage installed | 10 months after cage installed |

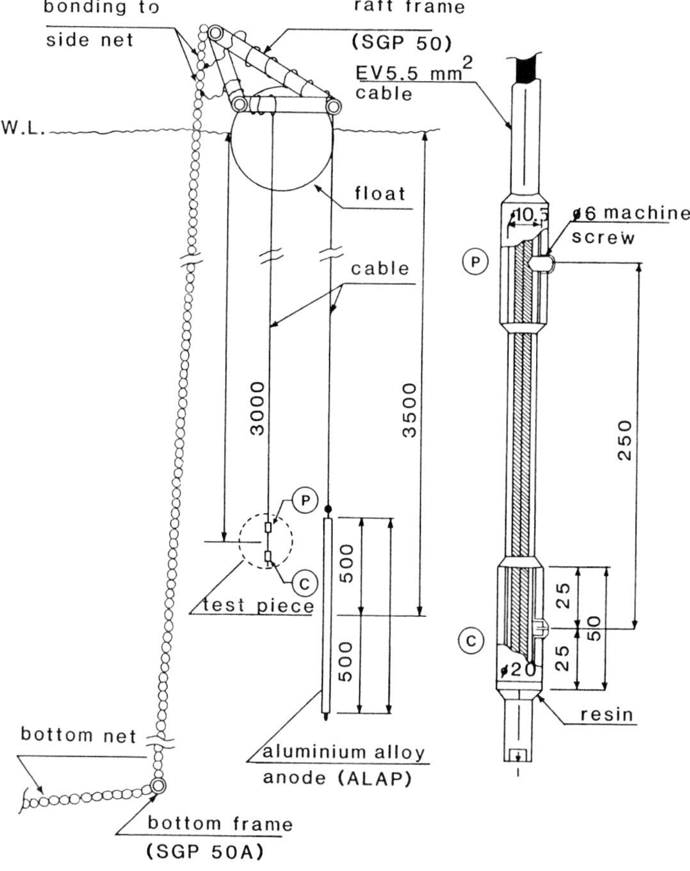


*Fig 3.8* Installation of cathodic protection system of aluminium alloy anode (p) and test piece (c) for wire netting cathodic protection (from Kuwa, 1983).

form is typically fabricated from polyvinyl chloride (PVC) coated galvanised steel wire which is woven into diamond mesh netting (Milne, 1972), whereas the plastic coated weldmesh is manufactured in a similar method to galvanised steel weldmesh and then coated in PVC. Comparisons of 25mm rigid mesh netting show that PVC coated chain link is similar in weight to galvanised weldmesh and has similar strength characteristics to galvanised chain link (*Table 3.3*). It is, however, considerably less resistant to fouling (see *Table 3.5*).

### 3.4.1.3 *Design criteria*

The function of the bag component of a cage is to hold the fish securely whilst permitting sufficient water exchange to replenish oxygen (and, in some cases, food) and remove potentially harmful toxic metabolites. It is also important that the cage volume remains relatively resistant to deformation by external forces, which would not only cause crowding and stress, but may also injure or cause mortalities amongst valuable stock.

Two principal types of force act on the cage bag; static and dynamic. Static loads act vertically and are imposed by the weight of the bag plus fouling and, in the case of net and small rigid mesh structures which are usually lifted during harvesting, by the biomass of fish. Static loads can be estimated from the area and density of netting or rigid mesh materials used, the weight of fouling, fish biomass and the quantities of ropes, weights of frame components used to strengthen and stiffen the structure.

However, whilst it is important to quantify static loads for design of the flotation and mooring systems, these are less critical in determining the design of the bag.

The more important dynamic forces impinging on the cage bag are caused by currents which act horizontally, although wind-induced surface waves

*Table 3.5* Seasonal weight increases for various fabrics due to marine fouling (from Milne, 1972).

| *Fabric* | *Wt kg m*$^{-2}$ | *July* | *Multiplication factor for weight increase* | |
|---|---|---|---|---|
| | | | *September* | *November* |
| Nylon | 0.23 | 2 | 85 | 108 |
| Ulstron | 0.34 | 2 | 64 | 110 |
| Courlene | 0.20 | 2 | 85 | 126 |
| Polythene: | | | | |
| (a) standard | 0.18 | 2 | 112 | 200 |
| (b) cupra-proofed | 0.18 | 1 | 44 | 94 |
| Netlon | 0.34 | 1 | 36 | 48 |
| Plastabond | 3.25 | 0.75 | 10 | 13 |
| Galvanised: | | | | |
| (a) chain-link | 2.03 | 0.3 | 0.5 | 2.75 |
| (b) weldmesh | 3.4 | 0.3 | 0.5 | 2.5 |

acting on the collar will, of course, induce some degree of vertical dynamic loading, whilst floating objects such as driftwood, jellyfish or water hyacinths can block portions of the mesh or netting, thereby increasing drag. Quantification of currents and the response of the materials can help predict how a particular design will perform in terms of water exchange and deformation of the bag and can aid in the design of frame members for rigid mesh materials and in rigging systems for netting. Data generated can also be used in the computation of moorings.

Water flowing through a mesh or netting panel imposes a load on the panel which in turn exerts additional dynamic loadings on the supporting frame, collar and mooring system. The load on a panel has been shown to be dependent upon the nature of the material used, the shape and size of the mesh, velocity of current and density of water such that:-

$$F_c = C_d \rho V^2 A/2$$ (Kawakami, 1964)

where $F_c$ = the force applied to the panel by the current (kg); $C_d$ = coefficient of drag of the material (dimensionless); $\rho$ = density of water; V = current velocity ($ms^{-1}$); and A = projected area of the mesh members ($m^2$). The latter parameter is calculated from the panel area, number of meshes and length and diameter of the mesh members. $C_d$ values have been experimentally determined using panels fabricated from a variety of clean and fouled materials (*Table 3.6*). For materials other than expanded metal meshes, $C_d$ values, which closely approximate those determined experimentally, can be calculated using the following formulae:-

$$C_d = 1 + 3.77\,(d/a) + 9.37\,(d/a)^2$$ for knotted materials,

and

$$C_d = 1 + 2.73\,(d/a) + 3.12\,(d/a)^2$$ for knotless materials

(Milne, 1970), where d = yarn diameter (mm), and a = bar length (mm). Note that $C_d$ values increase with bar width and consequently are greater for fouled than unfouled materials. The equations given above also show that $C_d$ values are considerably greater for knotted than knotless materials. The lowest $C_d$ value observed was determined from tests carried out by Woods Hole Engineering Associates (1984) on 90:10 Cu-Ni mesh, although for design purposes it is suggested that a more conservative value of 0.56, based on fouling trials, is used.

From experimental work, the above theoretical relationships have been found to be valid for panels of materials where the mesh members do not occlude more than 40% of the panel area, and for currents which flow perpendicular to the panel (Woods Hole Engineering Associates, 1984). There is also evidence that the $C_d$ values are independent of current velocity over the range typically encountered at fish farm sites (0.33–1.87$ms^{-1}$). However, because expanded metal meshes are non-symmetrical, the angle of an attack of the currents impinging on the mesh is important. In *Table 3.7*, the effect of current angle of $C_d$ is shown, and it can be seen that there is a greater than 3-fold variation. Interestingly, although mesh blockage is least at

*Table 3.6* $C_d$ values determined for perpendicular current forces acting on various netting[a] and rigid mesh materials before and after fouling[b]. Data from Milne (1970) and Woods Hole Engineering Associates (1984).

| *Material* | *Mesh type* | *Mesh size*[c] *(mm)* | *Unfouled yarn diameter (mm)* | $C_d$ | *Fouled yarn diameter (mm)* | $C_d$ |
|---|---|---|---|---|---|---|
| Nylon | Diamond | 50 | 2.3 | 1.42 | 10.2 | 3.99 |
| PP (Ulstron) | Diamond | 50 | 2.5 | 1.47 | 10.2 | 3.99 |
| PE (Courlene) | Diamond | 50 | 1.9 | 1.33 | 8.9 | 3.46 |
| PE | Square | 50 | 1.5 | 1.26 | 7.6 | 2.95 |
| PE (cupra-proofed) | Square | 50 | 1.5 | 1.26 | 5.1 | 2.13 |
| Netlon | Square | 50 | 3.3 | 1.19 | 7.6 | 1.48 |
| 90:10 Cu–Ni expanded metal | Hexagonal | 10 | 1.3 | 0.41 | — | — |
| Galvanised steel weldmesh | Square | 25 | 2.5 | 1.30 | 3.3 | 1.41 |
| Galvanised steel chainlink | Diamond | 25 | 2.0 | 1.24 | 3.8 | 1.48 |
| Plastabond (PVC | Diamond | 25 | 2.5 | 1.30 | 5.1 | 1.67 |
| coated chainlink) | Diamond | 76 | 2.5 | 1.09 | 6.4 | 1.25 |

a Refers to knotted netting
b Refers to fouling accumulated after two months' immersion in Scottish sea conditions
c Mesh size for netting taken to be 2 × bar length

*Table 3.7* The effects of current angle on the drag coefficient ($C_d$) of 10mm 90:10 Cu–Ni mesh. Current velocity = 0.267m $s^{-1}$ (modified from Woods Hole Engineering Associates, 1984).

| Angle | −50 | −40 | −30 | −20 | −10 | 0 | +10 | +20 | +30 | +40 | +50 |
|---|---|---|---|---|---|---|---|---|---|---|---|
| Cd | 0.23 | 0.41 | 0.29 | 0.48 | 0.51 | 0.47 | 0.73 | 0.62 | 0.63 | 0.58 | 0.59 |

an angle of −20° and greatest at −70°, these angles did not correspond with the minimum and maximum $C_d$ values. Gularte and Hugenin believe this to be due to the creation of complex flow regimes around the mesh members as a result of their configuration (Woods Hole Engineering Associates, *ibid*). Attack angle probably influences $C_d$ values for other materials, although to a lesser extent.

Using the equations presented above, it is possible to produce current velocity versus loading curves of the type shown in *Fig 3.9*, for different materials. However, whilst these relationships may be valid for rigid mesh materials and for small panels of netting mounted in a frame, they are unlikely to be representative of the actual forces which impinge on cage nets, which, unlike rigid mesh materials, undergo deformation in current flow. It is not uncommon for deformation to result in a loss of 25% of the cage bag volume (see Section 7.3). Tomi *et al* (1979), in a study of the effect of currents on a cage bag showed that the forces acting on the bag increased with current velocity in a fashion similar to that shown in *Fig 3.9*, resulting in a marked

(a) clean

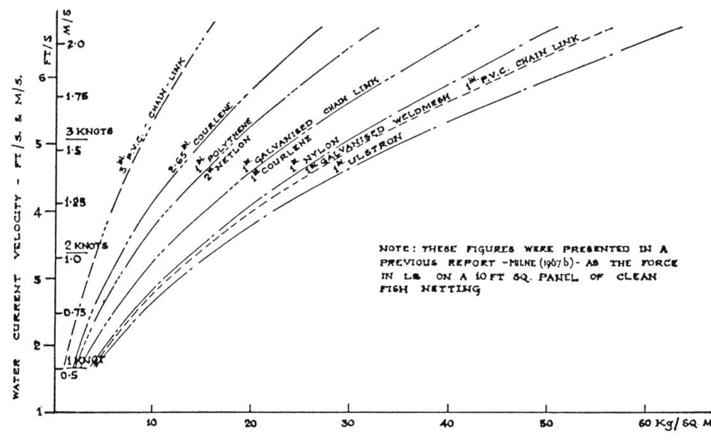

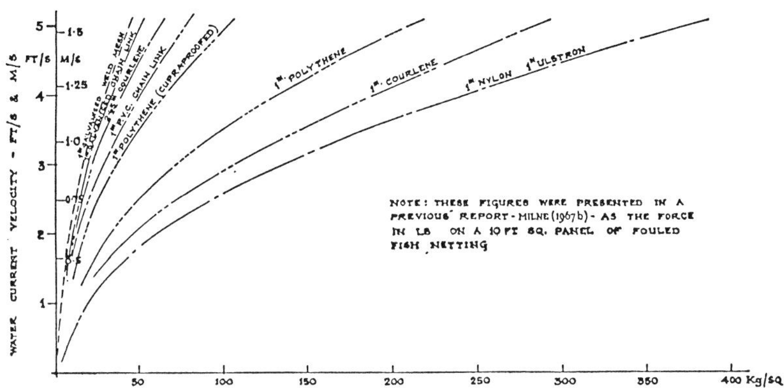


*Fig 3.9* Current forces (kg $m^{-2}$) on (a) clean and (b) fouled netting (from Milne, 1972).

decrease in net volume (*Fig 3.10*). However, whilst the effect of fastening weights to the bottom four corners of the bag resulted in a significant amelioration of the problem, horizontal forces acting on the cage bag increased by between 2 and 6-fold. Thus, the horizontal forces that may be expected to act on a net cage bag will be a fraction of those computed using the Kawakami (1964) and Milne (1970) equations, the actual value depending not only on material and current velocity, but also on the amount of weighting used to maintain bag shape.

The type of material, mesh size and degree of fouling also influence water exchange through the cage and are thus important determinants of stocking density. Studies carried out in Japan and the UK during the 1960s and 1970s have demonstrated that the transmission ratio, T (ratio of internal to external currents, $V_i$: $V_e$) falls with decreasing mesh size and with increased fouling (Hisoaka *et al*, 1966; Milne, 1970; Inoue, 1972; Edwards and Edelstein, 1976; Wee, 1979). In heavily fouled empty cages, fabricated from 13mm nylon mesh, Wee (*ibid*) found that T could be as low as 0.13. However, much of the above work was carried out with empty cages, over a limited range of $V_e$ values, and sometimes using current meters which were operating outside their design limits, and recent research has suggested that these studies oversimplify the situation. For example, Chacon Torres (1984) has demonstrated that the caged fish can greatly modify the rate and direction of water flow through a cage by their behaviour.

From a timed sequence of photographs of the dispersion of dye stuffs in empty and stocked $1m^3$ cages, he found that the exchange pattern changed

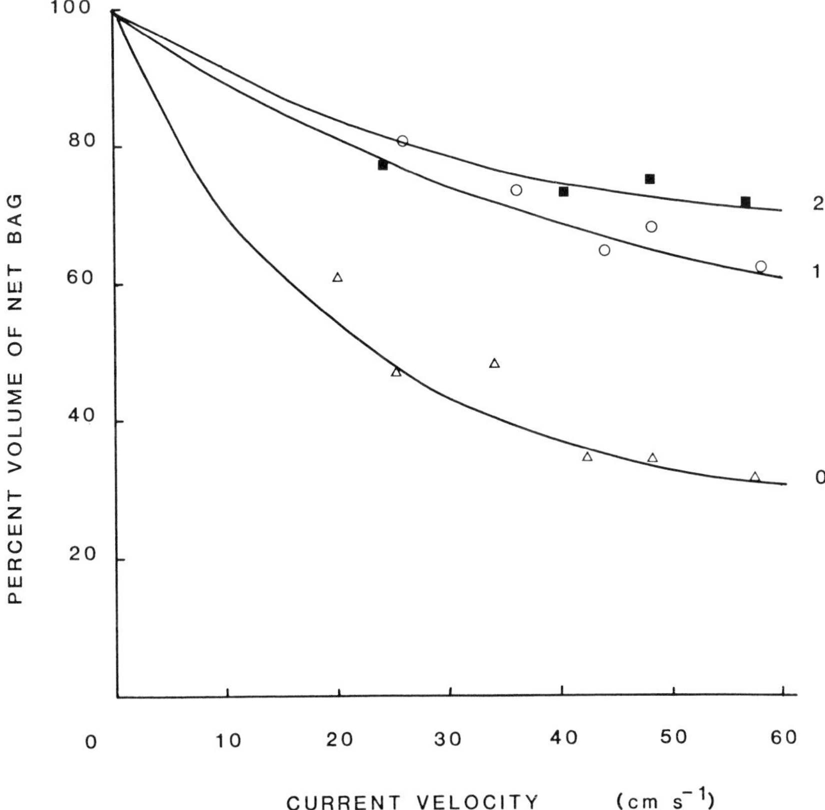


*Fig 3.10* Relationship between volume of cage bag and current velocity, with no weights (0), with 4 × 52 kg weights (1) and with 4 × 104 kg weights hanging from the bottom corners of the bag (modified from Tomi *et al*, 1979).

from unidirectional to multidirectional and estimated that the exchange rate increased by a factor of approximately 4 when rainbow trout were present. During feeding, the change in flow pattern was even more pronounced.

The behaviour of fish in a cage will vary with species and size and also be dependent upon environmental cues such as time of day, weather, current speed, salinity and temperature (see Chapter 2). Whilst T may be largely independent of fish behaviour in fast flowing conditions and in extremely large cages, behaviour is likely to be an important regulator of water exchange in most situations.

In summary, the values given in the literature for forces exerted on netting and for T may be of limited value, but the studies emphasise the importance of using as large a mesh size as possible and minimising the degree of fouling. The effects of weights on net deformation and horizontal forces acting on the netting are also well illustrated. Detailed calculations are unnecessary for the construction of small cages whilst most large net cage bags, which are usually fabricated by commercial netmakers, have evolved through a continual process of feed-back from fish farmers and technological innovations, rather than from information generated by experimenters. Experimental data, on the other hand, has proved invaluable for the design of large rigid mesh bags. For example, Hugenin and co-workers used data derived from laboratory and field trials to establish the range of forces likely to act on various expanded metal mesh materials, and employed these values to estimate the deformation of the panels and the forces that would be transferred to the frame members (Woods Hole Engineering Associates, 1984). Thus the optimum materials and dimensions for panels and frames were determined.

A great deal more research would have to be carried out before such an approach could be applied to the design of net cage bags.

### *3.4.1.4 Construction*

As rigid mesh materials are usually only available in a restricted range of panel sizes or roll widths, this can dictate the size, shape and design of the cage unless sophisticated engineering techniques are employed. Rigid mesh cages usually require some sort of supportive framework, otherwise they buckle and crack. Small designs can be assembled by simply nailing or binding sheets cut to the appropriate size to a previously fabricated frame. A top panel, usually hinged, is often included in rigid mesh designs. One deficiency of plastic-coated or galvanised materials is that the cut edges become prone to corrosion and must be treated with some sort of coating for protection. Similarly, bending sheets at right angles can accelerate corrosion not only by damaging any surface coating that is present, but also by inducing stress.

Small, simple frames can be constructed from a variety of woods, metals or synthetic materials (*Fig 3.11*), and the properties of many of these materials are discussed in Section 3.4.2.

Care must be taken over the choice of fasteners, not only to ensure that they are sufficiently strong, but also to avoid creation of galvanic cells. Plastic or nylon ties, some of which are purpose made, and synthetic twines are inert but are less suitable for use in larger cages than metal fasteners. Metal fasteners with similar electrode potentials to the metal mesh should be used. Hugenin and Ansuini employed nickel aluminium bronze fasteners ($E_o = -0.16$ to $-0.22$v) for 90:10 Cu-Ni sheets ($E_o = -0.15$ to $-0.30$v) as the electrode potentials were very similar (Woods Hole Engineering Associates, 1984). Stainless steel fasteners which were electrically isolated by plastic bushes ultimately proved to be unsatisfactory, due to inadvertent electrical contact. If such a system must be employed, each joint should at least be checked with a current meter prior to the cage being installed.

Although some larger rigid mesh cages rely on readily available galvanised steel or 90:10 Cu-Ni pipe for the framework, commercial development of modular systems has resulted in the use of specially designed and fabricated frame members. Hugenin and co-workers used a pultruded form of polyester and glass composite (fibreglass) which they found ideal: it was light, strong, highly resistant to weathering, and could be produced in an almost infinite variety of shapes and sizes. (Woods Hole Engineering Associates, *ibid*). Tests to determine tensile and flexural strength were carried out on a number of samples, some of which had been subject to various degrees of weathering, and showed that although the material did not significantly deteriorate, its performance was to some extent governed by whether it was wet or dry. Design specifications, therefore, had to be concerned with properties when wet.

Eventually, C-sectioned 5.1 × 1.4 × 3.2cm pultruded fibreglass was chosen as the most appropriate for construction of panel frames. Panels are constructed in a suitably equipped workshop by stretching the 90:10 Cu-Ni sheeting around the four frame members as shown in *Fig 3.12*, to form a standard panel 30cm wide, of variable length. Details of panel fabrication can be obtained from the manufacturer. Panels are joined together to form a rectangular or circular enclosure of variable size. Electrically-isolated galvanised steel reinforcements may be used at corners, joints, and at points where loads become concentrated.

Synthetic fibre net bags are usually designed with an area of freeboard (an area which protrudes above the water surface) to prevent fish jumping out. The height of freeboard is determined by species, some such as the silver carp, being tremendous leapers. However, the influence of wind forces on exposed netting should be borne in mind (see later). A compromise is to use a top net — also important for deterring predators — as well.

Synthetic fibre netting is sold in rolls which come in a variety of widths. Hence unless prefabricated bags are ordered from a commercial company, netting must be cut and assembled into cage bags on the farm. Prior to assembly, a large, reasonably level area, clear of twigs, bushes and debris must be available. The fabrication of simple square or rectangular cage bags from fine-mesh (2-5mm) mosquito netting (hapa), suitable for a lake-based tilapia hatchery (see Chapter 8) is comparatively simple, and an example is shown in *Fig 3.13*. The dimensions of the bag must be decided beforehand, and it is simplest and most economical if the total depth of the bag, including

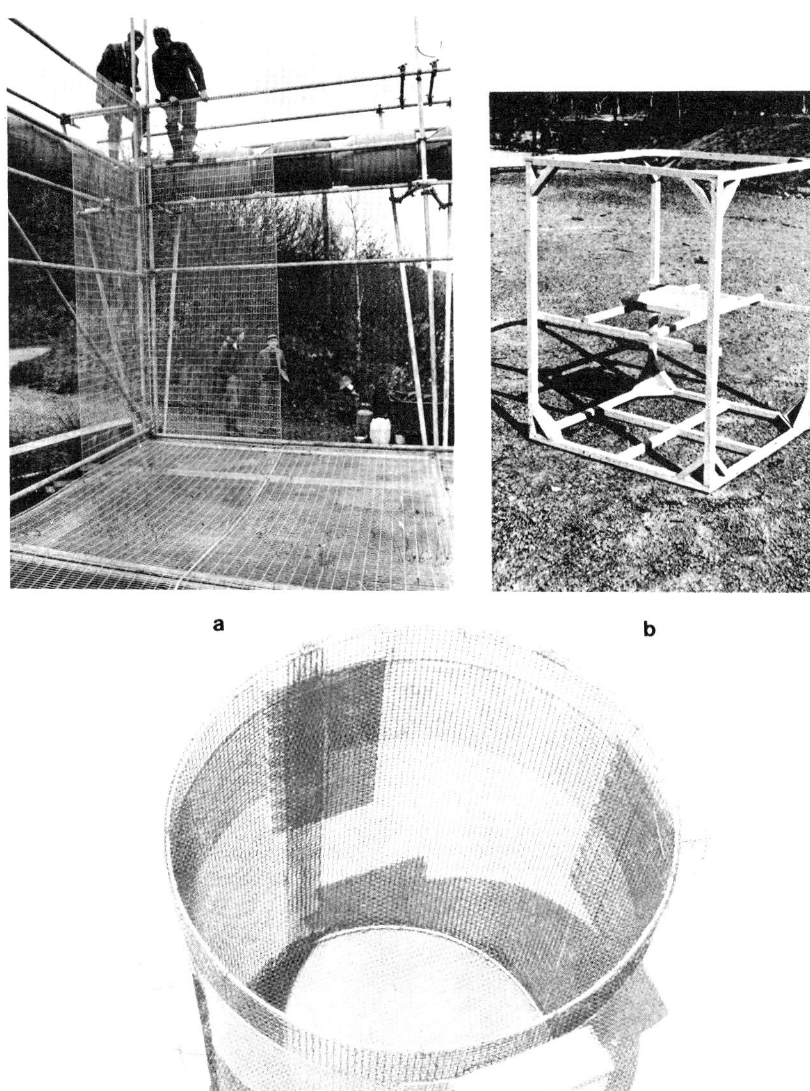

*Fig 3.11* Examples of rigid cage frames (a) 6 × 6 × 3.6m galvanised scaffolding cage for Atlantic salmon, Scotland (from Milne, 1979); (b) 1 × 1 × 1.3m mahogany cage frame with reinforced corners, for experimental culture of tilapias, Lake Kossou, Ivory Coast (from Coche, 1979); (c) $3m^3$ experimental cage for use in marine waters, fabricated from fibreglass hoopnet rings and vinyl coated steel mesh (from Swingle, 1971).

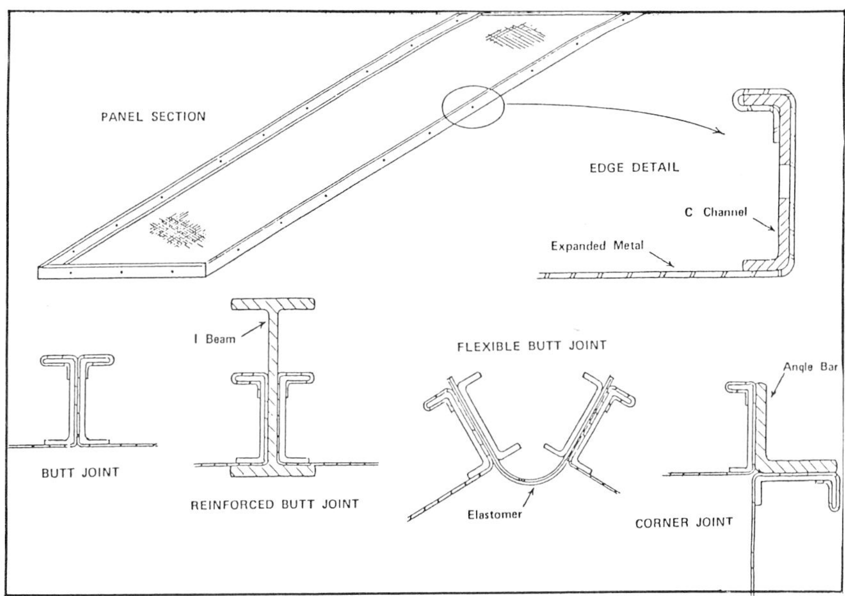

*Fig 3.12 (a)* Construction details of mesh-wrapped 90:10 Cu-Ni panels, and examples of ways in which they can be joined together to form units various shapes and sizes (from Woods Hole Engineering Associates, 1984). (b) 6.3 × 3.2 × 3.2m 90:10 Cu-Ni meshcage fabricated from panels (courtesy J Hugenin).

freeboard, and width of the base are a function of the diameter of the roll of material. For example, in *Fig 3.13*, the depth and width of the bag are twice the width of the roll. Double stitching of the edges in this case is recommended, and can be done using an ordinary domestic sewing machine fitted with strong nylon thread.

For larger mesh netting, cutting and assembly is a little more difficult.

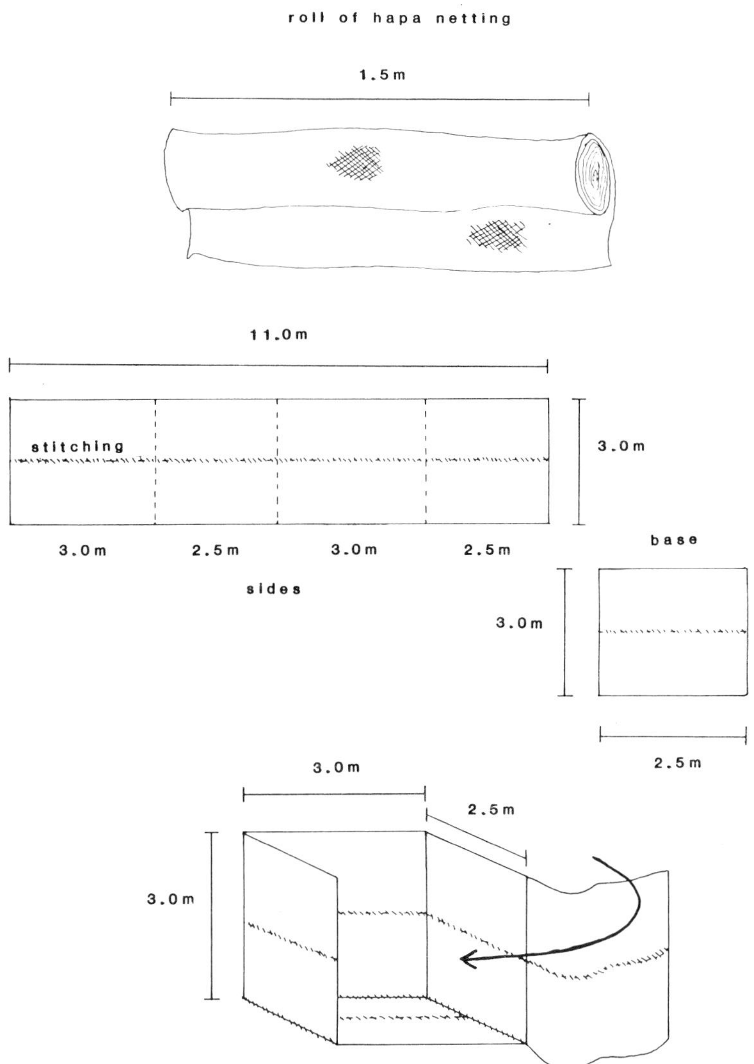


*Fig 3.13* Assembly of a simple hapa net cage. The sides are cut as one piece and are formed from two lengths of material stitched along the middle. The base is made separately (not to scale).

Most cage bags are fabricated from knotless netting which can be hung on the square or diamond. The former is reportedly easier and more economical for small mesh (<6mm bar length) sizes. However, if the netting is to be hung so that the meshes adopt a diamond shape, it is impossible to simply measure and cut lengths of material from a roll corresponding to the sizes of the panels which make up the cage bag without first of all deciding on the hanging ratio. Because of its shape, diamond mesh netting can be hung in a variety of ways which dictate the horizontal and vertical mesh openings. For fish cages, however, the netting is often hung with a hanging ratio (r) of 0.3 *ie* the horizontal opening of the mesh is equivalent to 70% of the stretched mesh value. For practical purposes, the following formula is used to calculate the length of netting:-

$$L_s = L_d/1-r$$

where $L_s$ = length of stretched netting (m), and $L_d$ = desired length (m). Thus, for a 4 x 3m cage bag, a total length of 14m of netting is desired. Assuming $r = 0.3$, $L_s = 14/0.7 = 20$m.

For the depth of netting:-

$$D_s = D_D/2r-r^2$$

where $D_D$ = desired depth (m) and $D_s$ = depth of stretched netting (m). An alternative to measuring the stretched netting is to count the number of meshes (M). The number of meshes can be computed using:-

$$M = D_D/(2r-r^2)ml$$

where ml = stretched mesh length (m).

Details of how to measure and cut netting for a circular cage bag are given in IDRC/SEAFDEC (1979).

After calculations of the dimensions of the side and bottom panels have been made, cutting from the roll should be planned so as to minimise waste. If possible, the four sides should be cut in one piece, as in *Fig 3.13*. An additional advantage of using a knotless material is that seams can be machine stitched, and hence are more resistant to tearing than the panels. However, machine stitching is impossible without specialised machinery and so unless the net bag is commercially fabricated, stitching of the sides to the bottom panel must be done by hand using a small net needle and twine which is similar to that of the netting yarn.

The extent and nature of the rigging required depends primarily upon the size of the cage bag, whether it is to be used in a fixed or floating type, and the design of the collar. For small bags less than 50m$^3$ volume, ropes should be attached to the outside of all edges, whilst for larger designs, rigging may also have to be attached at intervals along the panels to help strengthen and stiffen the structure and to facilitate lifting during harvesting and net changing. Usually 6-10mm PE or PP rope is used. There are many different ways to attach netting to a rope and these are discussed by Garner (1962). Usually the rope is simply lashed to the net using a twine which again is similar in diameter to that of the netting yarn. Each mesh is picked up by the needle and bound to the rope with a blanket stitch. Every 5-10cm, a clove hitch is tied to secure the rope firmly. For hapa netting, a twine of around 23tex is used, and a clove hitch tied every few centimetres.

Some authors recommend threading the rope directly through the meshes of the netting and securing it at each mesh with a clove hitch, whilst others suggest running a balch line through the netting in the same way, and attaching the rigging to the balch line (PPD, 1981; Alvarez, 1982). A balch line is an intermediate rope, typically 2-5mm in diameter, of low breaking strain, which reduces the effect of tension from the main rope on the netting, thus minimising the risk of tearing.

On the larger cage bags, ropes are stitched to the outside of the walls and floor of the bag every few metres, in the same way as described above. For floating cages, loops should be included not only at all corners, but also at metre intervals on all ropes, to facilitate lifting and net changing. The loops should either be knotted and whipped as shown in *Fig 3.14*, or spliced. Weights can be hung from the rigging which runs along the floor seam, in order to help the bag hang properly in the water or alternatively, a metal frame can be fabricated and lashed to the exterior of the floor seam. Design calculations of weights required for cage bags are generally so simplistic as to be of little practical use, and so the number and size of weights required is generally determined through trial and error. Alternative methods of rigging are given in the Appendix in IDRC/SEAFDEC (1979).

Bags fabricated for fixed cages should have at least 0.5m of free rope left on

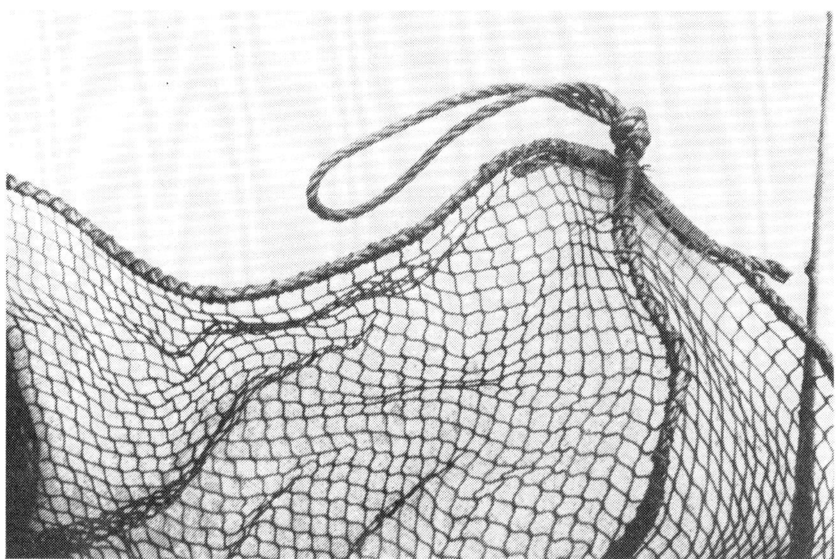

*Fig 3.14* Detail of corner of net bag illustrating how the rope is stitched to the netting along the top and corner seams of the bag, one stitch per mesh, and that there is a loop provided to facilitate attachment to the collar.

each rigging line to facilitate fixing to the frame.

Top nets, usually fabricated from large mesh knotted PE or nylon monofilament are cut to size and used to deter predators (see Section 7.9).

### 3.4.2 Cage collars and support systems

#### *3.4.2.1 Fixed cages*

Here the mesh bag is supported by posts driven into the substrate. In those parts of the world where this type of cage is used, bamboo is commonly employed for the supports. Bamboos are grasses which are widely distributed throughout the tropics, subtropics and mild temperate areas of the world and are particularly common in Southeast Asia, China, India and the Pacific region. The woody, hollow stems (culms) grow in branching clusters from thick underground rhizomes. They are fast-growing, the culms of many species reaching 15m + in 2-3 years. However, of the 1,000 or so species only around 20-30 are suitable for construction purposes, and include *Dendrocalamus spinosa*, *D. hamiltonia*, *D. merillianus*, *Bambusa blumeana*, *B. vulgaris* and *Giganthocloa gigas*.

The general anatomy of a typical bamboo culm is shown in *Fig 3.15*, and mechanical properties are summarised in *Table 3.8*. When compared with other materials on a weight for weight basis, bamboo is exceptionally resistant to bending and tension forces but, due to its unique structure, it is much weaker in terms of resistance to shear forces. In tests, it was found that once the ultimate bending force was overshot, failure occurred not because the tensile strength of the fibre was exceeded, but because of loss of cohesion between fibres as a result of the material's low resistance to the shear forces generated (Janssen, 1981). Thus, under excessive bending forces bamboo splits but does not snap. Hence, although bamboos may have certain disadvantages for use as supports in a fixed cage, they are easily split and readily fabricated into a cage mesh (see Section 3.3).

The properties of bamboo vary with species, moisture content, age and position along culm. Culms of between 3 and 5 years old have been shown to have the best mechanical properties and dry (12% moisture content) specimens are more resistant to bending forces than green. Thus, proper seasoning is important. Position along the culm has also been shown to influence resistance to shear and bending, the top sections being significantly weaker but stiffer.

Despite its many excellent properties, bamboo suffers from one principal disadvantage: it has a comparatively short useful working life. It is susceptible to destruction by boring insects and also eventually becomes waterlogged, thus losing its flotation properties (see later). Moreover it rots comparatively quickly. Evidence suggests that the section at the waterline degrades most rapidly through the combined action of weathering and continual immersion and drying out. Thus in fixed cages it is the exposed section which tends to fail first. The concensus of opinion is that bamboo has a useful working life of around 18-24 months in freshwater and 12-18 months in seawater (IDRC/SEAFDEC, 1979).

*Table 3.8* Summary of mechanical properties of bamboo, compared with timber, concrete and steel. Data from Chong (1977) and Janssen (1981).

| | *Bamboo* | *Softwood* | *Hardwood* | *Concrete* | *Steel* |
|---|---|---|---|---|---|
| Density (kg $m^{-3}$) | 600–800 | 450–650 | 400–750 | 2,400 | 7,800 |
| Ultimate tensile strength (N $mm^{-2}$) | 200–300 | 20–110 | | 4 | 270–700 |
| E-value (tensile[1] strength) (N $mm^{-2}$) | 17,400–10,300 | 5,000–9,000 | 8,000–18,000 | 28,600 | 210,000 |
| Ultimate short-term bending stress (N $mm^{-2}$) | 84 | — | — | — | — |
| E-value (bending stress) (N $mm^{-2}$) | 20,500 | — | — | — | — |
| Ultimate shear stress (N $mm^{-2}$) | 2.25 | 11 | 14 | — | — |

1 E-value = Modulus of elasticity = stress/strain (δ/E) ≡ ratio of load per unit area to deformation per unit length.

Hardwoods and softwoods are considerably more resistant to weathering and rotting but are, of course, much more costly. In the Philippines, a common alternative to bamboo is the 'anahaw' palm (*Livingstonia rotundifolia*). Although this material costs at least five times as much as bamboo, it has a lifespan of three years and, since it is considerably stronger, only a fraction of the number are required as supports. The latter fact has also proved advantageous at sites where storms and water hyacinth problems occur (see Section 7.5).

The posts in a fixed cage must fulfill the same functions as the collar in a floating cage *ie* they must be able to resist the static vertical forces imposed by the weight of the net and working staff, and the horizontal dynamic forces exerted by wind, waves, and currents. Unfortunately, there has been insufficient experimental work carried out to date to provide any but the most general guidelines.

In view of the high compressive strength of bamboo (Janssen, 1981) static forces are likely to be negligible and of the dynamic forces that are exerted on netting, Milne (1970) has shown that currents are the most important. Current forces will cause deformation of netting and so supporting posts must be used at regular intervals. The methods that have been developed for computing spacings of supports for net barriers (Milne, *ibid*) cannot be readily applied to cages due to their 3-dimensional configuration. However, fish farmers have found that with bamboo culms, spacings of around 1-2m are more than adequate under most circumstances.

The posts supporting the net bag must be buried sufficiently deep to resist moving as a result of the wind, wave and current forces acting on the structure. Milne's method (Milne, 1970, 1972) for net barrier pile calculations, applied to a typical fixed cage design using bamboo support posts at 1m intervals suggests that a depth of 2m is sufficient to support the structure securely, even in soft puddle clay. This figure, however, is likely to be extremely conservative when applied to cages, considering that the support given by horizontal frame members directly opposing the prevailing currents has been ignored. In practice, depths of 1-2m have been found to be satisfactory, although this, of course, will vary depending on the nature of the netting, the prevailing forces, the dimensions, spacing and type of posts used, and the cohesive properties of the soil.

Assuming a buried depth of 2m, a water depth of 6m and that the bamboo extends for 3m or more above the waterline, culms of 11m+ are usually required. Ideally the bamboo should be well-seasoned, aged 2 y+, and should have a diameter of around 10-12 cm at the base and 3cm at the tip. Prior to installation, the culms should be cleaned of any sharp projections which occur at the nodes and which could damage the netting, and the node diaphragms or internodes of the section to be submerged should be perforated to facilitate sinking. These tasks are best performed using simple tools such as machettes and chisels at a workbase purpose-built by the water's edge (*Fig 3.16*).

The bamboos are floated out to the site on a raft which also serves as a construction platform (*Fig 3.17*). A team of 2 or 3 men is sufficient to construct a small fixed cage in a few hours. Posts are driven into the mud by a twisting motion, often with someone in the water to aid accurate placement. The corner posts of one side are fixed first, and a line stretched between the two to ensure that the posts lie along a reasonably straight line, and after one

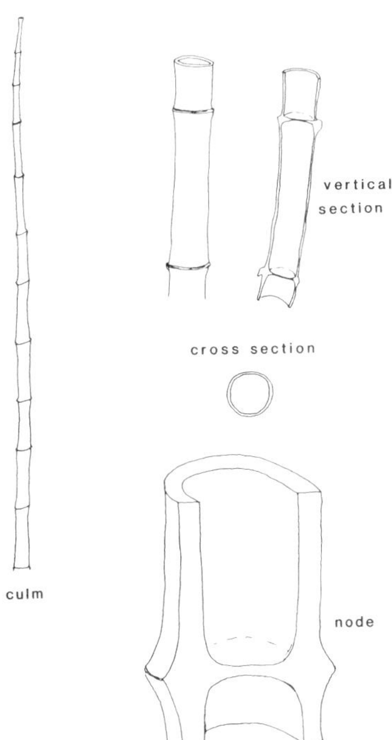


*Fig 3.15* Anatomy of a bamboo culm (redrawn from Janssen, 1981).

*Fig 3.16* Framework lashed together near the water's edge and adjacent to cage construction site, for cleaning and preparation of bamboos.

side has been finished, the others are completed in a similar manner. Horizontal supports, which utilise the top and middle sections of the culms, are lashed to the vertical posts — usually one on either side of a row — about 1.5-2.0m above the waterline and serve not only to strengthen the structure, but also as a narrow walkway. The completed framework should be around 0.5m longer on each side than the net bag dimensions to minimise the risk of abrasion. The cage net is then tied to the posts, the bottom rigging being secured by divers.

Where a shallow surface mud stratum is encountered, anchorage of the posts can be improved by driving bamboo pegs, 25mm diameter x 450mm long, through the internodes at the base of the culm to form two cross pieces, about 30cm apart (Alferez, 1982). Bracings to strengthen the structure may also be used at the corners and on the exposed side of the cage. Additional horizontals will stiffen the cage framework. If softwood or hardwood posts are used instead of bamboo, then notches can be cut into the trunk, and temporary rungs nailed in place so that a person can use his weight to help drive the posts into the substrate. Under difficult conditions, a raft equipped with scaffolding can prove useful (*Fig 3.18*).

### *3.4.2.2 Floating cages*

The diversity of floating fish cage designs was discussed in Chapter 2. The function of the collar is to support the cage bag in the water column and, particularly in the case of net bag designs, to help maintain the shape of the structure. Many cage collars also serve as work platforms. Both static and dynamic forces act on the collar. Static loads act vertically and comprise the weight of the bag, superstructure (including stanchions, freeboard and top nets, handrails, *etc*) and additional loads which may be expected during routine operations (*ie* weight of farm staff, feedbags and equipment such as graders, pumps, aerators, *etc*), and these must be quantified to design the flotation system. The following computation will serve as an example:-

Cage bag dimensions = 5 x 5 x 5m = $125m^3$

Material = nylon, raschel woven 24mm stretch mesh

Density = $0.24kg\ m^{-2}$

$\therefore$ Total weight of bag in air when clean and dry = 30kg.

Assume a 15-fold increase in weight in air due to fouling.

*Fig 3.17* Construction of fixed cage, training course, SEAFDEC, Binangonan, Philippines (a) corner posts are established first, and a line stretched between them to aid correct positioning of intermediates; (b) intermediate posts being driven into the substrate; (c) laterals are lashed in place to reinforce the structure; (d) completed cages.

Thus design weight of bag in air = 450kg.

Quantity of rope used for rigging = 90m.

Density = 5kg 100m$^{-1}$

∴ Total weight of rope in air when dry = 4.5kg.

Assume a 2-fold increase in weight in air due to fouling.

Thus design weight of rope = 9.0 kg

Maximum stocking density at time of harvest = 25kg m$^{-3}$ = 3125 kg

Assume 10% of fish may be supported by net during harvesting = 313 kg

Weight of superstructure and staff, equipment *etc* = 500 kg

∴ Total weight of cage in air = 1272 kg

Assuming a factor of safety at 1.25, the vertical design load in air = 1591 kg

In order to compute the weight of the cage in water, the densities of the different materials would have to be established, and in order that the cage floats, the static loads acting on the structure (*ie* the weight in water) must be counter-balanced by buoyancy forces. The buoyancy of the collar is dependent upon the upward force acting on those components which are wholly or partially immersed in the water, and is equal to the weight of water displaced. The buoyancy force can be calculated from data on the density of materials used to construct the collar:-

$$F_B = V_W Q_W - V_M Q_M$$

where $F_B$ = buoyant force (kg); $V_W$ and $V_M$ = volumes of water and the flotation material respectively (m$^3$) and $Q_W$ and $Q_M$ = the densities of water and the flotation materials (kg m$^{-3}$). For example, one cubic metre of water at 20°C weighs 1,000kg, and one cubic metre of bamboo weighs approximately 600kg (from *Table 3.8*). Thus, the buoyancy force acting on 1m$^3$ of bamboo in water = (1 × 1,000) — (1 × 600) = 400kg *ie* 1m$^3$ bamboo can support 400kg.

In some cage designs the flotation system is an integral part of the collar. The simplest type is contructed almost entirely from bamboo in which the bamboo not only serves as a rigid frame-work for maintaining the shape of the bag, but also supports the weight of the structure (see later). Other examples include gas and expanded polyurethane and polystyrene foam filled plastic and synthetic rubber pipes which form the basic framework of the collar on several of the more sophisticated commercially-produced cages. The alternative is to select dense materials such as steel and aluminium alloys for strength, and to add supplementary flotation such as steel drums or blocks of expanded synthetic polymers.

In theory it should be possible to design the flotation capabilities of a cage using data on densities of materials. In practice, however, it can prove

*Fig 3.18* Construction raft to help drive posts into difficult substrates.

difficult. The equation given above assumes an equal weight distribution around the collar and does not take account of situations where a considerable proportion of the vertical loading is acting on one side, or indeed at one point (*eg* at a corner) which frequently occurs on cages where collars serve as work platforms. The effect of cages being linked together is also ignored. However, for design purposes, it should be assumed that the cage is moored independently and that the staff, equipment *etc* are all standing on one side of the collar. The simplest way to calculate the required buoyancy is to assume that each side of the cage must be able to support not only its share of the loadings imposed by the net and superstructure but also the weight of staff and equipment. For example, if the weight of a cage in water is 1,000kg, and assuming that staff and equipment will weigh no more than 250kg, the buoyancy of each side of the cage should be:-

$$(1000 - 250)/4 + 250 = 438\text{kg}$$

For most materials, the specifications for quantifying the amounts required to ensure adequate flotation are readily available (*Table 3.8*). However, for materials such as bamboo, density is variable and depends upon species, age and method of seasoning. Moreover, because of its irregular and hollow shape, it is difficult to translate data on density per unit volume into flotation abilities per culm, although there are tables for estimating culm weight available from various tropical forestry research institutes (see also Appendix 3.1). The buoyancy of steel or plastic drums can be calculated from their weight and volume, using the formula given above.

Another difficulty which must be borne in mind if using wood or bamboo is the loss of buoyancy with time as a result of rotting and weathering. They can in part be protected by impregnation — preferably under pressure — with a preservative compound such as creosote or coaltar oil. Despite such measures, bamboo is unlikely to last for more than 2 years before it has to be replaced. For design purposes, however, any deterioration in flotation properties is usually ignored.

Some otherwise ideal materials such as plastic or steel drums rapidly accumulate fouling organisms under certain conditions, which reduces their flotation abilities and increases the drag on the mooring system. As a result maintenance costs can be high in terms of time spent on cleaning. Antifouling compounds can help retard the rate of fouling (see Section 7.6), whilst an alternative solution applicable to plastic and steel drums is to cover them in tough polythene bags which are simply changed as necessary. Steel and plastic drums may also be rotated every few months, thus exposing the fouling organisms to the air.

A number of flotation materials are either brittle or susceptible to the effects of weathering. One of the most commonly used materials is expanded polystyrene which unfortunately is not only brittle, but also yellows and becomes even less resistant to impact as a result of exposure to ultraviolet (UV) light (see *Fig 7.31*). One solution is to encase the floats in a material such as glass reinforced concrete or fibreglass, whilst in other designs the flotation system has been incorporated with the structural components of the collar and is thus protected from damage by impact (see earlier).

In view of the difficulties discussed above in quantifying precisely the materials required for adequate buoyancy it may seem tempting to over-engineer the flotation capabilities of the collar. Whilst cage builders do indeed play safe in this respect, often including a factor of safety in the computation (see earlier), a few words of caution should be issued here. Not only is this practice costly, it can also adversely affect the behaviour of the cage and the forces exerted on the collar and mooring system.

A body can be said to be in a state of stable equilibrium if, when it is displaced, forces are set up which tend to restore the body to equilibrium. By virtue of its shape, and because most of the mass lies close to or just above the waterline, a floating cage has a low centre of gravity. When a wave passes, the collar will incline through a small angle. Although the position of the centre of gravity of the cage, G, remains the same, the centre of buoyancy, B, — the centre of gravity of the displaced water — moves to a new position, $B_1$, as a result of the change in the shape of the volume of the water displaced by the cage (*Fig 3.19*). The intersection of the vertical line through $B_1$, with the displaced line which passed through the original centre of buoyancy is known as the metacentre (M). The magnitude of the restoring couple which restores the body to equilibrium is WGM sin $\theta$, where W is the mass of the cage (kg).

If the buoyancy of the cage collar is increased without greatly increasing the total mass, W, — which is possible if low density materials such as expended polyurethane or polystyrene are used — then the centre of gravity is raised.

The resultant decrease in draft causes the distance between the metacentre and the centre of gravity (the metacentric height) to increase as well, thus increasing the stability of the cage (see Stokoe, 1973, for review). However, the magnitude of the restoring couple also increases, so that the period of roll is reduced. From studies of the behaviour of ships it is known that increasing the metacentric height results in increased stresses on the structure whilst the resultant motion causes passengers to feel increasingly uncomfortable (Massey, 1979). Where cages are concerned, high surplus buoyancy is not only likely to cause the structure to move around a great deal, thus making it unpleasant or even dangerous for staff to work from, it is also likely to induce similar movement of the bag which may adversely affect the fish. Moreover, excessive movement of the collar will unnecessarily increase the forces acting on the structure. Linking cages together in a group may moderate the adverse effects of excess buoyancy.

Because wave and current forces acting on the collar are proportional to surface area (see later), any possible reductions in the dimensions of the flotation system are advisable, particularly if the cages are to be located at an exposed site. It follows, therefore, that the use of as low density materials as possible for supplementary flotation is desirable, although other factors such as resistance to fouling and weathering are of course, important.

The shape of the floats requires some consideration. At one exposed Scottish site, several cages incorporating air-filled, pillar-shaped aluminium floats, with drafts in excess of 2m, broke up during stormy weather as a result of excessive stresses on the joints between the floats and the collar. When the cages were displaced from equilibrium in rough seas, the buoyancy force on

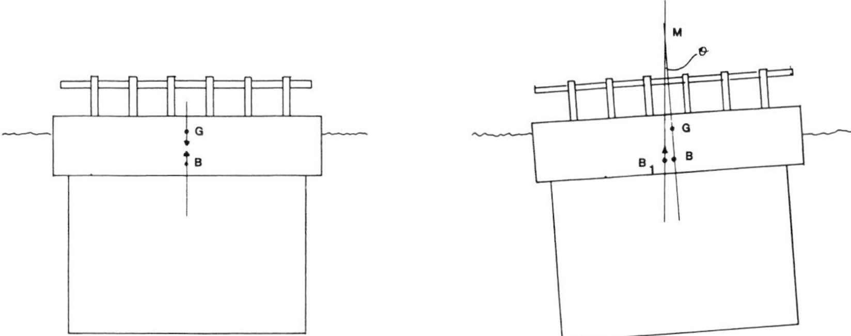


*Fig 3.19* Centres of buoyancy (B) and gravity (G) in a fish cage. When the cage tilts, B moves to a new position $B_1$ (see text).

the floats acted as an upward moment pivoted about the joints and eventually caused fatigue failure. Joints on floats on this type therefore need to be carefully designed.

The dynamic forces acting on a floating collar are principally horizontal forces produced by the wind, current and waves, although the latter does have a vertical component as well. Wind forces act on the superstructure — stanchions, handrails and freeboard netting, and those parts of the flotation/structural framework which are above the waterline. For design purposes, gust speeds are used to calculate wind forces, and the necessary long-term values required can often be obtained from the relevant meteorological institutes. Shellard (1965), in an analysis of data for the British Isles, states that 49-54m $s^{-1}$ gust speeds occur on average once every 50 years in coastal sea lochs, whilst 54-58m $s^{-1}$ gust speeds can be expected in offshore areas and the outer isles. However, from the experiences of Scottish fish farmers, it is suggested that these figures are likely to underestimate the gust speeds which can occur by 10% or more. Milne (1970) has suggested that the following equations are used to calculate wind forces:-

$$F = 0.0965\ A\ V^2 \text{ (Boven, 1968), and}$$

$$F = \tfrac{1}{2}\, k\rho V^2 \times 10^{-2} \text{ (Pankhurst and Holder, 1952)}$$

where F = wind force (kg); A = area ($m^2$); V = wind velocity (m $s^{-1}$); k = resistance coefficient = $1—\beta/2$; $\beta$ = blockage coefficient = $(1—d/l)^2$ d = mesh diameter (m) l = nominal mesh size (m); $\rho$ = density of air = 1.227kg $m^{-3}$. The former equation is used for solid components and the latter is used in the calculation of wind forces on netting and meshes.

Pankhurst and Holder's equation is likely to overestimate actual loadings since it was derived from empirical data on knotted materials. In *Fig 3.20*, the relationship between gust speed and wind force on various types of netting and meshes is illustrated. Assuming that 25mm nylon is used and that the freeboard height is around 1m, then wind forces at coastal marine sites are unlikely to exceed 42kg per m width of freeboard netting. However the forces produced by equivalent wind velocities on solid structures is around 280kg $m^{-2}$, thus emphasising the importance of designing buoyancy carefully, and reducing cage superstructure to a minimum.

Since cages are not usually sited in zones where breaking waves occur (see Chapter 4), it is the impact of the waves' kinetic rather than potential energy that is of concern here. In order to calculate the wave forces acting on the collar, the horizontal and vertical orbital velocities of the water particles must be known. These can be derived from information on the prevailing wave periods, significant wave heights and depth of water at the site (see Chapter 4), and by employing Stoke's theorem for waves of finite amplitude. The horizontal ($\mu$) and vertical ($\Omega$) components of wave particle orbital velocity can be calculated from:-

$$\mu = \frac{\pi H \cos h\,[2\pi\,(z + h)/L]}{t \sin h\,(2\pi h/L)} \cos\theta$$

$$\Omega = \frac{\pi H \sin h\, [2\pi\, (z + h)/L]}{t \sin h\, (2\pi h/L)} \sin \theta$$

(Muir Wood and Fleming, 1981), where H = wave height (m); t = wave period(s); L = wave length (m); h = depth of water (m); z = variation from mean water level (m) and $\theta$ = angle of wave relative to the structure. This equation applies to waves where $0.04 < h/L < 0.5$ (*ie* conditions prevailing at most sites.

In *Fig. 3.21* the maximum horizontal component of orbital velocity, $\mu$, is plotted against wave height for waves of restricted period ($2.5 \leq t \leq 8s$) in 6 and 15m water depth. According to Milne (1970), $\mu$ is unlikely to exceed 2m $s^{-1}$ in most fish farming conditions, and thus this value should be used for design purposes for marine cages. The forces generated by $\mu$ on exposed netting have been discussed by Milne (*ibid*), who proposed that the equation given above for calculating the drag forces exerted by currents on netting and mesh materials is applicable. From observations of fouling on exposed netting he suggested that approximately 20% of the surface should be considered as fouled and that in effect this results in a two-fold increase of the current forces calculated for clean netting.

However, most of the horizontal wave energy will be expended on the collar rather than on exposed freeboard netting. The force exerted on the collar can be described by an equation similar to that derived for the effect of current on netting and mesh materials:-

$$F = k\rho\mu^2 A$$

where F = force(N); $\rho$ = density of water (kg $m^{-3}$) and A = area of the cage collar perpendicular to the wave train ($m^2$); k is a dimensionless constant, similar to $C_d$ for netting and meshes, whose value will depend upon the nature of the collar (material, shape, construction, *etc*) and wave

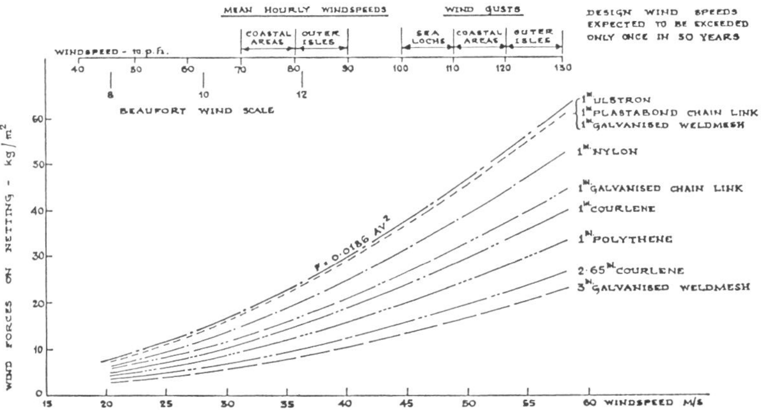


*Fig 3.20* Wind forces on netting for various wind speeds (from Milne, 1972).

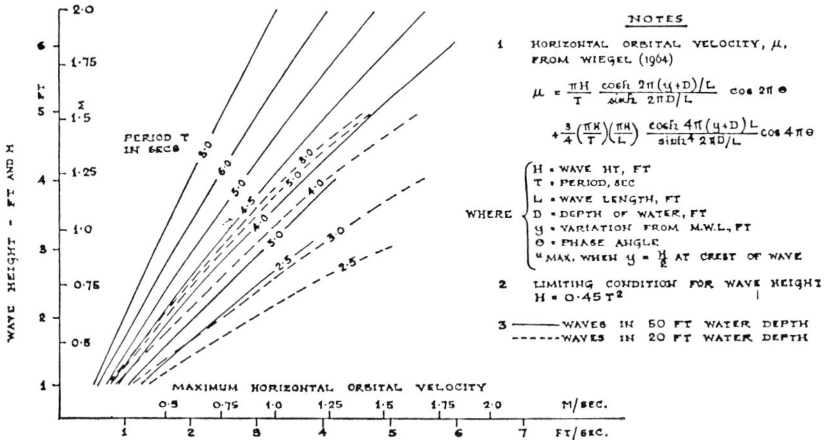


*Fig 3.21* Horizontal orbital velocities in 6m (20ft) and 15m (150ft) of water at various wave heights and periods (from Milne, 1972).

characteristics. Unfortunately k must be determined empirically and there are no values or rule-of-thumb estimates available. However, assuming a design μ value for marine cages of 2m $s^{-1}$, F will be considerably less than 400kg $m^{-2}$ which occurs when k = 1. The importance of minimising the surface area of the collar which lies perpendicular to the water surface is also illustrated in the above equation.

The vertical component of wave particle maximum orbital velocity, Ω, is around 82.5% of the maximum horizontal orbital velocity (Wiegel, 1964). Again, in the absence of any published experimental data, it is impossible to estimate the forces generated.

Currents will also exert a drag force on the collar which can be described using an equation identical to that given above. The current velocity at most coastal marine sites is unlikely to exceed 1.5m$s^{-1}$ (see Chapter 4). Since k is unknown, it is again impossible to give values for F. However, assuming the horizontal wave and current velocities given above for design purposes, the forces exerted by waves could exceed those generated by currents by a factor of almost 2, since the forces are dependent upon $V^2$.

In summary, although it is possible to estimate wind forces on the cage collar, it is impossible to quantify horizontal components of waves and currents — even using more sophisticated models derived from nautical engineering research — until a great deal more experimental work has been done. The k values have yet to be determined for different materials and collar designs and further study of the action of waves on freeboard netting and collar components is required, since wave particles follow orbital or elliptical trajectories, resulting in directional changes in the forces generated. The inertia of the net and dampening forces caused by waves generated from the movement of the collar, which interferes with the wind induced surface waves, will also modify the vertical movement of the collar. Moreover, flexible collars tend to deform, thus adopting a configuration of reduced resistance to prevailing forces, and the forces acting on a group of cages will be considerably smaller than those acting on the same number of cages moored individually (see Section 3.4.4).

Nevertheless, the simple equations presented here illustrate the importance of minimising the surface area of the collar in the plane perpendicular to the waves and currents, and demonstrate that at exposed sites the horizontal component of orbital wave velocity is likely to be greater than the forces generated by currents.

Some of the forces impinging on a collar will be reflected (*eg* a proportion of the wave energy), whilst the remainder will be absorbed. A proportion of the absorbed energy will be translated into kinetic and potential energy, thus causing the cage to move, the residual forces being either dissipated within the structure or transmitted to the adjoining cages, moorings and the water. A floating body, such as a fish cage, can have three rectilinear and three rotational motions. The rectilinear motions are:-

(i) heave : vertical motion
(ii) surge : horizontal motion along the longitudinal axis
(iii) sway : horizontal motion along the transverse axis

whilst the rotational motions are:-

(i) yaw : rotation about the vertical axis
(ii) roll : rotation about the longitudinal axis
(iii) sway : rotation about the transverse axis.

When a body is freely floating, heave, roll and pitch occur, largely as a result of wave action, and the displacement from equilibrium is corrected by gravity acting on the mass (see earlier). When moored, the body has three additional natural periods, surge, sway and yaw, with the elasticity and line weight of the mooring system being the restoring force (Wiegel, 1964). Whilst mooring has little effect on heave, roll or pitch of a cage or cage group moored from a single point, multiple point moorings will markedly reduce and modify these motions.

The energy dissipated within the collar is translated into deformation of the frame members and joints. Horizontal forces acting perpendicular to a moored square or rectangular cage will tend to induce bending along the edge and cause the collar to collapse (*Fig 3.22*), whilst horizontal forces acting at 45° will tend to produce distortion as shown in *Fig 3.22*. Although energy is transferred from one frame member to another at the joints, stress and fatigue are also produced since the joints have usually been designed to maintain the original angles under external loadings. The combination of vertical and horizontal forces will cause twisting of the individual structural members and indeed of the entire frame and further complicate the nature of the stresses acting at the joints.

The proportion of energy that is transmitted to the mooring system will depend upon the design and location of the mooring points (see later).

The overall robustness of a cage (defined here as its ability to function under adverse weather conditions) is thus governed by the following main criteria (Kerr *et al*,1980):-

(i) properties of the construction material and size and profile of the structural members;
(ii) strength of joints between structural members, especially when they occur at points of stress concentration;
(iii) degree and distribution of flexibility in joints and members;
(iv) design and location of mooring points to distribute resultant stresses.

All structures are flexible to some degree and for most cage types some flexibility in the collar is desirable, not only for the reasons given above, but also since it allows the cages to 'ride' the waves to a certain extent, thus reducing both the incident forces acting on the collar and the motion of the bag. Kerr *et al* (*ibid*) have further argued that since it is difficult and costly to design joints which allow some movement between neighbouring members, these should be kept as rigid as possible and the design should instead rely on the flexibility inherent in the frame members. Although this is the usual practice, some home-built designs have utilised simple flexible joints very successfully (*Fig 3.23*).

Flexibility is, of course, undesirable in the collar of rigid mesh designs and its effects can be minimised by mounting the mesh panels or framework at the points of least distortion (Hugenin and Ansuini, 1977).

To quantify the energy absorbed by the collar and the resultant deflection or deformation of the frame members and joints, techniques of structural analyses are required. Whilst it is beyond the scope of this book to deal with the subject, it is worth mentioning a little more about flexibility. An indication of the degree of deformation of a structural member that may occur can be derived from the modulus of elasticity (ratio of stress: strain). Mild steel has a modulus of elasticity of around $210 \times 10^3$N mm$^{-2}$, whilst wood and bamboo have values of 5-9 $\times$ $10^3$ and 10-17 $\times$ $10^3$N mm$^{-2}$ respectively (see *Table 3.8*). Steel is thus a much stiffer material and is much more resistant to deformation.

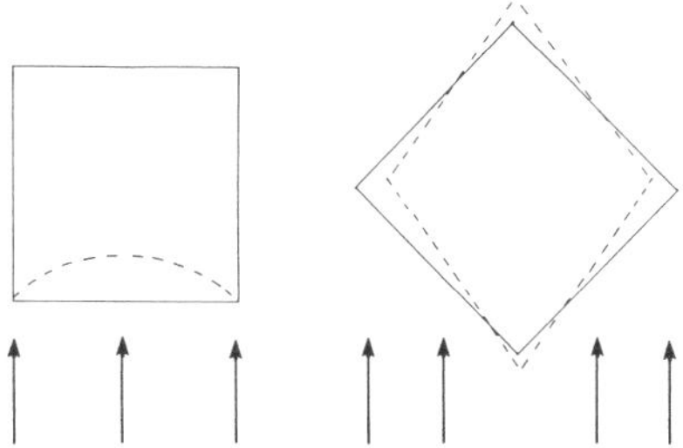

*Fig 3.22* Diagrammatic illustration of deformation of cage collars under external current and wind forces.

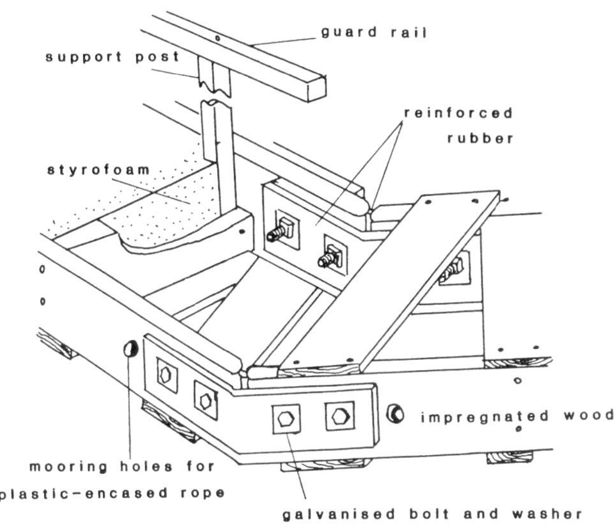


*Fig 3.23* Semi-flexible, home-made joint for hexagonal cage (redrawn from Sutterlin and Merrill, 1978).

Finally, it should be stressed that over-engineering of the cage is a costly and wasteful business.

In sheltered inland waters where current, wind and wave forces are unlikely to attain the same magnitude as at coastal locations relatively unsophisticated cages, constructed from materials of comparatively low structural strength have proven perfectly adequate. One of the simplest of these designs, widely used in the Philippines, is constructed almost entirely from bamboo. For a 3 × 3 × 3m cage, suitable for rearing tilapias, 20 properly seasoned bamboo culms and 100m of 4mm PE rope will suffice. The bamboos are prepared in the same way as described for fixed cages, and cut into lengths as shown in *Fig 3.24*. Using tyres to support each corner, 5 lengths cut from the culm bases are laid out to form each side of the collar. The corners are then secured using a combination of bamboo pegs and rope. The pegs, approximately 25cm long x 12mm in diameter and tapered to a point, should be cut from an internode section as near to the base of the culm as possible, and driven through the bamboos using a hammer (see *Fig 3.25*). The rigidity of each side is improved by lashing a section of split bamboo, outer face upwards, across the walkway. Stanchions, approximately 1m long, are firmly lashed to each corner and handrails run between them. Two struts are required at each corner to secure the stanchions firmly enough to support the cage bag.

The construction process is shown in *Fig 3.25*. Approximately 3-4 man-days are required to build each cage. The buoyancy of the collar is around 112kg — sufficient to support the weight of one man and the cage bag when it is in the water (see Appendix 3.1). Although it is impossible to

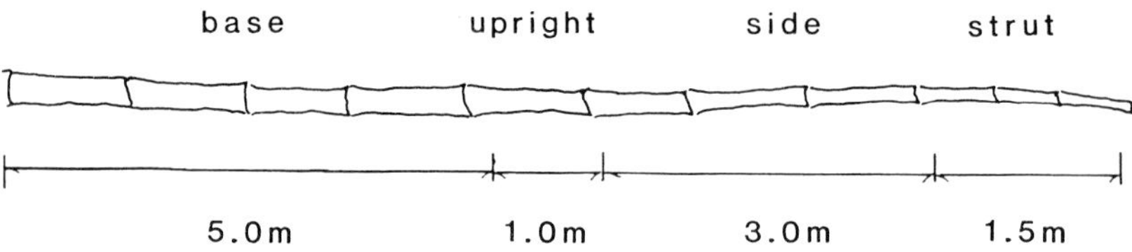


*Fig 3.24* Use of bamboo culm for floating cage construction.

*Fig 3.25* Floating bamboo collar cage construction (a) detail of corner, (b) uprights, sides and support struts being lashed in place; (c) the cage near completion, (d) the cage sitting in the water, ready to be towed to the site.

quantify the loads which the corner joints can withstand, Janssen (1981) has demonstrated that bamboo pegs are excellent connectors.

For larger cages it is recommended that plastic or steel drums replace the bamboos for flotation purposes and where the cages are to be used in slightly more exposed locations (*eg* sheltered coastal sites) lengths of wood and steel bolts should be substituted for bamboo. Hardwoods such as *Hopea odorata* (chengai pasir) and *Dryobalanops aromatica* (kapur) which are typically used for cage construction in Southeast Asia, are considerably denser and thus more resistant to bending and shear forces. Moreover they are less susceptible to rotting and damage by boring organisms. If regularly treated with preservative chemicals, Chua (1979) claims a hardwood collar will last up to 4 years.

The plan diagram of a typical wood and oil drum collar is shown in *Fig 3.26*. It is constructed from six 7m and three 8m hardwood beams, each with a cross-section of 15 × 7.5cm. The beams are joined by 12.5mm diameter mild steel bolts, 20cm long. Fifteen 200*l* steel drums secured to the frame by rope provide adequate flotation, and each compartment holds a 3 × 3 × 3m net. There are many variations on this design, details of which can be found in a number of the FAO/UNDP South China Sea Programme publications.

In Western Europe, the equivalent of these simple designs is the standard Kames-type cage, originally developed by Stuart Cannon, Kames Fish Farming Ltd, which has been extensively used in both inland and sheltered coastal areas (*Fig 3.27*). Although there are numerous variations of this design in existence, it typically consists of walkways which run along two sides of the cage, joined together at either end by timber cross-beams. The walkways are fabricated from larch — a comparatively dense (590kg $m^{-3}$) and strong softwood, which is usually pressure treated with wood preservative compounds to improve rot-resistance — and blocks of expanded polystyrene. Galvanised brackets secure the corners, and galvanised steel stanchions bolted to the walkways and crossbeams are used to support the wooden hand and net rails. Construction details are given in Berry (1981).

As discussed elsewhere in this book, there is a growing tendency to use larger cages and to try to rear fish in less sheltered locations. Both these trends have exposed weaknesses inherent in the types of cage design discussed above and have resulted in a number of solutions being employed. Some of these have involved no more than a few modifications. Materials with higher specifications have been introduced: galvanised steel or aluminium has been used to replace much of the timber, and larger and heavier brackets and bolts used at corners. Such materials are, of course, considerably more expensive, but arguably have a much longer life span — 10-12 years for a steel-framed design. In other cages the corner brackets have been re-designed to increase resistance to twisting, compression and extension forces (see *Fig 3.28*). Steel tubing should be joined by welding or with properly designed clamps. Instead of simply bolting the stanchions to the walkways, several designs have them built into the walkways and incorporate handrails fabricated from galvanised steel, so that the cage superstructure performs a greater structural role, improving opposition to bending and twisting.

Other designs have abandoned conventional shapes and materials, but have retained the principle of maintaining rigid joints and using structural members with some inherent elasticity (see earlier). By virtue of their shape, octagonal and hexagonal designs may absorb less energy than rectilinear collars (they present a smaller surface to prevailing forces) and distribute the loads over a larger number of joints so that the load per joint is reduced. However, 135° joints are weaker than right-angles, and must be carefully

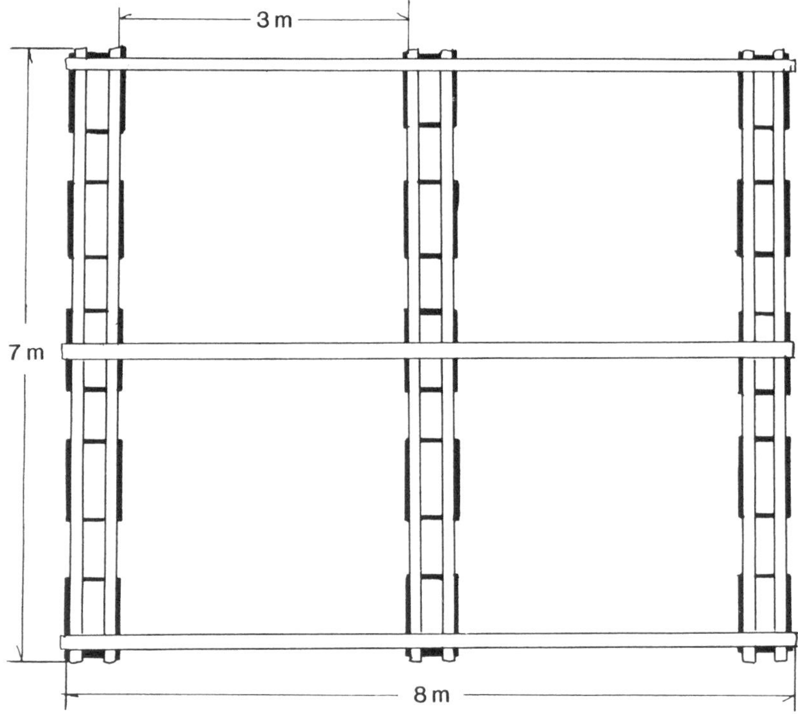


*Fig 3.26* Typical 4-cage raft constructed from oil drums and hardwood beams (redrawn from Mok, 1982).

*Fig 3.27* Kames-type cage.

designed with high specification materials. By avoiding all corners, circular collars are inherently more robust although they can also be more difficult to fabricate. For example, the circular floating cages developed by SEAFDEC for culturing milkfish are constructed from epoxyresin coated galvanised iron pipe and require a purpose-built frame for bending the pipes to specifications (Yu *et al*, 1979). Rigid PE pipe, which is light, strong and highly resistant to rotting and weathering, filled with high density expanded polyurethane is less problematic and is commonly used for circular collars in Western Europe. The base is typically fabricated from 16-20cm diameter pipe, whilst hollow sections serve as stanchions and handrails. Lengths of pipe are welded or bolted together on site to form circular or octagonal collars of various sizes. Sometimes the base incorporates an outer ring, connected by hollow-sectioned struts and saddle joints (*Fig 3.29*) which serves not only to strengthen the structure, but also increases buoyancy and improves the work area.

A number of commercial companies have approached the problems of farming in large cages at exposed sites by designing flexible collars which

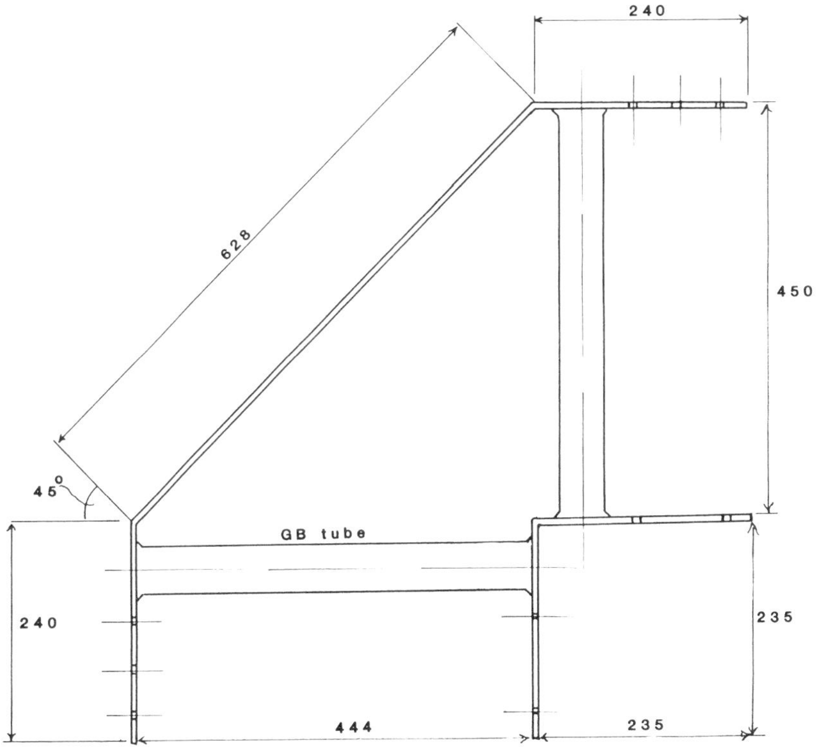


*Fig 3.28* Corner bracket for fish cage fabricated from 6mm galvanised steel plate and tubing. Dimensions in mm. Redrawn from Berry (1981).

*Fig 3.29* Saddle joint connecting inner and outer rings of a polyurethane foam-filled plastic pipe cage collar.

undergo pronounced deformation under the influence of external forces. This serves both to reduce the forces impinging on the collar, since it tends to adopt the shape which is least resistant to external forces, and to dissipate the absorbed energy through distortion of the structural members. The limits of flexibility should be set by the distortion of the bag that is admissable. The collars are usually 8-12 sided, and the structural members fabricated from synthetic rubber filled with an inert gas under pressure. Joints are rigid and made of steel or aluminium (see Section 7.12 for further discussion).

Discussion of the design and construction of floating collars has so far ignored the important question of the working environment provided for staff. Many small cages are designed so that the collar provides little more than flotation and so must be serviced from a pontoon or pier or by boat. For example, some of the cages typically used for culture of freshwater species in the United States are designed in this way (*eg* McLarney, 1984). Narrow collar designs, such as the simple flexible collar cages used in parts of the Philippines or Bolivia (*Fig 2.2f*), and the rigid designs widely used in Scandanavia are also designed without a work platform. Although the large, flexible cages described above do provide a narrow collar for staff to walk and carry out simple tasks, they are difficult to work from in rough weather. Moreover, on account of their size all heavy routine jobs such as grading and net changing must be carried out from boats, often using power-assisted gear.

The majority of fish cages have collars which serve as work platforms. Not only must they be sufficiently buoyant, they must also provide a work base which is sufficiently safe to work from. Most working collars have walkways that are at least 1m wide. Others, which have narrow walkways on several sides, may be connected to a central pontoon, often 3-4m wide, which facilitates the transport of items like feed by wheel-barrow. In some of the more recent designs, the pontoons are sufficiently wide and buoyant to support small trucks. Wooden slats, such as those used in Kames-type designs give an excellent non-slip surface, providing they are regularly cleaned and kept free from algae and other debris. Covering wooden decking with chicken wire or strips of scrap rubber conveyor belt material is also sometimes practised. The decking of steel or aluminium designs either has a non-slip finish, or is fabricated from an expanded metal mesh which not only provides good grip, but is also self-draining.

Handrails and stanchions are always fitted on the inside of walkways and in some designs, on the outside as well. Metal handrails, although strong, can be unpleasant in climatic extremes.

### 3.4.3 Cage linkages and groupings

Although most flexible collar and some small rigid collar cages are moored individually, it is common practice to group cages together, not only because it simplifies and minimises the costs of mooring arrangements, but also for management reasons. The number and arrangement of cages grouped together thus depends on:-

(i) the size of the farm;
(ii) the size and nature of the site;
(iii) the shape and design of the cage and linking system;
(iv) mooring constraints;
(v) environmental considerations

For a small production or artisanal unit, the first of the above considerations may be of little consequence. However, for larger farms, the farmer may wish to organise his production into convenient groupings of cages. For example, a salmon farmer may wish to keep one group of cages for grilse and another for second sea winter fish. The size, shape, depth and physical characteristics (*eg* degree of exposure) of the site may restrict the positioning of the cages and dictate the number that can be grouped together. Whilst square and rectangular cages can be assembled in a variety of configurations, hexagonal, octagonal and circular designs may be more limiting. Similarly, the linkage system on certain commercial designs is such that the cages can only be grouped together in a limited number of ways. Limitations may also be imposed by the availability of suitable anchorages at the site.

Environmental considerations are also extremely important. In a study of flow rates through a series of empty cages, fabricated from 2.4cm mesh net, moored together and oriented parallel to the current (0.7-1.9cm $s^{-1}$), Inoue (1972) showed that the transmission rate fell dramatically from 0.8 in the first cage to 0.59 in the second cage and 0.35 in the third cage. In trials conducted simultaneously, using 0.9cm mesh net cages, the results were even more dramatic. The transmission fell from 0.7 in the first cage to 0.35 in the second cage and 0.2 in the third cage. Although the presence of fish in the cages is likely to ameliorate flow rates, there can still be marked differences in water quality between cages on the outside and in the centre of a group. For example, in extensive cage culture trials conducted in Malaysia, production of fish in the central cage of a 3 × 3 group was significantly depressed compared with those in the outside cages (Awang Kechik *et al*, 1985). It is thus recommended that for most sites, although cage groupings may be 8-10 cages long, they should be no more than 2-3 wide.

As was stated above, the grouping together of cages markedly reduces the forces acting on the individual cages and modifies their behaviour, dampening both rectilinear and rotational motions. Linkages between cages should be designed so that the pitching motion is only moderately damped, whilst rolling and yawing and surge and sway are kept to a minimum. There are a number of reasons for this. Not only is it difficult and expensive to construct a joint which permits universal movement, it is also not desirable, either from the staff or from the fishes' point of view, to facilitate free movement of the cages. On the other hand, the more rigid the linkages, the greater the forces that are concentrated at these points. The most destructive forces impinging on the collar are waves, partly because they have both a vertical and horizontal component, and partly because they are cyclic in nature. They induce bending and shear forces whose magnitudes are a function of wave height and are greatest when the wavelength and the size of the largest rigid component of a structure are approximately the same — usually between 0.5 and 1.25 times the rigid component length (Kerr, 1981). Typical wavelength during coastal storm conditions are around 10-50m (see also Chapter 4.). Thus, it is best to construct either small cages with linkages which permit some degree of pitching and heaving or large cages with flexible collars.

The simplest type of linkages use rope or chain which is secured sufficiently tightly to reduce all motion, except in the vertical plane, to a minimum. Rope, however, tends to abrade fairly rapidly and so chain is preferable, although it may be necessary to protect timber collar members against abrasion. Shackles of appropriate size, or C-links, may be used to secure the chain. Rubber tyres are also lashed or bolted between cages and act as fenders (*Fig 3.30*).

There are several more sophisticated designs of linkages which are better suited to larger and heavier cages and more exposed coastal sites. These linkages are larger, made of stronger materials and are designed to distribute loads over a greater surface area or number of elements. In standard laboratory tests, they can often withstand loads as high as 10-20 tonnes. A typical design is shown in *Fig 3.31*. Both halves of the linkage have brackets welded to the walkways, which dovetail together. A large galvanised steel pin, secured at either end by split pins, is then used to join the sections together. Bushes made of nylon or a synthetic rubber compound help reduce noise and wear, and increase vertical dampening by a moderate amount, whilst allowing a certain amount of movement in other planes.

In the design and construction of linkages, due care should be given to the

*Fig 3.30* Disused car tyres as fenders between cages.

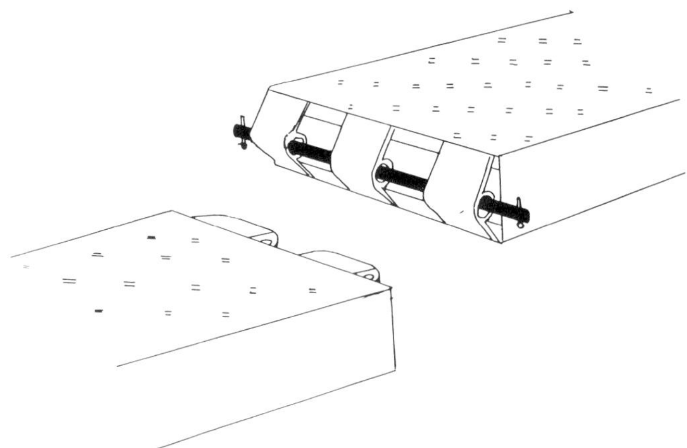

*Fig 3.31* Example of simple linkage between cages or cage platforms.

choice of materials, not only to satisfy loading criteria, but also to avoid galvanic action (see earlier). Linkages which facilitate the rapid removal of a cage from the group for repair or during harvesting or for re-siting, may also be desirable, although in some commercial designs it is impossible to separate individual cages since the central pontoon to which they are linked forms one side of the collar.

### 3.4.4 Mooring systems and access

Access to cages is normally by boat. However, where the cages are moored in a sheltered site close to shore as, for example, is common in the steeply-shelving fjords of Norway, pontoons may suffice. The pontoons must be secured so that they can accommodate tidal rise and fall and where the shoreline has a steep gradient or where access is via a jetty or pier, a ramp may be necessary (*Fig 3.32*). Commercial manufacturers have found it difficult and costly to construct ramps much longer than 15m.

Most mooring systems consist of lines and anchors, and their purpose is to secure the cages in a desired position. However, the moorings also influence the stresses acting on the cage members and the behaviour of the cages in rough weather, and can affect production, profitability and even staff safety. They are thus an extremely important part of a cage farming system, and so should be carefully designed. Indeed, the collar, net and mooring components of a cage system should be designed together, although in practice the cages are usually selected or built first, and the mooring system is very much designed as an afterthought. However, commercial fish cage manufacturers are increasingly willing to, or even insistent that, they design and install the mooring system, otherwise they may refuse to cover the cages against structural failure.

Mooring requirements should be determined by the design and type of cages used and the characteristics of the site. The procedure for carrying out such a computation would be to quantify the incident forces that are likely to act on the cages under worst possible weather conditions that could occur at the site, and then to evaluate the proportion of energy that is absorbed by the cage or group of cages. Part of this energy should be transferred from the cage to the mooring lines and anchors, the respective roles of the cages and mooring system in dissipation of the energy being determined by the specifications and group configuration of the cages and the costs of the mooring system.

Of course, as has been discussed above, it is impossible at present to apply this methodology to the design of cage moorings. However, it is possible to discuss many of the principles involved, to qualitatively evaluate different mooring systems, and to present some general guidelines. Consider, for example, a 5 × 5 × 5m cage. The bag is fabricated from 25mm stretched mesh nylon, and the collar consists of four blocks of wood-encased polystyrene, 5 × 0.4m. There is 1m of freeboard netting, supported by three 1 × 0.1m stanchions on each side of the collar. The forces that must be considered include the current forces acting on the bag and submerged portion of the collar, wave forces acting on the exposed portion of the collar, and wind forces acting on the exposed collar, freeboard netting and stanchions.

Consider, first of all, the horizontal forces acting on one side of the cage at various gust speeds and water current velocities. For design purposes, it is assumed that the net is fouled and that the collar is submerged to a depth of 30cm. Forces are calculated according to the equation given in Section 3.4.2.2 and *Figs 3.9* and *3.20*.

(a) *Currents*

| velocity ($m\ s^{-1}$) | force on netting ($kg\ m^{-2}$) | total force on netting (kg) | force on collar ($kg\ m^{-2}$) | total force on collar (kg) |
|---|---|---|---|---|
| 0.5 | 50 | 1,250 | ~250 | ~374 |
| 1.0 | 180 | 4,500 | ~1,000 | ~1,500 |
| 1.5 | 500 | 10,000 | ~2,250 | ~3,375 |

(b) *Wind*

| gust speeds ($m\ s^{-1}$) | force on netting ($kg\ m^{-2}$) | total force on freeboard (kg) | force on solids ($kg\ m^{-2}$) | total force on stanchions (kg) | total force on collar (kg) |
|---|---|---|---|---|---|
| 45 | 32 | 160 | 99 | 29 | 50 |
| 50 | 38 | 190 | 139 | 42 | 70 |
| 55 | 46 | 230 | 204 | 61 | 102 |

(c) *Waves*

Force ~400 $kg\ m^{-2}$ therefore total force~200 kg

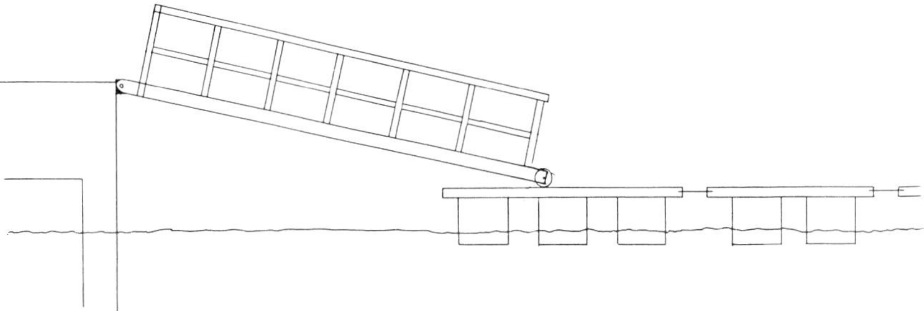

*Fig 3.32* Ramp from jetty to floating pontoons.

Thus, the horizontal forces on one side of the cage, under the conditions described above, will be in the range 1800 — 14,000 kg (*ie* current force on netting + current force on collar + wind force on freeboard + wind force on stanchions + wind force on collar + wave force on collar).

The errors associated with this calculation are, of course, enormous, and are based on the assumptions that:-

(i) the netting is knotted
(ii) there is no deformation of the cage bag
(iii) all of the current forces impinging on the collar are absorbed, and that
(iv) all the horizontal components of wave energy are absorbed by the collar.

Assuming a transmission factor of 0.7, the current forces exerted on the second net wall of cage by the water as it leaves will be:

| current velocity ($m\ s^{-1}$) | total force (kg) |
|---|---|
| 0.35 | 650 |
| 0.7 | 1,250 |
| 1.0 | 2,250 |

Other forces may be ignored, since it is assumed that all the wave and current forces acting on the collar are absorbed. Thus the total forces acting on the cage will be between 2.5 and 16 tonnes. The huge errors involved in such a computation give an extremely conservative estimate of the forces generated, and we may conclude no more than that under severe weather conditions a force of several tonnes may be exerted on the mooring system. Nevertheless, the exercise illustrates a number of important points:-

(i) Horizontal forces exerted by currents appear to be much greater than those exerted by wind or waves.
(ii) Because all forces are a function of velocity squared, the design velocities used in any computation can be crucial.
(iii) The effect of adding cages to a group will only marginally increase

the mooring specifications particularly when cages are being added to a large group.

However, this approach to estimating mooring requirements ignores shock loadings and vertical loadings, which in practice give farmers most problems. Hence, in view of the present state of knowledge, moorings are generally greatly overspecified. For particularly exposed locations it is best to consult other similar operations or a specialist mooring installation company.

There are two types of mooring system: multiple and single point. The former is more common and involves securing the cages in one particular orientation, whilst the latter allows the cages to move in a complete circle. Single point moorings tend to be used with square or rectangular rigid collar designs. They use less mooring cable and chain than multiple point moorings and the cages adopt a position of least resistance to the prevailing wind, wave and current forces. Although some would recommend their use at exposed sites, it must be borne in mind that there will be enormous forces generated where the mooring is attached to the cage as well as on the mooring line and anchors.

Cages moored from single point moorings distribute wastes over a considerably larger area than those secured by a multiple point system. If the areas covered by identical groups of cages moored by the two systems are compared, then there is a 20-40 fold increase in the area over which the single point moored cages dispenses waste, although wastes will not, of course, be evenly distributed over the entire area. Recent research has indicated that waste loadings to the sediments may play a crucial role in determining water quality and fish health, particularly at marine sites (see Chapter 5). However, the use of a single point mooring will also reduce the fraction of the site that is used for fish production. As a rule of thumb, as much as 20-30% of a site may be occupied by multiple point moored cages, the remaining space being used for moorings, whereas with single point moorings the cages only occupy 2-4% of the entire site area.

The choice of orientation of cages with multiple moorings depends upon the nature of the site and upon the type and group configuration of the cages. If particularly exposed, or if the prevailing currents are strong, then it is best to secure the cages in the position of least resistance to the prevailing wind and current forces. On the other hand, if the site is sheltered and water circulation is poor, then it is better to moor the cages so that water exchange is maximised. However, there may be restrictions on mooring orientation imposed by the size of the site or by availability of suitable mooring grounds.

Most methods of mooring involve the use of rope and chain to connect the cage or cage group to anchors or pegs. However, an alternative method is to drive long posts into the substrate and to attach the cage raft directly, either with ropes or with metal hoops or tyres, which permit tidal and wave-induced vertical movements of the cages. The number and dimensions of the posts required, and the depth to which they must be buried, in theory could be computed from estimates of forces acting on the cage system, and data on soil characteristics, following the methods described by Milne (1970). In practice, however, this is determined by experience. Stout hardwood posts, driven at least 2m into the substrate are recommended. Although sometimes used in sheltered and shallow inland and coastal sites with suitable substrates (Sodikin, 1977; Chua, 1979; Henderson, 1980) this method of mooring has not proved universally successful.

There are a variety of methods of using single and multiple point moorings. One or two heavy ground chains can be laid, to which the cages are attached via mooring lines. Alternatively the mooring lines can be run directly from the cages to the anchors. Where the mooring lines are secured to the cages, points of stress are formed and so it is important for multiple mooring systems in exposed locations that mooring lines are connected to the cages at a number of points. Joints, where stresses tend to accumulate or are transferred from one structural member to another, are frequently used as mooring attachment points. Anchors are usually deployed to resist the principal directions of force. Sometimes anchors may be installed on shore as well as at sea. Examples of various mooring systems are shown in *Fig 3.33*.

Mooring lines must, of course, be secured to cage collars via attachment points which are easily capable of resisting the forces generated. Thus, integral parts of the collar, rather than appendages fixed to the collar base or superstructure, are preferable. Where abrasion is expected, the line should be protected by encasing in plastic pipe.

Natural fibre rope is not particularly suitable as a mooring line, since it is easily abraded and prone to rotting (Klust, 1983). Steel cable, although immensely strong, is expensive and heavy (*Table 3.9*) and has little elasticity, although it is sometimes used for mooring cages to the land. Chain is

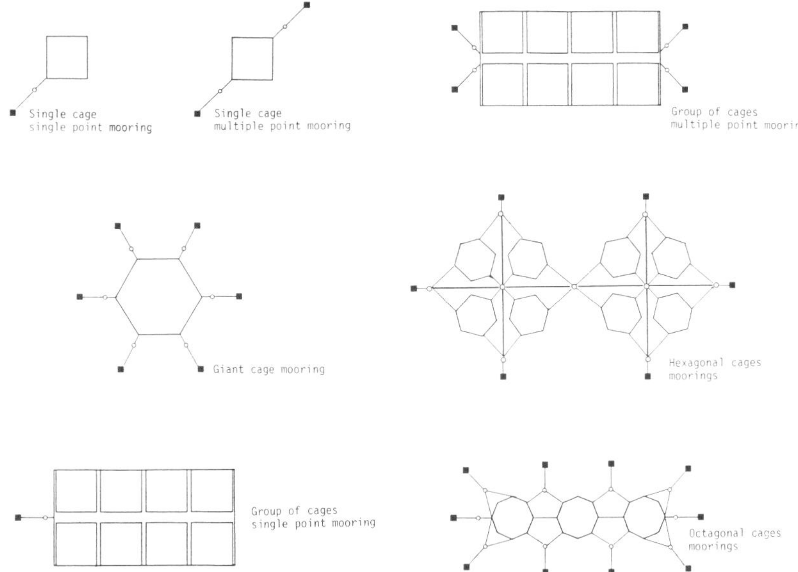


*Fig 3.33* Examples of various cage groupings and mooring systems.

extremely strong but for reasons which will be explained below, is usually used in conjunction with synthetic fibre rope to form the mooring line.

Synthetic fibre ropes are composed of nylon (PA), PE, PES, or PP. In the production of laid ropes, fibres are spun and twisted into rope yarn, several of which are then twisted into a strand. Three or four strands are twisted — the technical term here is 'laid' — into a rope, and for a cable laid rope, three or more of these ropes are twisted together (see Klust, 1983, for details). At each stage of production the direction of twist is reversed. Braided or plaited ropes differ from laid ropes in that the yarns are tied together by a process of interlacing, as opposed to twisting. There are a number of different types of braided rope which are described in detail by Klust (*ibid*).

For laid ropes of the same diameter, nylon and PES are considerably heavier than PP or PES (*Table 3.10*). However, nylon is much stronger than other ropes on a per unit weight or equivalent diameter basis. Braided ropes are lighter than laid ropes, but for equivalent weight or diameter, are generally weaker. Nylon is also more elastic than other materials. Klust (1983), in a comparison of three-strand laid ropes of equal diameter, found that at 20% of their breaking load nylon extended by about 20%, PE by between 7 and 11%, PP by between 6 and 9% and PES by around 3%. Recovery once the load is removed is extremely rapid for nylon rope. There is no marked difference between elongation characteristics of laid or braided ropes.

Mooring lines must perform two functions: they must both withstand and transmit forces. The loads imposed on a cage mooring system are principally dynamic and thus it is important that the lines used have a high breaking strength and can absorb much of the kinetic energy of rapidly changing forces, such as the cyclic loads imposed by waves and shock loads imposed by sudden gusts of wind, otherwise these forces will be transmitted directly to the anchors, thus necessitating the use of anchors with large holding abilities. One solution is to employ nylon rope. Although it can cost twice as much as PE or PP rope of equivalent strength, it has high extensibility and thus energy absorbing properties (*Table 3.11*). Braided ropes which cost more than laid ropes, have little advantage other than they are easier and more

*Table 3.9* Specifications of 7-strand galvanised steel cable. From Thomas *et al* (1967).

| *Diameter (mm)* | *Wt per unit length* ($kg\ 100m^{-1}$) | *Safe load*[1] *(kg)* |
|---|---|---|
| 3.2 | 4.8 | 227 |
| 4.8 | 11.2 | 635 |
| 6.4 | 18.6 | 1,043 |
| 7.9 | 31.3 | 1,724 |
| 9.5 | 43.9 | 2,268 |
| 12.7 | 75.9 | 3,856 |
| 25.4[2] | 296.8[2] | 14,300[2] |

1 Safe load is approximately 0.25 times breaking load.
2 = estimated.

*Table 3.10* Mass and breaking strength ($K_p$) of common braided and three strand laid synthetic fibre ropes (after Klust, 1983).

| *Nom* | *Nylon (PA)* | | *PES* | | *PP* | | *PE* | | *Braided nylon (PA)* | | | *Braided PES* | | |
|---|---|---|---|---|---|---|---|---|---|---|---|---|---|---|
| *dia (mm)* | *mass* | *Kp* | *mass* | *Kp* | *mass* | *Kp* | *mass* | *Kp* | *mass* | *Kp*[a] | *Kp*[b] | *mass* | *Kp*[a] | *Kp*[b] |
| 4 | 1.1 | 320 | 1.5 | 295 | — | — | 0.8 | 200 | 0.9 | 280 | 225 | 1.1 | 260 | 215 |
| 6 | 2.4 | 750 | 3.0 | 565 | 1.7 | 550 | 1.8 | 400 | 2.0 | 620 | 500 | 2.4 | 575 | 440 |
| 8 | 4.2 | 1,350 | 5.1 | 1,020 | 3.0 | 960 | 3.3 | 700 | 3.6 | 1,110 | 900 | 4.4 | 1,000 | 760 |
| 10 | 6.5 | 2,080 | 8.1 | 1,590 | 4.5 | 1,425 | 4.9 | 1,090 | 5.6 | 1,700 | 1,400 | 6.8 | 1,540 | 1,160 |
| 12 | 9.4 | 3,000 | 11.6 | 2,270 | 6.5 | 2,030 | 7.2 | 1,540 | 8.1 | 2,475 | 2,025 | 9.8 | 2,160 | 1,620 |
| 14 | 12.8 | 4,100 | 15.7 | 3,180 | 9.0 | 2,790 | 9.5 | 2,090 | 10.5 | 3,200 | 2,325 | 13.3 | 2,860 | 2,130 |
| 16 | 16.6 | 5,300 | 20.5 | 4,060 | 11.5 | 3,500 | 12.8 | 2,800 | 14.3 | 4,350 | 3,500 | 17.4 | 3,650 | 2,700 |
| 18 | 21.0 | 6,700 | 26.0 | 5,080 | 14.8 | 4,450 | 16.1 | 3,460 | 18.1 | 5,500 | 4,450 | 22.0 | 4,500 | 3,300 |
| 20 | 26.0 | 8,300 | 32.0 | 6,350 | 18.0 | 5,370 | 20.0 | 4,270 | 22.3 | 6,700 | 5,350 | 27.2 | 5,300 | 3,950 |
| 22 | 31.5 | 10,000 | 38.4 | 7,620 | 22.0 | 6,500 | 24.3 | 5,080 | 27.0 | 8,100 | 6,500 | 32.8 | 5,800 | 4,600 |
| 24 | 37.5 | 12,000 | 46.0 | 9,140 | 26.0 | 7,600 | 29.5 | 6,100 | 32.2 | 9,650 | 7,750 | 39.0 | 6,250 | 4,450 |
| 26 | — | — | — | — | — | — | 32.8 | 6,910 | — | — | — | — | — | — |
| 28 | 51.0 | 15,800 | 63.0 | 12,200 | 35.5 | 10,100 | 39.3 | 8,030 | — | — | — | — | — | — |
| 32 | 66.5 | 20,000 | 82.0 | 15,700 | 46.0 | 12,800 | 52.5 | 10,400 | — | — | — | — | — | — |
| 40 | 104.0 | 30,000 | 128.0 | 23,900 | 72.0 | 19,400 | 78.5 | 15,600 | — | — | — | — | — | — |

*NB* Kp = kilopond = 9.81N.
a = round braided rope.
b = solid braided rope.

*Table 3.11* Comparative properties of three-strand fibre ropes. The resistance to sunlight refers to synthetic fibres which are stabilised but not coloured (from Klust, 1983).

| *Ropes made of* | *PA cont. fil.* | *PES cont. fil.* | *PP monofil.* | *PP split fib.* | *PE monofil.* | *Manila* |
|---|---|---|---|---|---|---|
| *Group a* | | | | | | |
| Breaking length, dry, km | 30.6 | 19.2 | 28.8 | 28.8 | 21.2 | 10.8 |
| Wet strength in percentage of dry strength | 80 to 90 | 100 | 100 | 100 | 100 to 115 | 105 to 120 |
| Strength ratio to manila | 2.83 | 1.78 | 2.67 | 2.67 | 1.96 | 1 |
| *Group b* | | | | | | |
| Elongation, dry, at 30% of breaking strength, percentage | 23.3 | 5 | 12.7 | 8.5 | 9.9 to 15.1 | 4.7 |
| Elasticity | high | high | medium | medium | low, creep | low |
| Elongation ratio to manila at 30% of breaking strength | 4.9 | 1.1 | 2.7 | 1.8 | 2.1 to 3.2 | 1 |
| Toughness | very high | high | high | high | medium | low |
| *Group c* | | | | | | |
| Resistance to sustained load | very high | very high | high | high | medium | low |
| Resistance to repeated load | very high | very high | high | high | low | low |
| Resistance to shock loading | very high | high | high | high | medium | low |
| Flexing endurance | very high | very high | medium | low | low | low |
| *Group d* | | | | | | |
| Shrinkage in water | small, varying | none | none | none | none | great |
| Resistance to sunlight | medium | high | medium | medium | medium | medium |
| Resistance to rotting | very good | very good | very good | very good | very good | low |
| Resistance to ageing | good | good | good | good | good | medium |
| *Group e* | | | | | | |
| Surface feel | smooth | smooth | smooth, stiff | harsh | smooth, waxy | harsh |
| Fibre density, g $cm^{-3}$ | 1.14 | 1.38 | 0.91, floating | 0.91, floating | 0.96, floating | 1.35 |

pleasant to handle and do not kink.

Unless wooden pegs are used (see later) nylon ropes should not be directly attached to either shore or sea anchors, but instead should be connected via a section of chain. The chain serves to increase the effectiveness of the mooring system, partly due to its mass and shape which directly act as an efficient type of anchor (Wiegel, 1964) and partly due to its accentuation of the catenary shape of the mooring line which improves the holding power of the existing anchor by both reducing the angle between the mooring line and anchor (see below) and by increasing the energy absorbing properties of the mooring line. Moreover, a section of chain is necessary at the anchor since it is much more resistant than synthetic fibre rope to the prevailing high abrasion forces.

There are several types of chain available, including wrought iron and various types of carbon and alloy steels. Wrought iron is very variable in quality; the best has excellent corrosion resistance whilst the poorer grades are inferior in all respects. Some stainless steel chain is suitable for marine use, but is prohibitively expensive. Mild steel chain, with as low a carbon and manganese content as possible is widely recommended for cage anchorages. Specifications of one type are given in *Table 3.12*. Where buying secondhand or reconditioned chain, analysis of the composition is advisable. The length of chain is usually no more than one third of the total length of the mooring line, otherwise the vertical loading on the cage buoyancy would become excessive and necessitate the use of buoys. A fairly heavy grade of chain is recommended.

The total length of the mooring line to sea anchors should be at least three times the maximum depth of water at the site (see later) and where the rope joins the chain a galvanised, heavy duty thimble should be spliced into the rope. Galvanised shackles of the appropriate size — usually a fraction larger than the diameter of the rope and the anchor ring — should be used to connect the chain to the anchor and to the rope.

An alternative mooring line, composed almost entirely of chain is occasionally employed. Typically, 18-22mm chain, 2-3 times the maximum depth of water in length, is connected from the anchor to a float, 10m or so from the cage, and a section of rope — PES or nylon — used to link the float to the cages. The section of chain between the float and the anchor adopts a

*Table 3.12* Specifications of open link proof steel chain (from Thomas *et al*, 1967).

| *d*[1] (*mm*) | *wt per unit length* (*kg m*$^{-1}$) | *safe load*[2] (*kg*) | *d*[1] (*mm*) | *wt per unit length* (*kg m*$^{-1}$) | *safe load*[2] (*kg*) |
|---|---|---|---|---|---|
| 4.8 | 0.6 | 263 | 15.9 | 6.1 | 2,268 |
| 6.4 | 1.1 | 408 | 19.1 | 8.6 | 3,209 |
| 7.9 | 1.6 | 617 | 22.2 | 11.6 | 4,355 |
| 9.5 | 2.4 | 844 | 25.4 | 14.9 | 5,625 |
| 12.7 | 4.1 | 1,497 | 28.6 | 18.9 | 7,076 |
| | | | 30.2 | 23.2 | 8,709 |

1 d = diameter of link.
2 safe load is approximately 0.25 times working load.

pronounced catenary shape, the vertical and horizontal holding power depending on the mass of the chain and the angle between the chain and the anchor. The buoy minimises the vertical loading on the cages and must be sufficiently large to support the mass of the chain in the water and to resist the vertical component of the kinetic forces imposed by the cages on the mooring system. Under shock loads, the chain/buoy acts as a spring absorbing much of the energy that would otherwise be transmitted to the anchor. For detailed discussions of chain mooring lines and catenary action, see Wiegel (1964) and Berteaux (1976).

Land anchors are sometimes used where cages are sited inshore, close to land only marginally above sea level. They are convenient to install and maintain. Galvanised steel ring bolts, with 20-30cm long shanks, are either driven into rock, or a hole drilled and the bolt grouted in with cement. Wooden pegs, 15cm x 15cm in cross-section by 200cm long, are also sometimes used for marine anchors (Chua, 1979). No chain is used and instead the rope is passed through a hole drilled in the peg, and spliced to form a loop. The peg is then driven into the substrate using a guide pole (*Fig 3.34*). Where the substrate is rocky, an iron pin with an eye at one end may be used and the chain at the end of the mooring line secured to the eye via a shackle.

The simplest and cheapest type of marine anchor is the deadweight or block anchor which typically consists of a bag of sand or stones or a block of concrete or scrap metal. Consider a block anchor, sitting on the sea-bed and connected to a fish cage by a length of mooring line. The block will begin to move when the horizontal component of the force exerted by the cage and mooring line equals the frictional force between the block and substrate. The holding coefficient, k, of the anchor is defined as R/W, the horizontal force divided by the mass of the anchor. It can be shown that k depends upon the angle between the anchor and the cage and thus the ratio between water

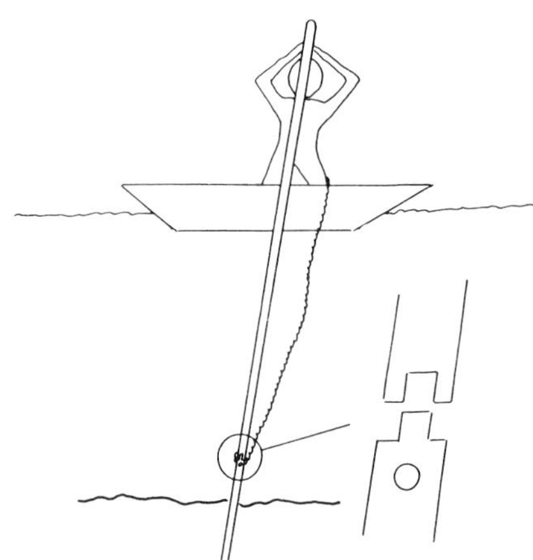

*Fig 3.34* Guide pole to drive anchor pegs into substrate (redrawn from IDRC/SEAFDEC, 1979).

*Table 3.13* The fixing coefficient, k, of sandbag anchors on different substrates and varying mooring cable length:water depth ratios (l:d) (from Nomura and Yamazaki, 1977).

| *Nature of substrate* | *l:d* 1 | 2 | 3 | 4 | 5 |
|---|---|---|---|---|---|
| Sand (vertically pulled) | 0.16 | 0.47 | 0.60 | 0.71 | 0.73 |
| Sand (horizontally pulled) | 0.21 | 0.59 | 0.65 | 0.68 | 0.74 |
| Mean value | 0.19 | 0.53 | 0.63 | 0.70 | 0.74 |
| Sandy mud (vertically pulled) | — | 0.33 | 0.41 | 0.49 | 0.61 |
| Sandy mud (horizontally pulled) | — | 0.31 | 0.31 | 0.43 | 0.62 |
| Mean value | 0.10 | 0.32 | 0.36 | 0.46 | 0.62 |
| Mud (vertically pulled) | 0.05 | 0.13 | 0.20 | 0.23 | 0.30 |
| Mud (horizontally pulled) | 0.05 | 0.33 | 0.34 | 0.48 | 0.52 |
| Mean value | 0.05 | 0.23 | 0.27 | 0.35 | 0.41 |

depth and line length — and the nature of the substrate. In *Table 3.13*, data on the k values of sandbag anchors are summarised.

Block anchors are inefficient *ie* have low holding power per unit installed weight. For example, a sandbag anchor weighing 100kg in water and installed on a sandy substrate, would have a holding power of between 19 and 27kg, depending upon the angle between the mooring line and the sandbag. The performance of the sandbag anchor is much poorer in mud. Note that there is a curvilinear relationship between the cable length: water depth ratio and the horizontal holding power which begins to level off at a k-value of 3-4. Similar relationships exists for other block anchors.

In practice, however, block anchors will be more efficient than this, since they tend to bed down into the substrate. Consider, for example, that a concrete block 1 × 1 × 0.4m, with a mass in water of 970kg, is sitting on a sandy substrate and is buried to a depth of 10cm. The cross-section area of the buried block opposed to the horizontal forces is 0.1m$^2$. Using data from Milne (1970) for the cohesive resistance of firm sandy soils to forces, gives a value of 548kg, thus increasing the holding power by a factor of almost two.

Concrete block anchors may be simply fabricated using wooden shuttering, tyres, or any other convenient object to form a mould. Steel rods for strengthening, and an eye bolt for a mooring attachment are usually incorporated. Once fabricated, the blocks can be dragged to the waters edge at low tide and floats attached, so that they may be floated out to the required location at high tide.

There are numerous types of embedding anchor (*Fig 3.35*). The holding power of an embedding anchor is related to its frictional resistance in soil, and so is dependent upon fluke area and soil penetration and the mechanical properties of the soil, rather than on the mass of the anchor. Anchor penetration is a function of fluke shape and the angle between the fluke and the shank whilst the frictional resistance of the soil is dependent upon soil cohesiveness and shear strength. As was demonstrated for block anchors, the angle between the anchor and the mooring cable is critical in determining holding power. In *Table 3.14*, the holding coefficients for various mooring cable length: water depth ratios are given for a 'kedge-type' anchor,

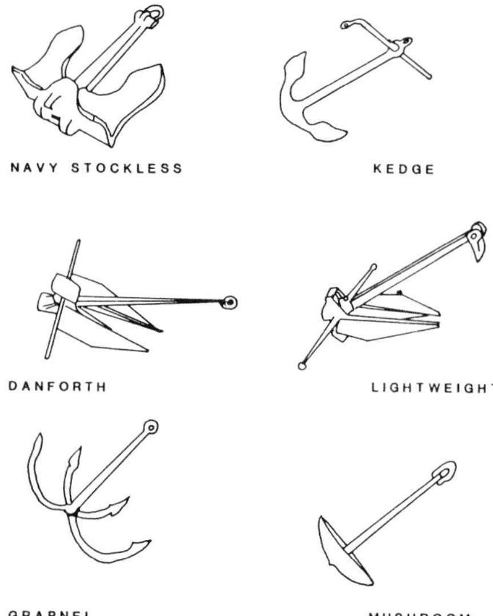


*Fig 3.35* Examples of various types of anchor.

illustrating differences between soil types and the effect of the angle.

Embedding anchors are thus very efficient *ie* they have a high holding power to mass ratio. Under optimum conditions (sandy mud soil, low angle between mooring line and anchor, *etc*) they are between 10 and 100 times as efficient as block anchors. They have, however, a number of disadvantages. They tend to be more expensive than block anchors in terms of cost per unit holding power and have to be bedded in properly. When dropped, embedding anchors do not immediately grip the substrate, but instead have to be moved for several metres. Moreover, if they drag, they can travel a considerable distance before bedding in again.

There are numerous other types of anchor, some combining the properties of block and embedding types (*eg* shaped concrete blocks) whilst others are designed for particular types of substrate (*eg* soft mud anchors). Anchor specifications can be obtained from manufacturers or chandlers, many of whom are increasingly becoming involved with supplying cage fish farmers,

*Table 3.14* The fixing coefficient, k, of kedge-type anchors on different substrates and varying cable length:water depth ratios (l:d) (from Nomura and Yamazaki, 1977).

| | *l:d* | | | | | |
|---|---|---|---|---|---|---|
| Substrate | 1.0 | 1.5 | 2.0 | 3.0 | 4.0 | 5.0 |
| Sand | 0.26 | 1.10 | 1.90 | 4.37 | — | 5.83 |
| Sandy mud | 0.23 | 1.90 | 3.27 | 4.40 | 5.50 | 5.15 |
| Mud | 0.11 | 0.60 | 1.99 | 3.29 | 5.11 | 6.46 |
| Mean | 0.20 | 1.20 | 2.39 | 4.02 | 5.31 | 5.81 |

and who are often willing to give advice.

Prior to choosing or installing anchors it is advisable to survey the lake or sea bed. When installing the moorings, the anchors should be positioned first, with excess mooring line which can then be adjusted to ensure that the cages are moored securely. Remember that embedding anchors must be dragged for a distance before bedding in. The position of the anchors can be accurately established using a Decca trisponder system, or by taking sitings with respect to local, easily visible land marks. Small buoys attached via light riser cables to the anchors may also be used. Mooring systems must be thoroughly checked at regular intervals and fouling should be removed from buoys and mooring lines. Shackles are likely to need replacing every 3-4 years.

*Appendix 3.1* Calculation of the buoyancy of a 3 × 3 × 3m bamboo cage (see Section 3.4.2.2).

The buoyancy of the bamboo collar can be computed from the volume and density of the bamboo used. Volume can be derived from:-

$$V = \pi nL[(0.5D)^2 - (0.5D - d)^2]$$

where V = volume ($m^3$); n = number of bamboos used; L = mean length of bamboos (m); D = mean outer diameter (m), and d = mean wall thickness (m). D and d are determined by taking a mean of several measurements.

For the 3 × 3 × 3m cage, twenty sections 4.5m long are used to construct the base. Assuming D = 0.12m and d = 0.008m, $V = 3.14 \times 20 \times 4.5 [(0.06)^2 - (0.052)^2] = 0.253m^3$. If the density of the bamboo used is 550kg $m^{-3}$, then the buoyancy = $(1{,}000 - 550) \times 0.253 = 114$kg, which would be sufficient to support the cage bag and one member of staff.

# 4 Site selection

## 4.1 Introduction

This chapter is concerned with the problems of selecting a suitable cage farm site in marine, freshwater and heated water environments. Current methods used to determine carrying capacity are discussed in Chapter 5.

The correct choice of site in any fish farming operation is vitally important, since it can greatly influence economic viability by determining capital outlay, and by affecting running costs, rate of production and mortality factors. Water based aquaculture systems suffer in comparison to land-based operations in that there is less room to make mistakes in choice of site. Poor land-based sites can, for example, sometimes be improved by drilling boreholes to increase water supply, or by introducing filters and sediment traps to remove suspended material. However there is little that can be done at a cage, pen or enclosure farm if the site is too exposed and water exchange is poor, or if water quality deteriorates. Thus it is extremely important to get site selection for cages right.

The criteria for selection of suitable sites for cage fish culture may be arranged into three categories (*Table 4.1*). The first is primarily concerned with the physico-chemical conditions which dictate whether a species of fish can thrive in an environment (temperature, salinity, oxygen, currents, pollution, algal blooms, exchange), the second lists the conditions that should be considered in order to site a cage structure successfully (weather, shelter, depth, substrate), whilst the third category reviews those factors which determine the possibilities of establishing a farm, and the profitability of the

*Table 4.1* Criteria for site selection of cages. Note that several factors are common to more than one category (see text).

| *Category 1* | | *Category 2* | *Category 3* |
|---|---|---|---|
| Temperature | | Depth | Legal aspects |
| Salinity | | Shelter | Access |
| Pollution | | Substrate | Security |
| Suspended solids | Currents | | Proximity to markets |
| Algal blooms | Fouling | | |
| Disease organisms | | | |
| Water exchange | | | |

venture (legal aspects, access, land-based facilities, security, economic and social considerations).

Although much of the data required must be collected by survey work and analyses of water samples, invaluable information can also be gained by talking to local people about prevailing weather conditions and the occurrence of toxic blooms or pollution. Consultations with them on the establishment of a cage fish farm may also help avoid problems of poaching and vandalism.

## 4.2 Environmental criteria for the cultured organism

### 4.2.1 Water quality

For a comprehensive treatment of water quality criteria relevant to cage culture, reference should be made to Kinne (1976), Alabaster and Lloyd (1980), and Poxton and Allouse (1982).

An ideal cage farm site must have good water quality, which means that it should not only be uncontaminated by toxic industrial pollutants, such as ammonia, nitrite, heavy metals and phenolic compounds, but also that the species' pH, temperature, oxygen and salinity requirements should be carefully considered. Advice on water quality assessment and methods for specific tests are detailed in a number of texts, such as Stirling *et al* (1985).

#### *4.2.1.1 Temperature and salinity*

Fish and other cultured aquatic organisms have no means of controlling their body temperature, which changes with that of the environment. A rise in temperature will increase the metabolism of the animal and cause a concomitant increase in oxygen consumption and activity, as well as in ammonia and carbon dioxide production.

Salinity is a measure of the amount of dissolved solids present in water (*Table 4.2*), and is usually expressed in parts per thousand (‰). Its relevance to aquaculture lies in its control of osmotic pressure, which can greatly affect the ionic balance of aquatic animals.

When selecting a site for cage culture, the species' optimum temperature and salinity conditions should be met, since even immediately outside these optima, behaviour, feeding, food conversion and growth can be adversely affected. Sub-optimal conditions can also contribute towards stress, leading to increased susceptibility to parasitic infections and reduced resistance to disease (Alabaster and Lloyd, 1980; Pickering, 1981; Shreck, 1982). Rapidly fluctuating temperatures and salinities are usually more harmful than gradual, seasonal changes, although some species are more tolerant than others. Optimal conditions for some of the more important cultured species are given in the relevant sections of Chapter 8.

In order to choose an appropriate site it is necessary to know what the temperature and salinity conditions are, and whether they are highly variable or not. It is thus important to have some understanding of the factors which control and influence temperature and salinity.

*Table 4.2* The major elements of 35‰ salinity seawater (after Kalle, 1971 and Poxton and Allouse, 1982).

| *Elements* | *g kg⁻¹* | [b]*Milli-equivalents kg⁻¹* |
|---|---|---|
| *Cations* | | |
| Sodium | 10.752 | 467.56 |
| Potassium | 0.375 | 10.10 |
| Magnesium | 1.295 | 106.50 |
| Calcium | 0.416 | 20.76 |
| Strontium | 0.008 | 0.18 |
| *Anions* | | |
| Chlorine | 19.345 | 545.59 |
| Bromine | 0.066 | 0.83 |
| Fluorine | 0.0013 | 0.07 |
| Sulphate | 2.701 | 56.23 |
| [a]Bicarbonate | 0.145 | — |
| Boric Acid | 0.027 | — |

[a] bicarbonate and carbonate will vary according to the pH of seawater.

[b] milli-equivalents $kg^{-1} = \frac{g\ kg^{-1}}{\text{atomic weight of ion}} \times$ ionic charge $\times$ 1,000.

There is a close correspondence between temperature regime and the amount of surface solar radiation received by a water body. The most important determinant of surface solar radiation is latitude, although cloud cover and shade caused by high surrounding hills or overhanging trees and shrubs may exert a short-term or localised influence (see Le Cren and Lowe McConnell, 1981, for review).

The longer wavelengths of the electromagnetic spectrum (>700nm) are the most important in the transfer of heat, and because of the high extinction coefficients associated with these near-infrared and infrared wavelengths, they are completely absorbed in the top metre or two of the water column. If it were not for mixing processes, then there would be an exponential fall in temperature with depth. The mixing of surface with deeper water requires energy, and the degree of mixing that occurs depends both on the energy inputs to the water body and on density differences between surface and underlying water, since the greater the difference, the greater the amount of energy that is required. In freshwaters, density differences are principally caused by temperature. Above about 4°C, water decreases in density with increasing temperature. In lentic (*ie* still water) freshwater systems such as lakes and reservoirs, a pattern of thermal stratification may often be observed where a layer of warm, less dense surface water (epilimnion) sits on top of a colder and denser deep layer (hypolimnion). Temperature differences within these two layers are small. Separating the two layers is a zone where temperature changes rapidly with depth, the thermocline (*Fig 4.1*).

In temperate regions, surface waters begin to heat up in the spring causing the onset of stratification which may last right through until the end of the summer. As the autumn advances, the surface waters cool down and the prevailing winds increase in strength and frequency until there is sufficient energy to turn the whole water body over, bringing up cooler water from below. During the winter, some lakes may freeze on the surface, resulting in a second period of stratification (*Fig 4.1*) which lasts until spring when the surface waters warm sufficiently to reduce density differences and facilitate wind-induced mixing.

Stratified lakes and reservoirs are typically deep and occur throughout temperate regions, and in some parts of the tropics. Water bodies which exhibit two periods of circulation per year (spring and autumn) are known as DIMICTIC, and tend to be continental in distribution (*ie* situated away from the warming influence of coastal air currents). MONOMICTIC lakes which turn over once in September or October are common in the UK and Western Europe. Both monomictic and dimictic lakes turn over completely. However, some extremely deep lakes, such as Lake Tanganyika in Africa, would require an immense amount of energy to turn over completely, and instead have a restricted or partial turn over. These are known as MEROMICTIC lakes.

Shallow and exposed lakes may not stratify at all, or stratify for short periods of time (days), and then turn over. POLYMICTIC lakes as they are called, occur throughout tropical and temperate regions. Lake and reservoir types and their distributions are summarised in *Table 4.3*.

Stratification in lotic (flowing water) systems is less pronounced than in lentic systems and is restricted to the slow-flowing lower reaches.

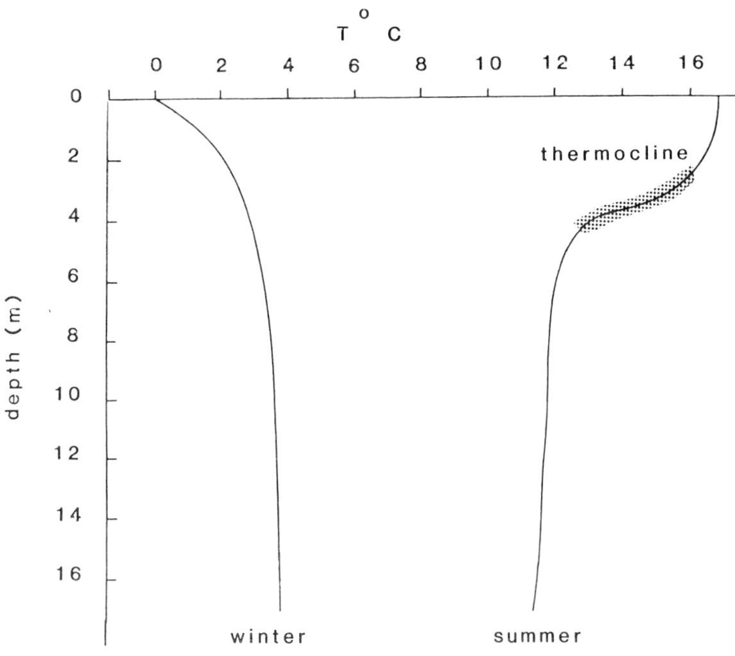


*Fig 4.1* Stratification patterns in a freshwater body.

*Table 4.3* Lentic water body types.

| *Type* | *Description* | *Distribution* |
|---|---|---|
| Holomictic | mix from top to bottom | universal |
| Dimictic | two circulation periods | continental mid latitudes and high altitude sub-tropics |
| Warm monomictic | one circulation period (autumn) | maritime mid-latitudes |
| Polymictic | Frequent stratification and circulation | shallow tropical and mid-latitude lakes |
| Meromictic | Only partially mixed | extremely deep lakes, universal |

In marine areas temperature variations with distance and depth are complicated by the effects of salinity. The salinity of seawater varies between about 32 and 40‰ and in open waters is determined by evaporation and precipitation. In offshore areas, where the stability of the water column is dependent upon tidal turbulence and water column depth (Pingree *et al*, 1978), stratification typically occurs in deep waters with low tidal velocities.

The temperature regime of marine coastal areas, which are more commonly used for cage fish culture may be greatly influenced by runoff from the land. In temperate regions, river discharges and surface freshwater runoff will be colder in winter and warmer in summer than the receiving coastal seas, whereas in tropical parts of the world, freshwater runoff is likely to be warmer than the sea during the dry season, but may be cooler during the wet season. Because the density of water is determined both by salinity and temperature (*Fig 4.2*), the mixing of terrestrial runoff with seawater requires energy. The degree to which the freshwater and seawater factions mix depends principally upon the volume of freshwater runoff and the mixing energy (tidal, wind) available.

An estuary may be defined as 'a semi-enclosed coastal body of water which has a free connection with the open sea and within which the seawater is measurably diluted with fresh water derived from land drainage' (Pritchard, 1967). Therefore, marked changes in temperature and salinity with depth might be expected in such an environment. However, because there are many types of estuary which vary in geomorphology there are a number of different stratification patterns. At one extreme, is the strongly-stratified estuary. Here, surface salinity differs from bottom salinity by several parts per thousand, and there is a strong salinity gradient (halocline) at mid-depth. There is a net flux of surface water seawards, and a net flux of deep water landwards, with some mixing of the two layers along the entire length of the estuary. However, where the river flow is large, then undiluted freshwater may flow out to open sea above the underlying layer of undiluted seawater, and no mixing between layers occurs. This type of strongly stratified estuary is called a salt-wedge estuary.

Another type of strongly stratified estuary is the fjord (see later). Here three layers may often be detected: a surface brackish water layer, an underlying layer of increasing salinity with depth, and a deep, stagnant layer

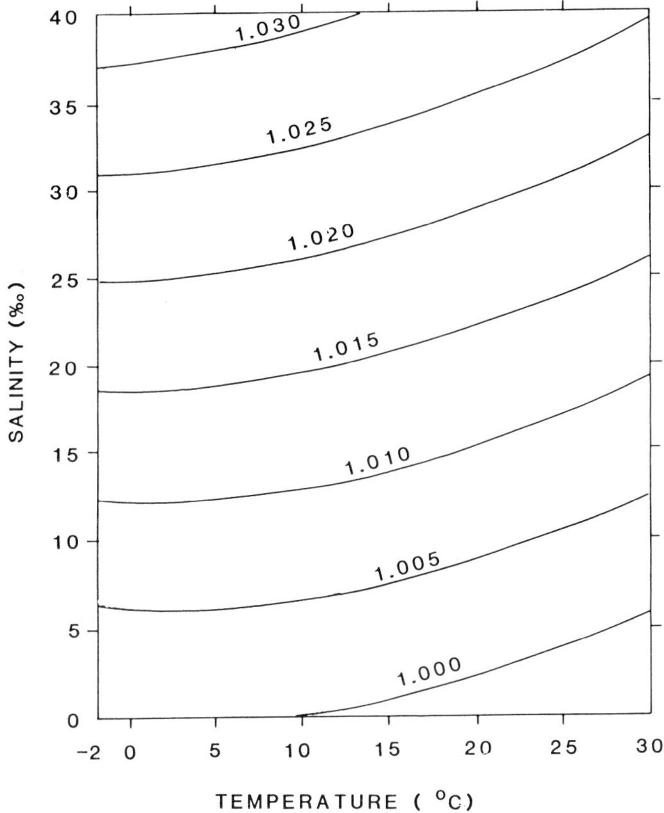


*Fig 4.2* Specific gravity of water at different temperatures and salinities. Lines of equal density are shown (redrawn from Kalle, 1971).

where salinity is constant to the bottom.

At the other extreme are the well-mixed estuaries where there is virtually no change in salinity with depth. In such estuaries, the transport of salt water landwards and fresh water seawards is achieved by horizontal eddy diffusion (Ketchum, 1983), and the net flux at all depths is seawards.

Intermediate in nature to the well-mixed and strongly stratified estuaries are the weakly-stratified estuaries, in which there is an increase in salinity of a few parts per thousand with depth. At the landward end of such estuaries is a zone of river water, undiluted with seawater, whilst at the mouth of the estuary we would find seawater undiluted with freshwater.

It should be pointed out, however, that the above description greatly oversimplifies stratification patterns in estuaries since within a single estuary changes in stratification with space and time occur. Detailed accounts of estuarine hydrography can be found in Dyer (1973) and Ketchum (1983). Some attempts have been made to model the stratification process in inshore coastal areas. Since density can be determined from temperature and salinity data, it is possible to calculate the amount of energy required to bring about complete mixing of the water column:-

$$\theta = 1/h \int_{-h}^{0} (\bar{\rho} - \rho) gz.dz \qquad \text{(Simpson, 1981)}$$

where $\theta$ = potential energy anomaly (J $m^{-3}$); h = water column depth (m), $\rho$ = density at depth z. Low values of $\theta$ (<10) correspond to a low stability or mixed water column, high values of $\theta$ (>10) are indicative of a stable or stratified water column, and values between 10 and 30 indicate a transitional or weakly stratified water column. Jones *et al* (1984) applied the model to data from the west coast of Scotland, and found a general agreement between water column depth and stability of the water column. Exceptions could be attributed to unusually stable weather conditions, resulting in increased heat input to surface water.

The potential for using thermal discharges from power stations for aquaculture purposes in temperate regions is widely recognised (almost 10% of publications relating to cage fish culture are concerned with rearing fish in thermal effluents) although the suitability of the water for cage fish farming very much depends upon the condition of the cooling water source and the design and operation of the power station. Water of varying salinities and quality are used to generate steam and drive turbines which in turn drive the generators, the waste steam being fed to a condenser and then discharged as thermal effluent. Some older stations, where water supplies are plentiful, use once through cooling systems whilst others recycle part of the water. Heated cooling water is either discharged directly into the receiving water body or cooled prior to discharge. Cooling methods involve either cooling ponds, some of which are equipped with fine-nozzle sprays to accelerate cooling by greatly increasing the water surface area, or cooling towers of various design. Heated effluent is eventually discharged via pipes or canals into the water body from which it was abstracted (see Langford, 1983, for review).

Cages may be sited in cooling ponds or canals or in the receiving water body. The temperature of the water available to caged fish depends upon geographic location (temperate/tropical), water supply (estuarine river/lake), the design of the system, site (pond/canal/receiving water body) and the proximity of the cages to the power station. Thermal effluent is usually between 6 and 15°C warmer than the intake (*ie* thermal power stations have a $\Delta$T of 6-15°C). When operating at full output $\Delta$T is constant, although because of changes in operating schedule due as much to political and economic pressures as to climate or hydrographic conditions, $\Delta$T can fluctuate widely. Changes in T of 10-15°C over a few days are not unknown (Langford, *ibid*). It is thus extremely important to determine the $\Delta$T pattern at a prospective thermally-heated site. Different species may have to be cultured at different times of the year.

### *4.2.1.2 Oxygen*

Oxygen is required by all higher organisms for the production of energy; energy that is required not only to fuel the essential functions of the organism, such as digestion and assimilation of food, and maintenance of osmotic balance, but also for activity. Oxygen requirements vary with species, stage of development and size, and are also affected by environmental factors, such as temperature. If the supply of oxygen to an animal deviates from the ideal, then feeding, food conversion, growth, and health can be adversely affected, and it is therefore of great importance to the aquaculturist that good oxygen conditions prevail at a site. Specific oxygen requirements for various cultured species will be discussed in Chapter 8. This section is principally concerned with the factors that determine environmental oxygen conditions.

The oxygen content of water is for the most part in a state of dynamic equilibrium with atmospheric oxygen, the gas passing in both directions across the air/water interface by diffusion. However, due to the low partial pressure of oxygen and its relatively poor solubility, the oxygen content of water is never more than 5% of the oxygen content of a similar volume of air. Solubility declines with increasing temperature and salinity. Thus a sample of seawater contains less dissolved oxygen (DO) than an equivalent volume of freshwater of the same temperature, and a sample of cold water will contain more DO than the equivalent volume of warm water, providing that the salinities are similar. An increase in altitude will result in a fall in partial pressure and a consequent reduction in the quantity of oxygen that the water can hold. These relationships are summarised in *Table 4.4*. Pressure corrections can be made by multiplying by a factor of P/760, where P is the observed barometric pressure in mm Hg, or, by P/1010, if the barometric pressure is recorded in mbar (Stirling *et al*, 1985).

At equilibrium, water is described as being 100% saturated. However, temporary departures from this state often occur, when the DO level may fall below or rise above saturation due to changes in oxygen supply and demand. A major influence on the pool of DO is the planktonic algae community. They require only light and a supply of nutrients in order to convert available carbon — usually dissolved carbon dioxide — into plant tissue by photosynthesis, releasing oxygen as a byproduct:-

$$6CO_2 + 6H_2O \xrightarrow{\text{light}} C_6H_{12}O_6 + 6O_2$$

During the day, there is a net production of oxygen, and at night when photosynthesis shuts down the algal community becomes a net oxygen consumer. If the algal populations grow sufficiently large, then due to the inefficiencies of air/water diffusion processes, a situation can arise where supersaturation of DO occurs during the day and sub-saturation conditions prevail at night, with late afternoon maxima and pre-dawn minima. In highly productive fish ponds, diurnal DO variations can be as great as 7-8 ppm (Boyd, 1979), severely stressing the fish. However, in water-based culture systems where algal populations are usually much smaller, diurnal variations in DO rarely exceed 2-3 ppm, although these may still prove stressful.

The environmental conditions which encourage large populations of algae (blooms) to develop typically occur during the warmer months in areas subject to high nutrient influxes. External sources such as sewage discharges

*Table 4.4* Solubility of oxygen in water (mg $l^{-1}$) at different temperatures and salinities when exposed to water-saturated air at a total pressure of 760mm Hg (= 1.01 bar) (from Stirling *et al*, 1985).

| *Temp* °C | *Salinity ‰* 0 | 5 | 10 | 15 | 20 | 25 | 30 | 35 |
|---|---|---|---|---|---|---|---|---|
| 0 | 14.6 | 14.1 | 13.6 | 13.2 | 12.7 | 12.3 | 11.9 | 11.5 |
| 2 | 13.8 | 13.3 | 12.9 | 12.5 | 12.1 | 11.6 | 11.3 | 10.9 |
| 4 | 13.1 | 12.7 | 12.2 | 11.8 | 11.5 | 11.1 | 10.7 | 10.3 |
| 6 | 12.5 | 12.1 | 11.6 | 11.25 | 10.9 | 10.5 | 10.2 | 9.8 |
| 8 | 11.8 | 11.45 | 11.1 | 10.7 | 10.4 | 10.1 | 9.7 | 9.4 |
| 10 | 11.3 | 10.9 | 10.6 | 10.2 | 9.9 | 9.6 | 9.3 | 9.0 |
| 12 | 10.8 | 10.45 | 10.1 | 9.8 | 9.5 | 9.2 | 8.9 | 8.6 |
| 14 | 10.3 | 9.95 | 9.7 | 9.4 | 9.1 | 8.8 | 8.6 | 8.2 |
| 16 | 9.9 | 9.55 | 9.3 | 9.0 | 8.7 | 8.5 | 8.2 | 7.9 |
| 18 | 9.5 | 9.15 | 8.9 | 8.6 | 8.4 | 8.1 | 7.9 | 7.6 |
| 20 | 9.1 | 8.8 | 8.6 | 8.3 | 8.1 | 7.8 | 7.6 | 7.3 |
| 22 | 8.7 | 8.6 | 8.3 | 8.1 | 7.9 | 7.7 | 7.5 | 7.2 |
| 24 | 8.4 | 8.3 | 8.1 | 7.8 | 7.6 | 7.4 | 7.1 | 6.9 |
| 26 | 8.1 | 8.0 | 7.7 | 7.5 | 7.3 | 7.1 | 6.8 | 6.6 |
| 28 | 7.8 | 7.7 | 7.5 | 7.3 | 7.0 | 6.8 | 6.6 | 6.4 |
| 30 | 7.6 | 7.4 | 7.2 | 7.0 | 6.8 | 6.6 | 6.4 | 6.1 |
| 32 | 7.3 | 7.2 | 7.0 | 6.9 | 6.6 | 6.3 | 6.1 | 5.9 |
| 34 | 7.1 | 7.0 | 6.9 | 6.7 | 6.4 | 6.2 | 6.0 | 5.8 |
| 36 | 6.9 | 6.8 | 6.7 | 6.5 | 6.2 | 6.1 | 5.9 | 5.7 |
| 38 | 6.7 | 6.6 | 6.5 | 6.4 | 6.1 | 5.9 | 5.7 | 5.6 |
| 40 | 6.5 | 6.5 | 6.3 | 6.2 | 6.0 | 5.7 | 5.6 | 5.5 |

and agricultural runoff may be important contributors. However, a sudden upwelling of nutrient-rich water from the deeper parts of the water body during the breakdown of stratification may also stimulate blooms. Certain types of water body such as enclosed areas, with poor flushing rates (see Section 4.2.2) tend to be more subject to blooms. Serious problems can occur when algal blooms die due to sudden changes in climatic conditions affecting light and temperature, or exhaustion of some essential nutrient and during the subsequent decomposition, microbial respiration may remove much, or even all of the DO resulting in fish kills (Grave, 1981).

Another community which may make demands on the pool of DO is the benthos, and it is suspected that the high particulate waste loadings often associated with intensive cage fish farm sites may greatly increase the removal of DO by the benthic microbial and invertebrate communities from the overlying water, which in turn may reduce the DO content in and around the cages (see Chapter 5). Investigations into the importance of oxygen consumption by sedimented wastes are, however, still at an early stage.

Supersaturation of dissolved gases ($O_2$, N) in thermal power station effluents can arise from pressurisation of water by pumping or from flowing over spillways, raising saturated water from depth to the surface, or from heating saturated water (Sylvester, 1975). High dissolved gas levels have caused extensive mortalities of a number of species held in cages at various power stations through gas bubble disease (Marcello and Strawn, 1973;

Ciesluk, 1974). Submerging cages to reduce exposure to gas supersaturation has proved an effective means of reducing mortalities (Chamberlain and Strawn, 1977), since gas saturation decreases by about 10% per metre depth due to increased hydrostatic pressure.

In summary, sites which are strongly stratified for much of the year, and/or where algal blooms tend to develop are likely to have periodic poor oxygen conditions and should be avoided if possible. Sites which have good bottom currents and which therefore disperse sedimenting wastes are also preferable. (*NB* This cannot, of course, apply to lentic inland water bodies where currents are generally low). The actual supply of DO to the caged fish is dependent not only upon the DO concentration in the water, but also upon water exchange through the net cage. Currents and fouling are important in these respects, and will be discussed later.

### *4.2.1.3 pH*

pH is a measure of the hydrogen ion activity ($^{\alpha}H^+$) of a solution, and is determined by the equation:-

$$pH = -\log {}^{\alpha}H^+$$

It is important to aquaculturists because extreme values of pH can directly damage gill surfaces, leading to death (McDonald, 1983). However pH is also important because it affects the toxicity of several common pollutants (ammonia, cyanide) and heavy metals (aluminium).

pH is expressed on a scale ranging from 0 (acidic) to 14 (alkaline). A pH value of 7 is considered to be neutral. Note that because it is a logarithmic scale, a change in pH of one unit means a ten-fold increase in the hydrogen ion concentration.

Seawater is usually alkaline, with pH values in the range 7.5-8.5, due to the chemical composition having more anions than cations. It is also well buffered, which means that it is comparatively resistant to changes in pH caused by the addition of alkaline or acidic compounds. Thus the pH of seawater does not show as marked seasonal or diurnal changes due to photosynthetic $CO_2$ production/uptake, as inland waters frequently do.

Freshwaters, because of the great variability in their ionic composition, may have pH values that lie anywhere between pH3 and pH11. Acidic waters are often associated with highland lakes, or with lakes subject to acid rain precipitation such as those that occur in certain areas of Northern Europe or North America. Alkaline waters typically occur where the geology of the watershed is dominated by alkaline sedimentary rocks, such as limestone or chalk. Lakes with exceptionally high pH values are often found in desert areas, where inflows are less than outflows, resulting in an accumulation of salts.

The ideal pH for most species seems to be in the range 6 to 8.5 and thus pH is not a problem at most marine sites. Some care must be taken with freshwater sites, since there can be marked seasonal and diurnal changes, particularly in low-buffered waters. Intensive operations will stimulate phytoplankton production which can, due to increased photosynthetic $O_2$ production, lead to elevated pH values particularly in summer. Ammonia toxicity may become a problem at such times (see later). In areas prone to acid rain precipitation, a sudden fall in pH can often occur in January — March, due to surface runoff of acidic snow-melt waters into lakes. Problems of low pH waters are discussed more fully in Chapter 7.

### *4.2.1.4 Turbidity*

Turbid conditions are caused by organic or inorganic solids suspended in the water column as a result of soil erosion, mining wastes, sewage and sewage effluent discharges, pulp and paper mill wastes, as well as a number of other industrial effluents (Alabaster and Lloyd, 1980). Some of these suspended solids may have toxic properties (*eg* the salts of various metals) whilst others, such as many of the organic wastes, can cause $O_2$ depletion in the water column during microbial breakdown. However, these properties will not be considered here. Planktonic algae, although also classified as suspended organic material, will be considered elsewhere (see Section 4.2.2).

The quantity and quality of suspended material present in the water column at any particular moment in time is largely under the control of water movement, which transports, fractionates and modifies the properties of the solids. Due to the effects of gravity, large, dense particles are more easily sedimented than small, less dense particles. However, water currents can prevent particles from sedimenting, and resuspend already sedimented materials. The chemical nature of the water, and salinity in particular, may also influence turbidity through its effect on floculation and thus sedimentation. This is particularly important in sediment transport in estuaries. Reviews of these processes may be found in Postma (1967) and McCave (1979).

Although suspended solids can cause a number of problems in aquatic systems (Alabaster and Lloyd, 1980), it is principally their direct effects on the caged fish that are of concern to the cage fish farmer. At sufficiently high levels suspended solids will cause gill damage inducing the gill epithelial tissues to proliferate and thicken (Ellis, 1944; Eller, 1975; Raghavan *et al*, 1979). If damage is sufficiently severe, the fish will die. Mortality rate varies with species, and also with the nature of the suspended material. Ellis (1944) has argued that the larger the particles, the greater their hardness and angularity, and therefore the greater the possibility of injury to gill tissues. Mortalities often occur after the period of exposure, thus complicating diagnosis (Alabaster and Lloyd, 1980).

The presence of suspended solids in the water has also been implicated in diseases such as 'fin-rot' (*Myxobacteria*) (Herbert and Merkens, 1961; Herbert and Richards, 1963), and in poor growth performance of fish. There is some evidence that the latter problem is in part due to the effect of turbidity on visibility, thus increasing food losses and impairing growth (Sigler *et al*, 1984).

It seems that at turbidity levels below about 100 mg $l^{-1}$ there is little effect on most fish species. Above this value the picture becomes complicated, particularly by factors such as exposure time. Although cage fish farmers would be best advised to avoid sites where such turbidity levels occur, this may not always be possible in riverine sites where several thousand mg $l^{-1}$ suspended solids can occur during flood periods (Alabaster and Lloyd, 1980). It should not be forgotten either, that cage fish farms are themselves a source of suspended solids.

### *4.2.1.5 Pollution*

Pollution has been defined by Holdgate (1979) as: 'The introduction by Man into the environment of substances or energy liable to cause hazards to human health, harm to living resources and ecological systems, damage to structure or amenity, or interference with legitimate uses of the environment.'

Thus, with respect to cage fish farming, a pollutant could be something which damaged the cage structure, adversely affected the cultured fish or its food, or accumulated in the fish to such a degree that it proved toxic to man when ingested. The number of pollutants entering aquatic ecosystems is enormous — Mason (1981) estimates at least 1,500 in freshwaters — and it is beyond the scope of this book to discuss or list even those most relevant to cage fish farming. Readers who are interested are advised to consult Alabaster and Lloyd (1982), Poxton and Allouse (1982) and Liewes (1984) as starting points. The principal categories of aquatic pollutant are given in *Table 4.5*, and whilst it would require extensive sampling and sophisticated methods of laboratory analysis to detect many of these compounds, risks may be significantly reduced by siting cages as far away as possible from large industrial conurbations. Unfortunately, avoidance of pollution is not as easy as it once was, since in many parts of the world competition for resources has forced fish farming to develop alongside industry or in heavily populated areas. In Japan, for example, one of the major hazards faced by yellowtail

*Table 4.5* Categories of aquatic pollutants (from Mason, 1981).

| |
|---|
| Acids and alkalis |
| Anions (*eg* sulphide, sulphite, cyanide) |
| Detergents |
| Domestic sewage and farm manures |
| Food processing wastes (including processes taking place on the farm) |
| Gases (*eg* chlorine, ammonia) |
| Heat |
| Metals (*eg* cadmium, zinc, lead) |
| Nutrients (especially phosphates and nitrates) |
| Oil and oil dispersants |
| Organic toxic wastes (*eg* formaldehydes, phenols) |
| Pathogens |
| Pesticides |
| Polychlorinated biphenyls |
| Radionuclides |

farmers is industrial pollution, which is now virtually ubiquitous around the southern Japanese coast (Tabira, 1980). In Laguna de Bay in the Philippines, the cage and pen industry has grown up alongside the industrial development and rapid population explosion which has occurred in the area. Until the Napindan Channel barrage scheme was implemented in 1983 the Pasig River, which connects the lake to the sea, used to reverse its direction of flow during the dry season, thus dumping considerable quantities of toxic wastes and domestic sewage from the city of Manila into the lake, and these were held responsible for a number of fish kills (Beveridge, 1984b).

However, geographically isolated areas can also be at risk from pollution. Oil spills from tankers using routes close to coastal fish farms have caused a number of problems, and even more insidious is the increasing risk of reductions in pH due to acid rainfall in some of the more remote lakes of Northern Europe and North America.

Thermal effluents from power stations may contain biocides such as chlorine which are used to control fouling, corrosion inhibiters and heavy metals (*eg* copper, zinc, cadmium, nickel and iron) (see Langford, 1983, for review) and care must be taken to ascertain the levels of these pollutants during the site selection process.

### 4.2.2 Phytoplankton blooms

The term 'phytoplankton bloom' refers to the periodic occurrence of large populations of planktonic algae in fresh and marine waters, which can happen whenever the appropriate conditions prevail: high light levels, high nutrient levels, warm water temperatures, and a combination of favourable hydrographic conditions. Blooms can adversely affect fish, not only through their physical presence in the water, which can damage or clog gills, but also through their influence on the pool of DO. Several phytoplankton species are sources of a musty flavour in fish, whilst other species are known to produce toxins which can kill a wide range of aquatic organisms, or which can accumulate in their tissues and prove fatal when ingested by man. Whilst phytoplankton blooms are discussed in general in this Section, emphasis will be placed on the factors which control the development of toxic blooms. The effects of phytoplankton on DO are discussed in Section 4.2.1.2.

At any moment in time the phytoplankton community in a natural aquatic environment will consist of tens of different species of algae. Since each species differs in its temperature, light or nutrient requirements, then under any given set of environmental conditions one species will tend to dominate. As conditions fluctuate during the year, due to changing light levels, temperature, nutrient availability and not least through the influence of the algae themselves, so the dominant species may change. This process is known as succession (see Round, 1981, for review) and is much less marked in the tropics, where environmental conditions are more stable.

Changes in the species composition of the phytoplankton community may also take place due to the influx of algae from other areas, through the action of coastal currents or rivers.

During bloom conditions the algal community, typically dominated by one species, becomes extremely large, and densities in excess of $10^6$ cells $ml^{-1}$ can occur. In theory, blooms of any algal species could occur; in practice, blooms of certain species seem to occur more frequently than others. In freshwaters the most important groups are the diatoms and Cyanobacteria (blue-green algae).

Freshwater diatom blooms tend to occur in temperate regions most frequently during the spring and occasionally during the autumn months, when they thrive in the increased light levels, yet low temperatures (Hutchinson, 1967). In fact, diatom blooms are usually comparatively small, and are termed blooms because they are present in relatively large numbers compared to the small numbers of other planktonic algal species present during the spring and autumn periods. Such blooms are scarcely discernable to the naked eye. However, blooms of diatoms may cause gill damage because of their silica-impregnated cell coats (see later).

Blooms of Cyanobacteria are more common than diatom blooms in inland waters, occurring in both tropical and temperate countries. In tropical water bodies Cyanobacteria blooms may be a permanent feature (*eg Spirulina* bloom in L. Chad), whereas in temperate countries transient blooms of Cyanobacteria typically occur from mid to late summer. Cyanobacteria blooms are the result of a combination of factors. Firstly, there must be sufficient light and nutrients present to stimulate the growth of the population and hence such blooms occur most commonly in productive lakes where nutrient loading is high. However, the formation of the bloom is also due to the accumulation of the algae at the surface of the water column (*Fig 4.3*). Many species of Cyanobacteria accumulate gas vacuoles in their cells

*Fig 4.3* Dying bloom of freshwater Cyanobacteria in a fish cage (courtesy J A Stewart).

during photosynthesis in dim light conditions (*ie* deep water), causing the cells or colonies to float towards the surface. If this occurs during a period of calm weather, then there may be insufficient turbulence in the surface waters to counteract the behaviour of the algae and so a bloom is formed (see Reynolds and Walsby, 1975; Reynolds, 1984, for reviews). If a gentle wind begins to blow, then a thick scum of Cyanobacteria may appear along the shore (*Fig 4.3*).

The more common genera of Cyanobacteria which form blooms are summarised in *Table 4.6*. Amongst these are several species belonging to the genera *Oscillatoria* and *Anabaena* which can cause problems of off-flavour in farmed fish (Lovell and Sackey, 1973; Tabachek and Yurkowski, 1976; Persson, 1980). Of possibly greater concern to fish farms, however, are toxin-forming bloom species which belong to the genera *Microcystis*, *Anabaena*, *Aphanizomenon* and *Oscillatoria*. Although populations of toxin-producing Cyanobacteria have been discovered throughout tropical and temperate inland waters (Carmichael, 1982), a recent review by Skulberg *et al* (1984) has shown that the factors which determine toxicity are complex and poorly understood. Algal toxins are of two basic types: alkaloid neurotoxins and protein or peptide hepatotoxins. However, not all species within a genus, and not all populations of a particular species may be toxin-producing, and even within a water body some bloom patches may be toxin-producing whilst nearby patches may not. The toxin produced by a particular species may vary from strain to strain (*Table 4.7*), thus making it impossible to identify a toxic bloom or the specific toxin involved without carrying out laboratory tests.

Whilst freshwater Cyanobacteria toxins are amongst some of the most potent which occur in nature (*Fig 4.4*) their role in reported fish kills is far from clear. Direct immersion of rainbow trout for 10 days in cultures of a known toxic strain of *Microcystis aeruginosa* resulted in no mortalities (Phillips *et al* 1985a) and the implications from this experiment seem to be that toxic effects can only occur if the fish is directly exposed to the toxin. There is some evidence that this may happen during the breakdown of a toxic bloom (Prescott, 1948), but is even more likely to occur if the alga is ingested directly.

From a cage fish farmer's point of view, it would therefore seem best to avoid sites where toxin-producing Cyanobacteria may exist, if possible; despite the fact that fish mortalities caused by such algae seem to be rare at cage farms. Extensive and semi-intensive cage operations may seem to be particularly at risk, since the fish rely wholly or in part on cropping the available phytoplankton. However, these less intensive methods of cage culture are almost entirely restricted to the tropics and subtropics, where little is known about the distribution of toxin-producing Cyanobacteria, or their effect on economically important phytoplankton-feeders, such as silver carp or tilapia. As far as we know, toxic species tend to occur in highly productive environments, which although ideal for extensive cage culture, are less suited to intensive operations.

A number of marine algal groups form blooms, including the diatoms, Cyanobacteria, prymnesiophytes and dinoflagellates. Marine diatom blooms are common in both temperate and sub-tropical coastal waters, and usually occur during the cooler months of the year. One diatom species, *Chaetoceros convolutus* has a number of prominent spines and has been responsible for heavy losses, due to the combined effect of interference with gill function and loss of blood from injury, at a number of North American salmon farms (Kennedy, 1978). Sockeye salmon (*Oncorhynchus nerka*) seem to be particularly susceptible to damage. However, only certain sites seem to have suffered losses, and it has been suggested by Kennedy ( *ibid*) that by careful site selection the problem can be avoided.

Marine species of Cyanobacteria which form blooms are uncommon, but include the genera *Trichodesmium* and *Noduleria* (Round, 1981). Although a few tropical and sub-tropical species are toxic, such as some members of the genera *Lyngba* and *Oscillatoria* (Moore, 1982), they seldom form blooms and have rarely been associated with fish kills.

*Table 4.6* Genera of bloom-forming Cyanobacteria in inland waters.

| *Order* | *Family* | *Genus* |
|---|---|---|
| Chroococcales | Chroococcaceae | *Coelosphaerium* |
| | | *Gomphosphaeria* |
| | | *Microcystis* |
| Nostocales | Oscillatoriaceae | *Oscillatoria* |
| | | *Spiralina* |
| | | *Trichodesmium* |
| | Nostocaceae | *Anabaenopsis* |
| | | *Aphanizomenon* |
| | Rivulariaceae | *Gleotrichia* |

From Reynolds and Walsby (1975).

*Table 4.7* Cyanobacterial toxins (modified from Skulberg *et al*, 1984).

| *Species* | *Toxin name* | *Structure* |
|---|---|---|
| *Lyngba majuscula* | lyngbatoxin A | alkaloid |
| | debromoaphysiatoxin | phenolic |
| *Schizothrix calcicola* | debromoaphysiatoxin | phenolic |
| *Oscilatoria nigroviridis* | oscillatoxin A | phenolic |
| *Nodularia spumigena* | nodularia toxin | unknown |
| *Microcystis aeruginosa* | microsystin | peptide |
| | microsystin type-c | peptide |
| *Anabaena flos-aquae* | anatoxin-a | alkaloid |
| | anatoxin-b | unknown |
| | anatoxin-c | peptide |
| | anatoxin-d | unknown |
| | anatoxin-a(S) | unknown |
| *Aphanizonenon flos-aquae* | aphantoxin | alkaloid |
| *Oscillatoria agardhii* | oscillatoria toxin | unknown |
| *Oscillatoria agardhii var. isothrix* | oscillatoria toxin | unknown |
| *Oscillatoria rubescens* | oscillatoria toxin | unknown |

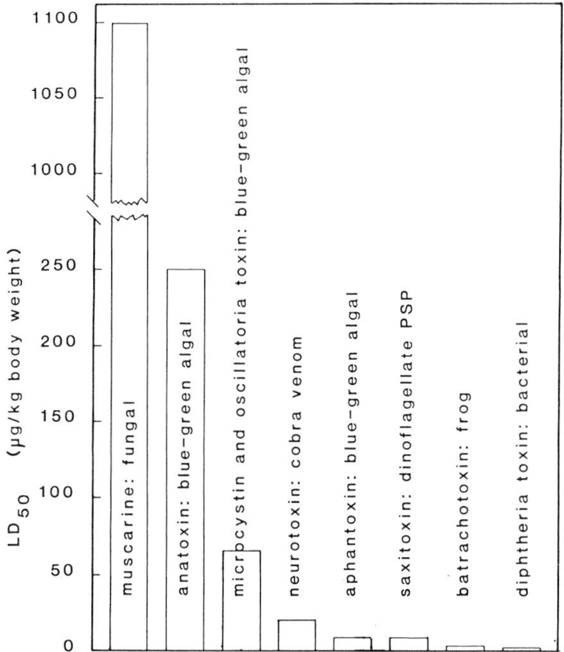


*Fig 4.4* Toxicities of various animal, fungal, bacterial and blue-green algal toxins. $LD_{50}$ represents the dose, in g purified toxin per kg body weight, required to kill 50% of a given population of laboratory mice or rats by intraperitoneal injection (redrawn from Skulberg *et al*, 1984).

*Prymnesium parvum* is a euryhaline, toxin-producing alga which has caused widespread fish mortalities in brackish water ponds in Israel (Hepher and Pruginin, 1981), although it does not seem to have been responsible for fish deaths elsewhere. Details of the toxicity of this species can be found in Shilo (1982).

The most important group of toxin-producing marine algae from the fish farmers' point of view are undoubtedly the dinoflagellates, which during bloom conditions cause the so-called 'red tides'. However, of the 1,200 species, many of which cause red tides, only a dozen or so have been shown to produce substances which are toxic, and only half of these have been implicated in fish kills (*Table 4.8*). Nevertheless, a number of incidences involving several species of fish (salmonids, gadoids, eels) have occurred in fish farms in Western Europe in the past decade (Tangen, 1977; Taylor and Seliger, 1979; Parker, 1981; Jones *et al*, 1982; Mortensen, 1985). In many cases, *Gyrodinium aureolum* has been responsible, although a mysterious toxic microflagellate, known as 'Flagellate X' has been implicated in a number of fish kills at salmon farms in Scotland and Ireland.

Roberts *et al* (1983) have described the sequence of events leading to death in *S. gairdneri* following exposure to toxic *Gyrodinium*: increased opercular rate, hyperactivity followed by inactivity, and death within 1 to 24 h. The gills of the dead fish displayed marked histological changes characterised by necrotic degeneration and disintegration of the lamellar epithelium (*Fig 4.5*). The structure of the toxin still remains to be determined.

Other reports of dinoflagellates causing deaths at fish farms have come from Japan (Doi *et al*, 1981; Nishimura, 1982).

Toxic dinoflagellate blooms tend to occur in warm water, and are therefore restricted to the summer months in temperate regions and are rare during cold water upwelling episodes in tropical coasts (Reyssac and Roux, 1972). Gowen *et al* (1985) have suggested that there are three ways in which dinoflagellate or 'Flagellate X' blooms could occur at a coastal site:-

(i) The bloom could develop at the site from species already present ('Flagellate X' or other toxic microflagellate);
(ii) The bloom could develop at the site from species transported into the farm area ('Flagellate X' or other toxic micro-flagellate);
(iii) The bloom could develop outside the site and be transported into the farm area (dinoflagellate).

The risk of a type (i) or type (ii) bloom occurring depends very much on the nature of the site. If water exchange at the site is in days and if nutrient levels (especially dissolved $NO_3$) are low, then the risk is low. Risks of type (i) and (ii) bloom occurring are much higher in weakly stratified fjordic bays, with long (several weeks) exchange periods and if there are also high levels of dissolved $NO_3$ present, even in deep water, then the risk is increased further.

For sites at risk from type (iii) blooms, environmental conditions at the site are less important than proximity to areas where dinoflagellate blooms might occur. Recent studies of Northern European waters have demonstrated that dinoflagellates thrive in shelf sea-front boundaries (*ie* areas where tidally-mixed inshore waters and thermally stratified offshore waters meet) (Pingree *et al*, 1978), and that certain locations, by virtue of their proximity to these fronts and the prevailing currents in the area, are particularly susceptible to blooms (Gowen *et al*, 1985).

It is thus apparent that some coastal sites, which are otherwise highly suitable for cage fish farming, may be at risk from toxic dinoflagellate blooms. One simple way to avoid such sites is to inquire from local people and the relevant authorities about the occurrence of red tides in the area, since it seems that toxic dinoflagellate blooms are recurrent events. Coastal shelf sea fronts may also be detected by satellite. However, despite such precautions, the risk of a toxic algal bloom cannot altogether be dismissed, since the cage farm will inevitably change the nature of the site by adding wastes to the water, and research in Japan has shown that fish farm wastes can stimulate dinoflagellate blooms (Nishimura, 1982). It would be best, therefore, to avoid nutrient-rich marine sites and sites where the exchange period is longer than a few days.

Based upon the above criteria, Gowen (1984) has identified a number of high risk sites on the west coast of Scotland. This approach might also be

*Table 4.8* Some of the known toxic dinoflagellates. Interested readers should also consult Proc. 3rd Int. Conf. Toxic Dinoflagellates, 8-12 June 1985, New Brunswick (in press), for most recent information.

| *Dinoflagellate* | *Usual distribution* | *Poison* |
|---|---|---|
| *Gonyaulax catenella* | North Pacific coasts, California to Japan, Chile, South Africa | Causes PSP[a] Structure determined |
| *Gonyaulax tamarensis*[b] | Coasts of New England, Canada, countries along North Sea | Causes PSP Structure determined |
| *Goynaulax acatenella* | Coast of British Columbia | Causes PSP Poison not isolated |
| *Gonyaulax monilata* | Gulf of Mexico | Toxic to fish, but not to warm blooded animals Poison not isolated |
| *Gonyaulax polyedra* | Coast of Southern California, North Sea | Poison reported but not verified |
| *Gymnodinium breve* | Gulf of Mexico | Toxic to fish, chicks and mice Partially purified |
| *Gymnodinium venficium* | English Channel | Toxic to fish and mice |
| *Gyrodinium aureolum* | North Sea | Toxic to fish and mice and possibly shellfish |
| *Pyrodinium bahamense var compressa* | South China Sea | Toxic to fish and shellfish |
| *Pyrodinium phoneus* | North Sea | Causes PSP Poison not isolated |
| *Exuviaella mariae-lebouriae*[c] | Japan | Causes degeneration of liver and kidney tissue Partially purified |
| *Chattonella antiqua* | Japan | Toxic to fish. Causes damage to gills and probably interferes with gas exchange |

[a]PSP = paralytic shellfish poisoning.
b = *G. excavata*.
c = *Prorocentrum maximum* var. *mariae-lebouriae*.
*Source:* Schantz (1982), Jones *et al* (1982), Shimada *et al*, 1983, McLean, 1984.

applicable to fjordic marine sites in other parts of the world (*eg* Canada, New Zealand, Norway, Chile).

### 4.2.3 Disease

The worst types of site from a disease risk point of view are (i) those in which pathogenic or potentially pathogenic organisms exist prior to establishment of the farm, and (ii) those in which disease organisms are likely to thrive following the establishment of a farm.

Organically polluted water bodies seem to harbour more disease agents than unpolluted water bodies. 'Red-boil disease' in the estuarine grouper (*Epinephelus salmoides*), for example, is produced by the bacterium *Vibrio parahaemolyticus*, and is contracted following skin damage caused by handling (Wong *et al*, 1979; Chua and Teng, 1980). The organism has been found in exceptionally high numbers in sewage-polluted waters, such as in the Straits of Penang. *Vibrio*-induced mortalities of up to 90% of caged stocks have been recorded from here, whilst almost no deaths were caused at a clean water site on another part of the island (Chua and Teng, 1978). Similar differences between the two sites in terms of *E. coli* numbers were also recorded (Chua and Teng, 1980).

Haemorrhagic septicemia is a disease of cultured freshwater carps (*Labeo rohita*, *Cyprinus carpio*) which can be caused by *Aeromonas punctata* (Toor *et al*, 1983) and its presence in freshwaters is related to trophic state (Rippey and Cabelli, 1980). There is evidence, too, of a link between trophic state and pathogenic fungal infections in fish (Grimaldi *et al*, 1973). The occurrence of other fish pathogens also seems to be related to trophic conditions. For example, Chua (1979) has observed that the ectoparasitic marine isopod *Nerocilia* which attacks the rabbit fish (*Siganus* spp.) is also more prevalent in organically enriched marine waters.

Potentially pathogenic organisms, therefore, are often associated with water bodies where a source of infection, such as an untreated sewage discharge, is present or where there is a good supply of organic nutrients that the organisms can utilise. There is now mounting evidence that some long-established cage farm sites may fall into this category, and as such, are a bad disease risk.

Both wild fish populations and organisms which are intermediary hosts in a particular fish parasite's life cycle can also cause disease problems for the fish farmer. The introduction of large numbers of caged fish to a system can have a dramatic effect on disease agents, although the mechanisms involved in the

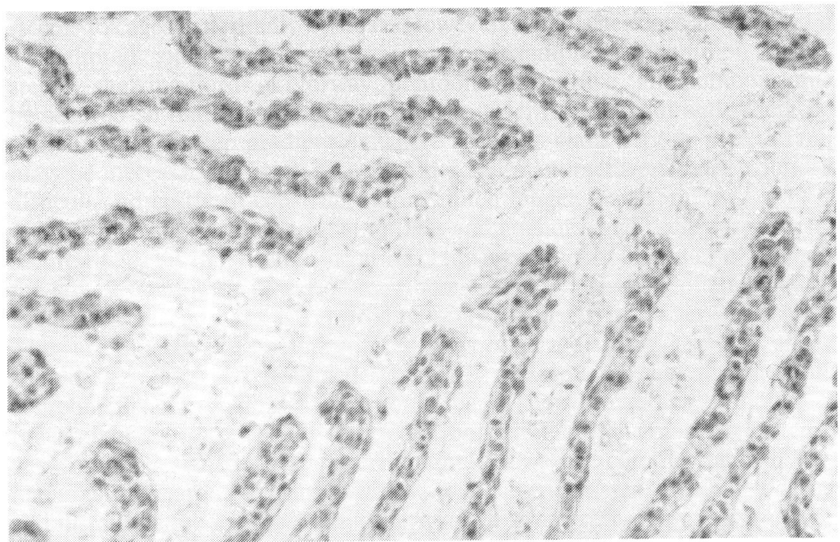

*Fig 4.5* Section of gill affected by the marine dinoflagellate, *Gyrodinium aureolum*, showing distorted secondary gill lamellae. Cytoplasmic and nuclear debris is evident between the lamellae (H & E, x 325) (courtesy Professor R J Roberts).

transmission of parasites from wild to cage fish and *vice versa* are poorly understood (McGuigan and Sommerville, 1985). In several cases in the UK caged fish have become severely infested with the cestodes *Triaenophorus nodulosus* and *Diphyllobothrium* spp. resulting in heavy mortalities and the eventual closure of one farm (Wootten, 1979; Jarrams *et al*, 1980). Those infections were traced to the wild fish populations which were subsequently found to be carrying the parasites.

Data from Matheson (1979) showed that Atlantic salmon parr raised in cages in a freshwater Scottish loch became heavily infected with *D. ditremum* and *D. dendriticium* within two months of being introduced to the site. Although the parasites were not isolated from the wild brown trout (*Salmo trutta*) in the loch, few specimens had been examined. In another Scottish loch, a survey of the wild brown trout failed to uncover the presence of any endoparasites, and as a result, one month later cages of rainbow trout, which were also examined and found to be free from parasites, were introduced. However, two months later the cage fish were found to be heavily infected with *Diphyllobothrium* spp. (Sommerville, unpublished data). It is believed that infestation may have been precipitated by the influx of gulls to the area and exacerbated by the inadequate use of feeds which caused the caged fish to ingest copepods, the intermediary hosts of these cestodes (Phillips *et al*, 1983).

*Cryptocotyle lingua* is a marine digenean the metacercariae of which encyst in the skin of fish, causing disfigurement. The first host of this parasite is the common whelk (*Littorina littorea*) which is found on rocky coasts, and consequently, this parasite is a problem primarily encountered by salmon farms which are sited close to such areas (Needham, 1980).

Disease risks, therefore, can be minimised by avoiding sites where there are parasites or other disease agents present in the wild fish, intermediary hosts, or environment which could be transmitted to the caged fish. Russian workers suggest that diplostomatosis in caged trout and carp may be avoided by siting the cages 80-100m from fringing macrophyte beds where the snail *Lymnea stagnalis*, the intermediate host of *Diplostomum spathaceum* resides (Martyshev, 1983).

Unfortunately disease outbreaks still occur through the introduction of diseased stock to the farm, or the attraction of birds and other opportunistic predators to the area (Beveridge, 1984b).

### 4.2.4 Water exchange

Good water exchange, or flushing, at a site is essential for intensive cage culture operations in minimising the build up of wastes and all the attendant problems that this can cause. Water exchange is dependent upon currents, although the picture can be complicated by the effects of salinity, temperature and topography.

Some marine cage sites are in relatively open coastal waters, such as straits, or in the shelter of a group of coastal islands or a coastal indentation. In such locations, water exchange is likely to be rapid and unrestricted, occurring in a period of days rather than weeks. Applying the derivations of Edwards and Edelsten (1976), the volume of water in the bay, V, can be calculated from:-

$V = AD$, where V is in $m^3$, $A$ = surface area ($m^2$), and $D$ = mean depth (m).

The flushing time, T, can be estimated by dividing V by the incoming flow of water, F, such that:-

$T = V/F$

The hourly mean flow of water, F, is dependent upon tidal currents. Hence:

$F = AH/12.5$, where $H$ = mean tidal height (m)

Thus, $T = 12.5\ D/H$

However, many cage farms, particularly in Scotland and Norway, are situated in enclosed sea lochs (fjords) or bays, where the pattern of water exchange is much more complicated. Fjords were formed in mountainous regions of the high latitudes during the Pleistocene era, when the movement and pressure of ice sheets resulted in overdeepening and widening of pre-existing river valleys, leaving rock sills or bars near the mouth (Dyer, 1973). The resultant form is typically rectangular in cross-section, with a small width:depth ratio. Fjords may extend as far as 150km inland, often with one or more inflowing rivers. Although the above equation can be used to estimate the flushing rate of a fjord, it assumes that the fjord is well-mixed, and ignores the effects of 'short-circuiting'. The flushing of a fjord is dependent upon the movement of water at the sea end caused by the tides, and at the landward end due to freshwater inflows (see *Fig 4.6*) In reality, the idealised picture shown in *Fig 4.6* is greatly complicated by (i) the size and topography of the basin, (ii) the number, location and depth of sills, (iii) the

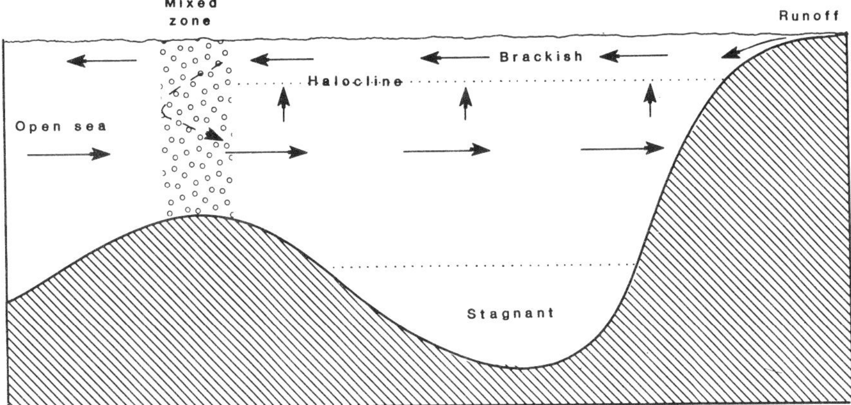


*Fig 4.6* An idealised fjordic circulation. The surface, brackish layer discharges to the sea. A compensation current flows in over the sill to replace water which is entrained (vertical arrows) into the surface outflow. Strong mixing may occur at the sill and there may be recirculation of brackish water into the compensation current (modified from Edwards and Edelsten, 1976).

magnitude and location of the freshwater inflows, and (iv) the tidal range in the area.

In some fjords, part of the water in the incoming flood tide may be recirculated ebb water, leading to a short-circuiting of the flushing system and an increase in flushing time. This is most likely to happen in fjords where there may be a number of sills restricting water movement, or in fjord systems with a long and complicated topography. Sites at the head of such fjords may be particularly affected.

The mouths of some fjords, such as those on the Pacific coast of North America have sills which extend to 150m or more, whilst others, such as those on the Norwegian, Scottish or New Zealand coastline are much more shallow, and sometimes have sills of 10m or less in depth (Dyer, 1973; Edwards and Edelsten, 1976). The presence of such shallow sills will not only increase the exchange time of the whole basin through the effects of short-circuiting, it may also greatly limit the flushing of deep layers of water (*Fig 4.6*).

The effect of increasing freshwater flow into a fjord system will be to decrease the overall exchange time (Dyer, 1973), although because of salinity/density effects (see Section 4.2.1.1), this will have a proportionally greater effect on surface rather than deep water exchange. An example of this is given by Edwards and Edelsten (1976), who found that the flushing time of Loch Etive, a fjord on the west coast of Scotland, was 8 days at a depth of 2m, whereas at 14m, it was 28 days.

The above points concerning the influence of sills and freshwater inflows illustrate that although currents at a site may appear to be ideal (10-60cm $s^{-1}$) for cage fish culture, water exchange can still be low, leading to poorer-than-expected water quality in surface waters, and a build-up of wastes on the bottom.

In freshwater lentic sites, the exchange time, or flushing rate (volumes per year) depends upon the volume of the water body, V, and the total volume of water flowing out of the water body each year, $Q_o$, such that:-

$\rho = Q_o/V$ (Dillon and Rigler, 1975).

This formula is appropriate for cages sited in the main body of a lake or reservoir but is likely to lead to an over-optimistic estimate if applied to a farm sited in a bay within a lake. Although the flushing rate of such a bay can in theory be computed using the formula above, substituting the volume of the bay for V, and the total annual outflowing water from the bay for $Q_o$, in practice the problems involved in estimating the latter component make such a computation very difficult. As a general rule, however, it is strongly advised that large-scale intensive units are not sited in enclosed bays.

It should also be noted that improved exchange at a site (*ie* an over-pessimistic prediction of $\rho$) can also occur, when cages are located near outflows.

The exchange time or flushing rate of inland waters ranges from a matter of seconds for riverine locations, to several years for many lakes. As stated in the introduction to this section, intensive cage fish farms are best sited in areas with good water exchange. In marine areas, this means at sites where there are good bottom as well as surface currents, and where the exchange period is in days rather than weeks. In fresh water, lakes and reservoirs where flushing occurs in weeks and months, rather than years, are best. In rivers, currents are generally too strong for intensive cage culture (see Chapter 2).

Semi-intensive and extensive cage culture operations, however, depend upon the supply of plankton/detritus, and thus a high rate of primary production is necessary. As is demonstrated in Chapter 5, primary productivity and flushing rate are inversely correlated, and thus sites with poor water exchange may be preferred for these types of culture.

### 4.2.5 Fouling

The fouling of net cages decreases the specified mesh size of the net bag, whilst increasing its surface area. The reduction in mesh size restricts the flow of water through the cages and thus reduces the rate of DO supply and waste metabolite removal which can adversely affect the fish. The increased resistance to water flow will often cause deformation of the net bag and thus a decrease in cage volume, as well as imposing additional stresses on the cage structure and moorings. The additional weight of fouling on cage bags can lead to net failure, and makes net changing difficult and time consuming.

According to Lovegrove (1979) the United States Navy estimate that throughout the world there are some 200 species of marine fouling plants and animals. The few studies that have been carried out on fouling of marine cages and pens suggest that the diversity of organisms colonising them can be large and that one or two of these organisms (*eg* the mollusc *Martesia striata*) can damage wooden structures by boring. Cheah and Chua (1979) counted more than 34 species of algae (cyanophytes, rhodophytes, chlorophytes), coelenterates, polyzoans, annelids, arthropods, molluscs, and simple chordates clinging to net cages sited in the Straits of Penang after immersion for only two months. No increase in diversity was found during the subsequent two months of the study although the weight of attached organisms more than doubled.

A number of studies have shown that colonisation begins by the formation of a primary film of detritus and algae (Horbund and Freiberger, 1970; Milne, 1970; Santhanam *et al*, 1983). However, the subsequent size, diversity and development pattern of the fouling community varies with materials used and the environmental conditions at the site. In Scotland, Milne (1970) tested the rate of fouling of mesh and netting panels of different material, and found that the galvanised panels developed much less fouling than the synthetic fibre netting panels (*Table 3.5*), whilst in Tuticorin, India, Santhanam *et al* (1983) observed that different organisms colonised the bamboo, oil drum and polyethylene netting parts of cages.

The size and diversity of the fouling community also seems to be greatly affected by temperature and the productivity of the environment. Thus, in warm, organically enriched waters such as the Straits of Penang and Hong Kong Harbour, the rate of fouling is high (Cheah and Chua, 1979; Tseng and Yuen, 1979). The rate of fouling is also reportedly higher in cages sited in thermal effluents (Chamberlain and Strawn, 1977). There is evidence too, that fouling is most rapid in areas of slow currents. In Singapore, for example, fouling is only reported as a problem where flood tide currents are less than 25cm $s^{-1}$ (PPD, 1981).

The range of fouling organisms and the rate of fouling decreases with salinity, and so fouling is generally not a problem at freshwater sites, although some in African lakes, the boring larval stage of the mayfly *Povilla adusta* is present, and can cause considerable damage to wooden cage structures (Coche, 1979) (*Fig 4.7*). Santhanam *et al* (1983) found many fewer fouling animals on net cages and pens sited in brackish than marine waters. Observations of fouling at a freshwater cage site in Scotland have shown that algae are the principal organisms responsible, although during late April/early May, large numbers of *Hydra* sp. are present. In their studies of

*Fig 4.7* Damage caused by *Povilla adusta* larvae to a submerged cage structure made of softwood after 13 months of continual immersion (from Coche, 1979).

fouling of tilapia cages in Laguna de Bay, Pantastico and Baldia (1981) also found algae to be the principal fouling organisms present, but that growth was highest in the upper portions of the net cages where light levels were greatest. The algae were also continuously cropped by the caged fish.

In summary, fouling is principally a problem at marine sites with low current velocities, and is greatly modified by temperature and nutrient status. Either such sites should be avoided, or management practices adapted to take account of the problem (see Section 7.6).

## 4.3 Environmental criteria for cages

### 4.3.1 Weather

The weather can determine the suitability of a particular site or area for cage fish culture through its influence both on the cage structure and on the enclosed fish. Of particular concern are violent storms and conditions of extreme cold.

The storms of tropical latitudes may be classified into four types, according to intensity:-

(i) Tropical cyclones (wind strengths $\geqslant$ force 12, 33m $s^{-1}$)

(ii) Severe tropical storms (wind strengths = force 10 - 11, 24 - 32m $s^{-1}$)

(iii) Moderate tropical storms (wind strengths = force 8 - 9, 17 - 23m $s^{-1}$)

(iv) Tropical depressions (wind speeds $<$ force 8, 17m $s^{-1}$)

The most violent of these, the tropical cyclones, have torrential rain accompanying the hurricane force winds and occur principally between latitudes 5° and 30°. There are a variety of regional names. In the southwest Pacific and Bay of Bengal, they are known as 'cyclones'; those which occur over the North Atlantic, West Indies and northeast Pacific are known as 'hurricanes', whilst those which occur in the China Sea and northwest Pacific are called 'typhoons'.

Although the precise factors which determine the origin of cyclones are not yet understood, a great deal is known about how and why they tend to occur. They develop from disturbances in the Inter-tropical Convergence Zone (ITCZ) (*Fig 4.8*) and tend to follow a westward track curving polewards in a clockwise manner in the northern hemisphere and anticlockwise in the southern hemisphere. As they travel they intensify and accelerate until they are typically 600-800km in diameter, although direction and speed can be unpredictable. The vast amounts of energy that fuel their progress are derived from heat released in the condensation of water held in the moist air sucked upwards from the ocean surface (Couper, 1983). There is more energy released when water vapour is condensed from warm air, since warm air holds more moisture. Thus, cyclonic development is greatest over the western areas of oceans during late summer and autumn when sea temperatures are highest (*Fig 4.9*). As a cyclone travels across a large land mass, it becomes starved of moisture and decays rapidly. Cyclones are thus classed as maritime phenomena.

Tropical cyclone paths are plotted in *Fig 4.9*. Most occur in the West Pacific and Indian Ocean (*Fig 4.10*). In many of the typhoon-affected areas the development of marine fish culture has been hampered. In the Philippines, for example, there is no commercial cage or pen culture of fishes in marine waters, due in part to the adverse coastal climatic conditions. In other parts of the world, such as Japan, cage fish farmers have learned to live with the problem and accept occasional losses. Breakwaters and submersible marine cages (see Chapter 7) can be used to minimise typhoon damage, although even the latter may be lost during severe typhoons (Milne, 1974). Inland waters adjacent to coastal areas are also sometimes hit by typhoons (*Fig 4.11*), causing enormous damage and loss of stock. Risks are so great in Laguna de Bay that several of the recently established lake-based tilapia hatcheries have reverted to land-based systems (Beveridge, 1984a).

Other areas subject to extreme climatic conditions which can make cage culture a risky business include those affected by ice packs or icebergs, which not only damage cages, but can also cause super-chilling of water resulting in fish deaths (see later). The principal marine areas affected are those within the influence of the Arctic seas, and include the coastlines of northern and eastern Canada as far south as Nova Scotia, northern and western Alaska, Greenland, northern and eastern Iceland and the Baltic. In many of those areas, cage culture is virtually impossible due to the almost year-round stranglehold of the ice and prolonged periods of low surface temperatures, whilst in other more marginally affected areas (*eg* New Brunswick, Baltic) risks posed by icebergs or abnormally low temperatures can adversely affect insurance premiums. In the latter type of site there is some interest in the possibilities offered by submersible designs. Occasional superchilling of seawater can occur outside the above areas. In northern Norway in 1982, one fish farmer lost 20,000 Atlantic salmon smolts after the temperature dropped to − 1.2°C (Anon, 1982a).

Temperate continental inland lakes and reservoirs often freeze during winter. However, although some damage may be inflicted on the cages themselves, there is some evidence to suggest that the movement of fish can keep the actual cage environment ice-free (Paetsch, 1977) (see also Chapter 7).

### 4.3.2 Shelter

Few structures other than oil rigs can withstand the impact of the open sea. Winds blowing across its surface tear at any structure projecting above the water, whilst waves attack any object on the water/air interface. Shelter from these forces, and in particular waves, is therefore a prime consideration in site selection.

An idealised sinusoidal wave is shown in *Fig 4.12*. In the aquatic environment there are many kinds of waves which differ in origin, form and velocity (*Fig 4.13*). Although those which are of paramount importance in siting cages are those generated by the wind.

Wind-driven waves are propagated by the frictional drag of wind blowing

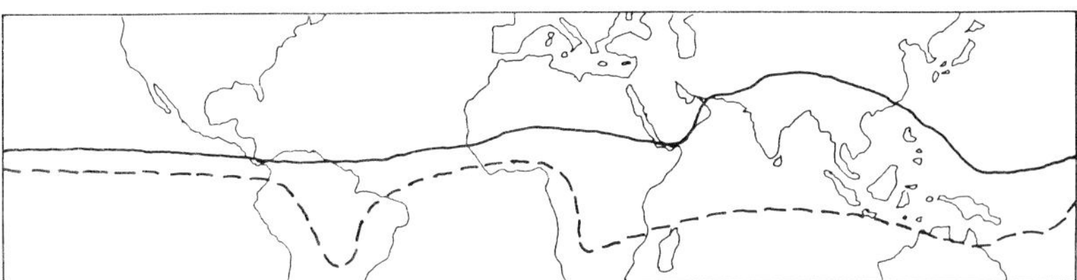

*Fig 4.8* Mean positions of the intertropical convergence in January (solid line) and July (dashed line) (redrawn from Barrett, 1974).

across a stretch of water and transferring energy to the fluid. The size of wind-generated waves is determined by (i) wind velocity, (ii) the duration of time that the wind blows, and (iii) the distance of open, unobstructed water across which the wind blows (fetch) (Bascom, 1964). The resultant size of the waves is also affected by waves which were already present when the wind started blowing. At the upwind end of the fetch, waves are poorly developed with small wave heights, and short periods of oscillation. However, they develop with distance until they reach their maximum possible dimensions which occurs when the waves reach the same velocity as the wind. In practice, however, only waves generated by comparatively low velocity winds ever reach their full potential (*Table 4.9*). When a wave moves out of the generating area it is modified and becomes a swell. The shorter waves lose their energy quickly and die out with the result that the swell becomes more uniform and longer crested whilst the wave height gradually decays.

The important point about wind-generated waves is that wave height increases with wind velocity (*Table 4.9*), and wave energy increases proportionally to the square of wave height. When considering the suitability of a site from the point of view of shelter, it is the worst possible conditions that are of concern *ie* the highest waves that are likely to occur at the site. A number of authors give instructions on how to predict wave heights at a particular location, given the appropriate meteorological and topographical data (Darbyshire and Draper, 1963; Milne, 1972; Muir Wood and Fleming, 1981; US Army Corps of Engineers, 1984) and the following section has been adapted from these papers.

In order to predict wave characteristics at a particular site it is necessary to have information on the long-term frequency and direction of surface wind speeds. This can often be obtained from national meteorological offices, such as the British Meteorological Office, or the Norwegian Meteorological Institute, who may provide information in the form of a windrose, in which the percentage frequency with which winds of various strengths occur are plotted as vectors for the eight primary direction sectors (*Fig 4.14*). The total length of the vectors comprising each windrose, calms included, equals 100%. Note that the length of each vector in *Fig 4.14* is different, and that in this example, the prevailing winds are from the southwest and west. Each vector is subdivided into four segments of unequal lengths, illustrating the proportion of winds of different strengths blowing from each direction.

Alternatively, there have been several reviews of wind speed data for

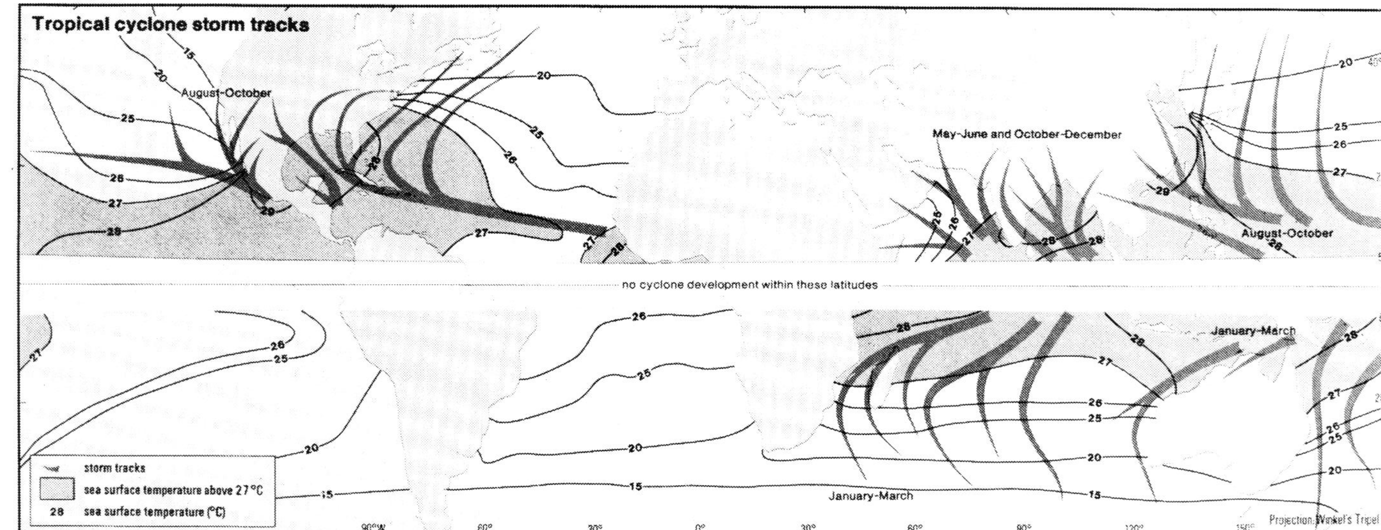


*Fig 4.9* Map to show sea surface isotherms during the warmest season, and paths typically taken by tropical cyclones. For a tropical cyclone to develop, sea surface temperatures must be at least 26.5°C. The Earth's rotation is also instrumental and storms never develop within 5° of the equator. Coast lines and estuaries where the tidal range exceeds 4m are particularly at risk (copyright Times Books 1983, reprinted with permission from Times Books Ltd.)

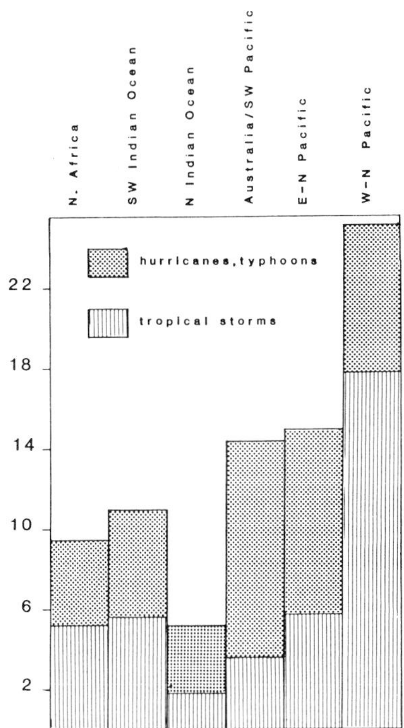


*Fig 4.10* Average number of hurricanes (wind force 12) and tropical storms (wind force 8-11) per year. (redrawn from Couper, 1983).

*Fig 4.11* Typhoon damage, fishpens and cages, Laguna de Bay, Philippines.

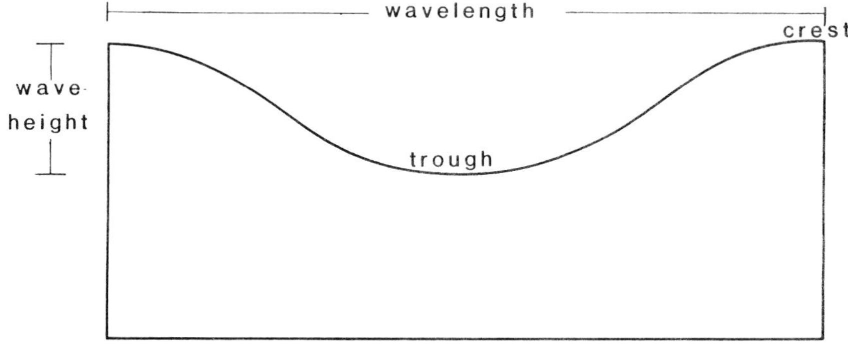


CREST : the high point of a wave

TROUGH : the low point of a wave

WAVE HEIGHT : the vertical distance between trough and crest

WAVELENGTH : the distance between two corresponding points on successive waves

WAVE PERIOD : the time taken for a wave crest to travel a distance equal to one wavelength

*Fig 4.12* Wave terminology.

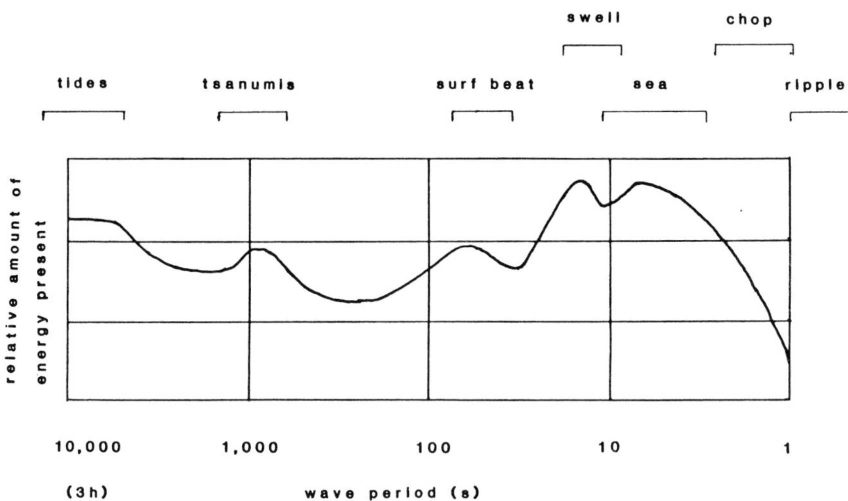


*Fig 4.13* Period and relative energy of different ocean waves (redrawn from Bascom, 1964).

different parts of the world which could be consulted. In a review of the British records, Shellard (1965) gave highest mean hourly wind speeds and gust speeds for different areas of the British Isles which would on average only be exceeded twice every hundred years. Note that it is the mean hourly speeds that are used to predict wave characteristics.

Much of the above data is based on ship-board observations and has a considerable degree of bias that may be corrected by using the expression:-

$W = 2.16\ W_S^{0.777}$ (Cardone, 1969, in US Army Corps of Engineers, 1984)

where $W_S$ = the ship reported windspeed in knots, and $W$ = corrected windspeed.

If neither of these two sources has information relevant to the site being considered, then 'guesstimates' of highest mean hourly wind velocities based on nearest-to-site information and modified by any local observations that can be made, may have to be employed. Correction factors for various alternative sources of data (*eg* wind data collected at elevations other than 10m; wind speeds recorded when there are large air-sea temperature differences) are detailed in the Shore Protection Manual (US Army Corps of Engineers, *ibid*).

The wind speed is then adjusted to wind stress factor ($U_A$) to account for drag, using the formula:-

$U_A = 0.71\ U^{1.23}$ m s$^{-1}$

(*NB* $W \equiv U$).

The next piece of information that is required is the fetch lengths at the site corresponding to the directions of the strongest prevailing winds. A fetch is defined as an uninterrupted region in which the wind speed and direction are reasonably constant. For any given wind direction it is recommended that the fetch length be determined as a mean of nine measurements, 3° apart, taken from a map (see *Fig 4.15*). Unlike the open expanses of the oceans, fetches in inland waters are limited by surrounding land masses, and fetches which are long in comparison to width occur frequently. In such cases, fetch width may

*Table 4.9* Conditions in fully developed seas (modified from Bascom, 1964).

| *Wind velocity (knotts)* | *($ms^{-1}$)* | *Fetch length (nautical miles)* | *(km)* | *Time (hours)* | *Average height (m)* | *$H_3$ significant height (m)* | *$H_{10}$ Average of the highest 10% (m)* | *Period where most of the energy is concentrated (S)* |
|---|---|---|---|---|---|---|---|---|
| 10 | 5.1 | 10 | 18.5 | 2.4 | 0.27 | 0.43 | 0.55 | 4 |
| 15 | 7.7 | 34 | 63.0 | 6 | 0.76 | 1.07 | 1.52 | 6 |
| 20 | 10.3 | 75 | 138.9 | 10 | 1.52 | 2.44 | 3.05 | 8 |
| 25 | 12.9 | 160 | 296.3 | 16 | 2.74 | 4.27 | 5.49 | 10 |
| 30 | 15.4 | 280 | 518.6 | 23 | 4.27 | 6.71 | 8.53 | 12 |
| 40 | 20.6 | 710 | 1,314.9 | 42 | 8.53 | 13.41 | 17.37 | 16 |
| 50 | 25.7 | 1,420 | 2,629.8 | 69 | 14.63 | 23.77 | 30.18 | 20 |

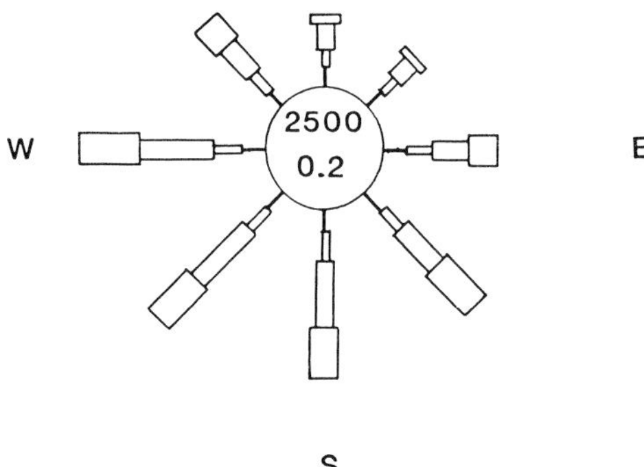


*Fig 4.14* A typical windrose. The values in the centre of the rose give the total number of observations and percentage of calms. (See text).

have a limiting effect on the generation of waves. This was first recognised by Saville (1954) who proposed a method for determining the *effective fetch*, $F_E$, for relatively uniform fetch widths, such as occur in inland waters or coastal fjords. Recently, however, Saville's and other researchers' methods have been shown to greatly overestimate the restrictive effect of fetch width on wave heights and until studies currently being undertaken in the United States are completed, it is recommended that fetch is estimated using the above procedure.

The final piece of information required to predict wave characteristics at a particular site is the depth of water along the fetch, since for a given set of fetch and wind data, wave heights will be smaller and wave periods shorter if the waves are generated in shallow rather than deep water, due to friction exerted by the sea bed, and percolation in the permeable sediments (U.S. Army Corps of Engineers, 1984). In *Figs 4.16* and *4.17*, shallow water (3m, 7.5m, 10.5m, 15m) and deep water (>15m) wave forecasting curves are plotted, assuming constant depths over the fetch. Milne (1970) has recommended that inshore and sheltered coastal sites use shallow water curves and that exposed coastal sites use deep water values. The wave height is estimated by tracing a line from the wind stress factor ordinate until the appropriate fetch length ordinate is intersected.

This method of calculating wave height assumes that the fetch length and not the wind duration is limiting (see above), although *Figs 4.16* and *4.17* have plots of wind duration included. The predicted wave height actually refers to the *significant wave height*. A glance at any water body will tell you that the distances between waves is highly irregular and that wave heights are also variable due to the presence of a large number of wave components at any particular time. The practical solution adopted by oceanographers has been to deal with waves statistically. The significant wave height refers to the mean height of the highest 33% of the waves (Sverdrup and Munk, 1947), which is around 40-60% higher than the overall mean wave height (see *Table 4.9*).

One other assumption upon which the calculation of significant wave height is based is that the atmospheric pressure system is stable. However, in rapidly moving pressure systems, such as those which can occur during tropical cyclones, the situation becomes complex. As an illustration of the modifying effect this can have on the significant wave height, Muir Wood and Fleming (1981) cite Bretschneider's example of a 12m $s^{-1}$ advancing cyclone producing a 50% increase in wave height.

Problems may also be experienced with swell. Coastal swell typically has wavelengths in the region of several hundred metres and periods of 7-15 seconds, and the source of origin may be several thousand kilometres distant. Sites exposed to large tracts of open ocean may periodically experience heavy swell which can be highly damaging. In Scotland, some sites exposed to the Atlantic Ocean have experienced heavy swell which has caused descaling of

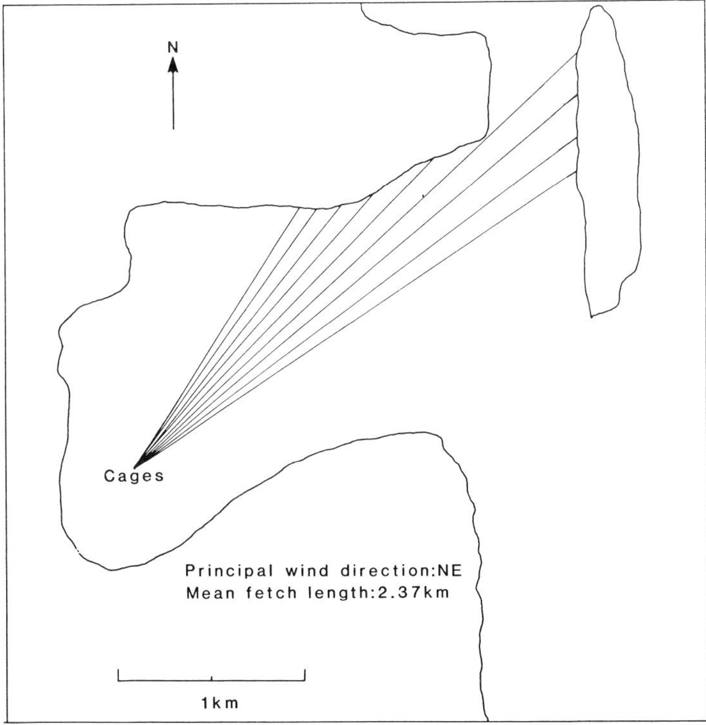


*Fig 4.15* Map of cage site. Nine radials, with 3° difference between them are drawn for the purposes of calculating fetch length along principal wind direction.

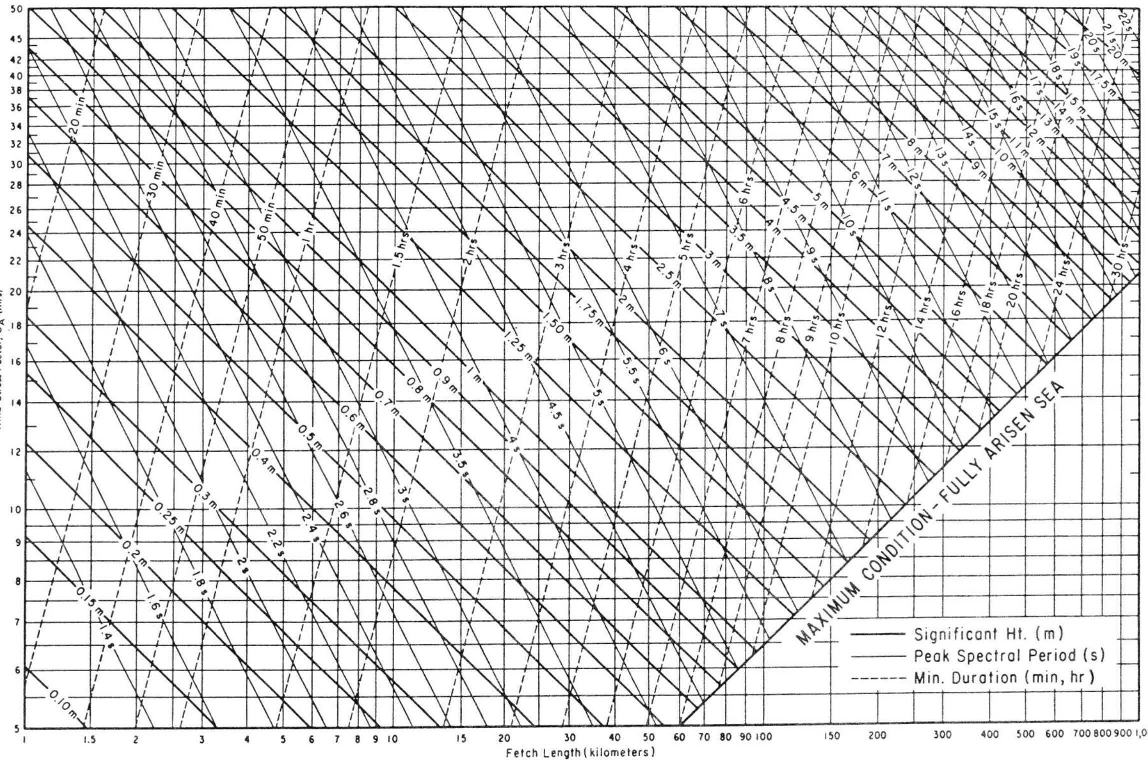


*Fig 4.16* Nomogram for deep water significant wave predictions as functions of windspeed, fetch length and wind duration (from US Army, 1984).

fish and consequent high mortalities, through excessive motion of the cage bag.

In summary, the above methods can be used to estimate significant wave heights at prospective sites. Although cages and moorings can be designed and built to withstand a range of weather conditions, as a rule of thumb it is assumed that most floating cage structures are no stronger than other types of coastal structure, and therefore unless specially designed will be unable to withstand significant wave heights greater than 1-1.5m (Moller, 1979).

### 4.3.3 CURRENTS

Good water exchange through cages is essential both for replenishment of oxygen consumed by fish and removal of waste metabolites. In extensive culture situations, water currents are also essential for the supply of food. However, excessive currents will impose additional dynamic loadings to the cage, supporting structure and moorings, may adversely affect fish behaviour and contribute towards food losses from semi-intensive and intensive operations. High flow rates have also been implicated in skeletal deformities of cage-reared carps (see later).

In all but a few coastal regions of the world, such as in the Mediterranean or Baltic, tidal currents are the predominant source of surface water currents. Attractive forces, exerted by the moon and sun on the earth produce tidal waves, which have extremely long wavelengths (half the circumference of the earth) and periods of oscillation (12h 25min). The crest and trough of the wave are termed high and low tide, and the wave height is known as the tidal range.

Associated with the rise and fall of the tide are horizontal motions of water, or tidal currents. Maximum current speeds occur at the middle of the rise (flood) and fall (ebb), and minimum current speeds with slack water at the time of high and low tides. The effect of the Earth's rotation produces a rotational tidal current which in the northern hemisphere, where semi-diurnal tidal patterns predominate, is principally clockwise. Around the British coastline, for example, an eastward flowing tide at high water, will be flowing westwards six hours later. In the southern hemisphere, where mixed and diurnal tides prevail (*Fig 4.18*), the direction of rotation varies much more.

However, the pattern of tidal currents is greatly influenced by local topography and other factors, such as surface runoff from the land and prevailing winds, resulting in marked changes in current over short distances and over depths of only a few metres. Take for example the idealised bays shown in *Fig 4.19*. On the north coast (a), the flood stream starts after hour six, reaches a maximum at hour 9, and decreases to a minimum, slack water, at hour 12. During the ebb, the tidal current will run out of the inlet at a maximum speed at hour 3, and reach slack water at hour 6. Rotation is thus in a clockwise direction. Although the tidal pattern in the basin on the southern coastline is identical, the tidal currents show a counterclockwise rotation.

It is possible to estimate tidal currents in some types of coastal areas. For

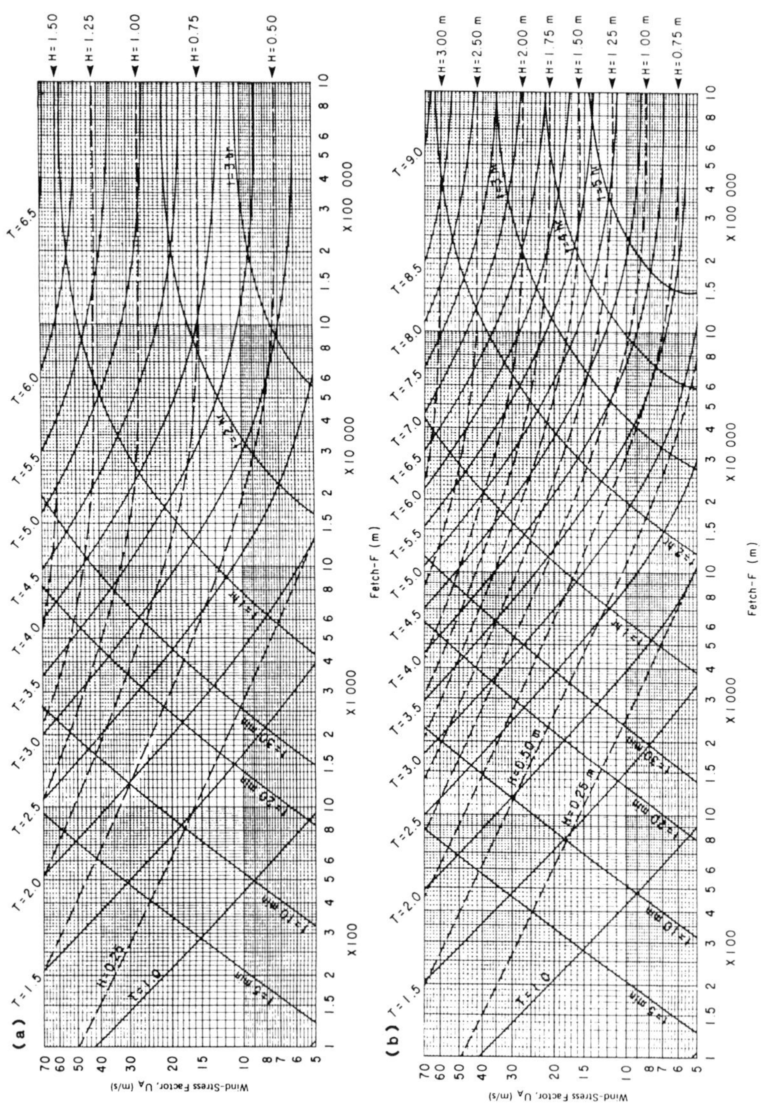
(a)
H=1.50
H=1.25
H=1.00
H=0.75
H=0.50
H=0.25
T=1.5
T=2.0
T=2.5
T=3.0
T=3.5
T=4.0
T=4.5
T=5.0
T=5.5
T=6.0
T=6.5
T=1.0
t=5 min
t=10 min
t=20 min
t=30 min
t=1 hr
t=2 hr
t=3 hr
X 100
X 1 000
X 10 000
X 100 000
Fetch-F (m)
Wind-Stress Factor, U_A (m/s)
(b)
H=3.00 m
H=2.50 m
H=2.00 m
H=1.75 m
H=1.50 m
H=1.25 m
H=1.00 m
H=0.75 m
H=0.50 m
H=0.25 m
T=1.5
T=2.0
T=2.5
T=3.0
T=3.5
T=4.0
T=4.5
T=5.0
T=5.5
T=6.0
T=6.5
T=7.0
T=7.5
T=8.0
T=8.5
T=9.0
T=1.0
t=5 min
t=10 min
t=20 min
t=30 min
t=1 hr
t=2 hr
t=3 hr
t=4 hr
t=5 hr
X 100
X 1 000
X 10 000
X 100 000
Fetch-F (m)
Wind-Stress Factor, U_A (m/s)

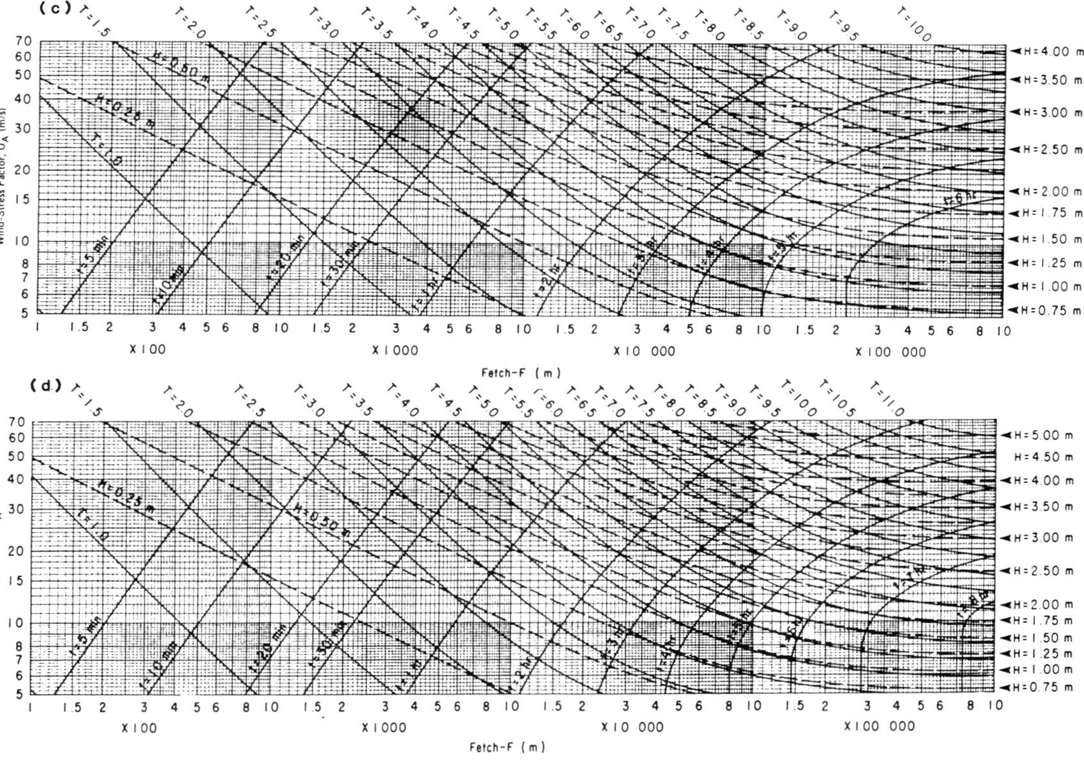


*Fig 4.17* Nomograms for shallow water wave predictions as functions of windspeed, fetch length and wind duration (a = 3.0m; b = 7.5m; c = 10.5m; d = 15.0m) (from US Army. 1984).

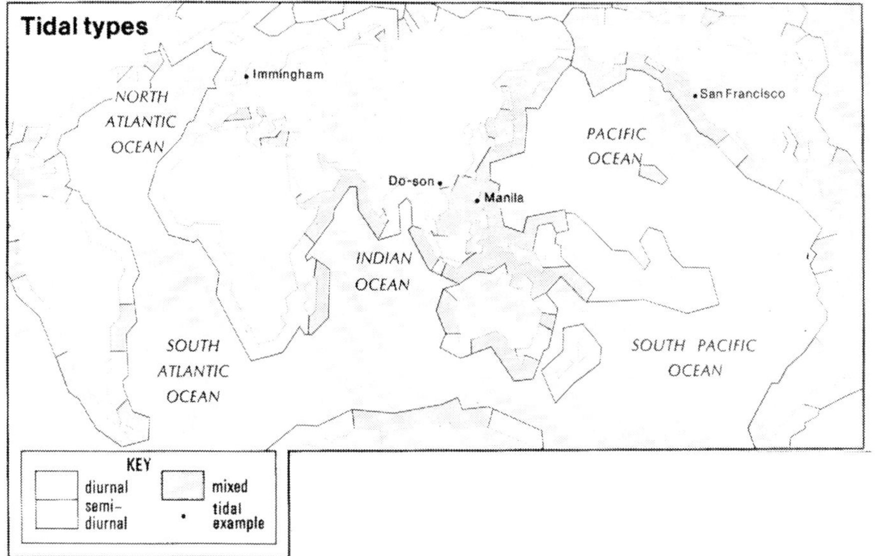


*Fig 4.18* Tidal types (copyright Times Books, 1983. Reprinted with permission from Times Books Ltd.)

example, maximum current speeds during ebb and flood tides in a fjord can be calculated from:-

Vmax = (AH/B) $10^{-4}$m $s^{-1}$ (Edwards and Edelsten, 1976).

where A = surface area of the fjord landward of the site, B = cross-sectional area of the fjord at the site, and H = range of the tide, metres.
However, in real life this formula is of limited value, because of local influences. Rapid changes in depth or coastal contours can result in eddies with areas of dead current in their centres. The presence of a sill at the mouth of a fjord will cause an acceleration of the tidal currents in the region of the sill and greatly complicate the movement of water in the inner part of the inlet whilst in yet other types of bay, resonance effects may greatly increase the tidal range, and thus increase maximum tidal current velocities. A significant fall in barometric pressure may also cause an increase in tidal range, and thus an increase in currents.

A detailed account of tidal currents is beyond the scope of this book, the above examples serving only to illustrate how complicated the picture is. Further details concerning tides and tidal currents can be found in Pond and Pickford (1978) and Muir Wood and Fleming (1981). The important point to remember is that prospective marine cage farm operations must survey the currents at a proposed site. The methodology for such a survey has been discussed by several authors (Edwards and Edelsten, 1976; Landless and Edwards, 1976, 1977).

Current velocities in coastal marine areas typically range from 0 cm $s^{-1}$ at slack water, to figures in excess of 250 cm $s^{-1}$ in some sites at full flood and ebb tides. In the ideal cage site, periods of slack water should be minimal. It is impossible to recommend a specific maximum current velocity as optimum for a particular species, because of the influences of cage design and stocking density. As current velocities increase, the capital costs of constructing cages and installing moorings which can withstand such velocities escalate. Production will also increase with tidal current velocity, since stocking rate can be increased. However, beyond a certain point deformation of the cage structure will cause an unacceptable reduction in volume of flexible net structures (Okabayashi, 1958) and fish may have to start expending large amounts of energy to maintain station thus adversely affecting production.

In practice, most researchers have found ebb and flood tidal currents in the range 10-60cm $s^{-1}$ to be satisfactory, and do not recommend choosing a site where such currents exceed 100cm $s^{-1}$ (Braaten and Saetre, 1973; Chen, 1979;

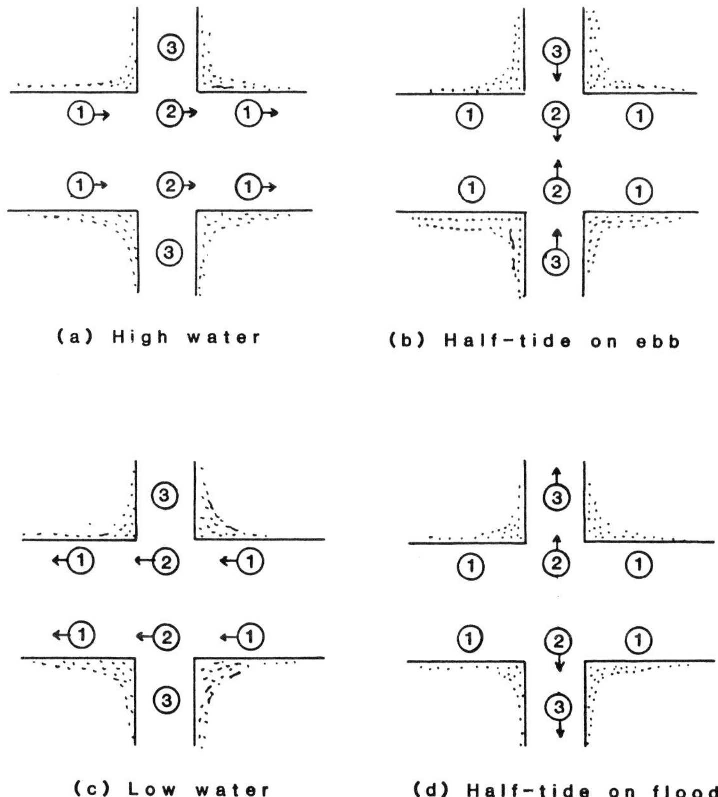


*Fig 4.19* Rotational tidal currents at the mouths of two inlets separated by a wide channel connected to the open sea. The tide in the inlets is assumed to be approximately in phase with the tide immediately off the mouth. Current vectors : (1) = offshore, (2) = off mouth to inlets (3) within inlets.

Kerr *et al*, 1980; Chua and Teng, 1980; PPD, 1981; Kafuku and Ikenoue, 1983).

Inland waters have currents which are different both in origin and pattern. In lotic water bodies (*ie* flowing water systems, such as streams and rivers) water is derived from three principal sources; surface runoff, through flow, and groundwater, and is subject to two major forces, gravity and friction. As water flows along a channel there is a loss of energy through frictional forces exerted by the banks and the stream bed. The simplest, and best known equation which describes this relationship, is the Manning equation (Manning, 1891):-

$$V = 1/n \times R^{2/3}\ S^{1/2}$$

where V = velocity (m $s^{-1}$), n = roughness coefficient, R = hydraulic radius (cross-sectional area divided by wetted perimeter, m) and S = slope.

The radius, R, and slope, S, can be obtained from field survey measurements. Roughness, n, is dependent upon the size and shape of the grains forming the wetted perimeter of the stream, and other characteristics such as the quantity and type of vegetation present. Although there are exceptions, such as when sand is formed into dunes or ripples, in general the smaller the grain size, the lower the 'n' value will be. Most streams have values of between 0.025 and 0.075 although tables of roughness values can be obtained from many standard hydraulics textbooks (*eg* Chow, 1959; Richards, 1982).

Lotic systems differ markedly in their characteristics, depending on how and where they were formed, and on whether or not they have been modified by man. Upper sections characteristically have high current velocities, low flows and turbulent flow patterns caused by the prevailing stony or gravel substrates, whereas lower sections tend to have low current velocities, high flows, and laminar flow patterns due to the typically fine-grained sand or mud substrate. As discussed earlier, cage culture in lotic systems is principally practised in Southeast Asian countries, where fish are reared in traditional extensive/semi-intensive situations. In Kampuchea, slow-flowing reaches of the Mekong and Tonle Sap rivers have cages (Lafont and Savoeun, 1951; Pantulu, 1979), whilst in Indonesia, rivers and irrigation canals, which for much of the year are also fairly slow-flowing, are used (*Fig 2.2(i)*). In Bandung, for example, both branches of the River Tjibunut, which are used to culture carps, are about 5m broad by about 0.35m deep with typical current velocities of around 10cm $s^{-1}$ (from Vaas and Sachlan, 1957), although, as in many other tropical rivers, current velocities greatly exceed this during monsoon floods. However, these cages can withstand such conditions because of their unusually stout construction, and method of anchorage.

The pattern of current velocities differs between inland and marine sites. Instead of the usual four or eight periods of fast flowing currents per day associated with peak ebb and flood conditions, currents in inland waters are more constant and change with the weather (precipitation, wind). Thus, high rates of water exchange through cages may be tolerable in marine situations because feeding during peak flow conditions can be suspended. However, the relatively constant flow of water through cages in lotic sites means that the feeding regime cannot be adjusted in this way. Feed loss is a problem in lotic sites which has been discussed by several authors (Collins, 1975) and is one of the main reasons why commercial intensive cage culture in fast-flowing rivers or streams is unlikely to prove viable.

Current velocities in thermal effluent canals can also be highly variable. For example, at La Casella oil-fired power station situated on the Po River, the flow in the discharge canal can vary from $1 \mathrm{m\ s}^{-1}$ to nearly zero, depending upon the water level in the river (Bronzi and Ghittino, 1981).

At the Kozienice power station in Poland, carps reared for one year in cages held in the thermal effluent canals suffered severe body flexures, which Backiel *et al* (1984) attributed to the prevailing currents ($2.6\text{-}12.3 \mathrm{cm\ s}^{-1}$). No other reports of similar problems have been reported elsewhere.

Surface current velocities in lentic water bodies, such as lakes and reservoirs, are usually much lower ($0.2\text{-}2.0 \mathrm{cm\ s}^{-1}$) than in rivers or marine sites. The currents are principally of wind-driven origin, although draw-down from outflows and discharges from inflowing rivers and streams can have localised effects on velocities whilst other factors, such as seiches can modify whole lake patterns (Csanady, 1975; Smith, 1975).

However, current velocities in lentic water bodies, being usually fairly uniform, rarely influence site selection. The exception to this is in some extensive culture operations, where cages can be sited near the outflow and can thus take advantage of the increased current velocities and supplies of planktonic food (*Fig 4.20*).

### 4.3.4 Depth

Some cages, such as the rigid wood and bamboo constructions used in traditional fish culture in Indonesia, and the experimental fixed cages used to grow prawns in the Philippines, are designed to sit in shallow water with their bottoms in contact with the substrate. The depth of water in which these cages are sited is not critical, provided that they are covered, or largely covered, by water for most of the culture period. Too great a fall in water level could seriously reduce cage volume, thus significantly increasing stocking density and adversely affecting water quality.

Fixed cages are usually used in shallow areas of lakes and reservoirs or rivers, where the depth does not exceed 8m or so, since it is difficult to find sufficiently strong supporting posts which are any greater in length. Floating cages may in theory be used in virtually any depth of water, although the cost and other associated problems concerned with mooring increase with depth.

For most types of cage culture, the cages should be sited in sufficient depth to maximise the exchange of water, and yet to keep the bottom of the cages well clear of the substrate. As discussed in Chapter 3, fishes often augment the passive, externally generated currents which flow through their cages by drawing in water during normal swimming and feeding activities. Some of the internally generated currents, particularly during feeding, pull water into the cage through the bottom panel (Chacon Torres, 1984) and as the cage

*Fig 4.20* Extensive culture of bighead carp, Selatar Reservoir, Singapore.

bottom approaches the substrate, this flow will become increasingly impeded. There is also a great deal of evidence, from intensive culture situations at least, that wastes can build up under cages resulting in localised oxygen depletion and a build up of potentially toxic compounds (Shelkov *et al*, 1981; Merican and Phillips, 1985). Although the extent and degree of deoxygenation depends very much on how the operation is managed and the nature of the site, it is wise to keep the cages well clear of the bottom. Under some circumstances noxious gases such as $H_2S$ are released and these have been attributed by some authors as a possible causative agent of gill damage in caged Norwegian salmon stocks (Braaten *et al*, 1983; Lieffrig, 1985), and recent reports from both marine and freshwater sites have shown that there can be a build up of potentially disease-causing micro-organisms in the sediments under established cages (Rosenthal, 1985).

Although the relationships between cage culture, sedimentation of wastes, water quality and disease remain poorly understood, it is best to avoid unnecessary risk by holding the fish at least 4-5m above the sediments. This is not always practical, particularly in lotic sites or ponds where cages are being used. However, in lentic freshwater sites, suitable depths can readily be found using simple survey equipment (*eg* plumb line, echo sounder) or a chart, and by taking annual fluctuations in water level into account. In some parts of the world, annual water level fluctuations are very large, particularly in the case of reservoirs or lakes where water is abstracted for irrigation purposes. In water bodies of this type, fixed cages are inadvisable because although the rigging can be untied and the net bags lowered to a new position so as to maintain cage volume, this is only possible if the drawdown is <4m per annum or so. (Remember that the posts are usually ~8m, and cage bags >3m). Floating structures have the advantage, of course, that they can be moved to deeper water if and when necessary.

In marine situations, tidal fluctuations must be taken into account in any depth calculations. Spring tides of maximum rise and fall occur every two weeks at new and full moons, and can range from little more than a half a metre in some parts of the world, to over 10m in the Bay of Fundy, Newfoundland. For the fish farmer's purposes the tidal range for a particular site can be calculated from nearest-to-site values given in tidal tables augmented by site observations of several tidal cycles. Observations should be made during relatively stable and calm weather as strong offshore or onshore winds can cause correspondingly smaller or larger than expected tides. A difference in barometric pressure of 34 millibars from the average will result in a height difference of 0.3m (*ie* 10mm per mb) (Muir Wood and Fleming, 1981).

### 4.3.5 SUBSTRATE

The substrate at a site can range in type from rocky to soft mud, and may have a considerable influence on the choice of cage design. In freshwater sites, where fixed cages are commonly used, it can prove difficult, if not impossible, to drive stakes into hard, rocky ground, even when enlisting the help of a pile driving platform (*Fig 3.18*), and ultimately floating cages may have to be used. For marine sites, it may be better to choose sites with rocky substrates since it indicates good current scour, with reduced risk of waste build-up. Mooring cages in this type of site however, can be a problem (see Section 3.4.4).

## 4.4 Site facilities and management

### 4.4.1 LEGAL REQUIREMENTS

The law with respect to aquaculture varies enormously from one part of the world to another with the result that whilst it is relatively easy to set up a fish farm in some countries, it is almost impossible in others. The principal reason for this is that most countries have not yet drafted or passed specific legislation dealing with aquaculture, relying instead on amendments to existing laws. In this section the types of conditions and controls that prospective cage fish farmers are likely to meet are briefly reviewed.

Cage fish farms sit in the water column and are moored to the lake or sea bottom and/or the shore. In some cases, the cages may also be attached to the land above the high water mark via an access walkway. In most countries the land below the low water tide level is owned by the state, and so permission to moor a structure must be obtained from the appropriate government department. In the UK, for example, the Crown owns the sea bed and anyone wishing to build a cage fish farm must apply for a lease from the District Valuer of the Crown Estate Commissioners, which if granted will cost a nominal £20 per annum. Since most land above the high water mark is privately owned it follows that the fish farmer must buy, lease or obtain permission from the lawful owner if he wishes to use the land for access or mooring.

Some countries insist that the prospective cage fish farmer also obtains a licence in order to culture fish in cages, and in so doing, often lay down restrictions concerning site, species, size and type of development. Few countries have allocated specific areas for cage culture developments, the exceptions being Hong Kong, Singapore and the Philippines (PPD, 1982; Tseng, 1983; Beveridge, 1984b). In Hong Kong and Singapore, the legislation applies to marine cage farms, whilst in the Philippines, laws have been passed to try and control cage and pen farming developments in Laguna Lake (Agbayani, 1983) (*Fig 4.21*). Other countries have tried to restrict the type of site used, or have laid down legislation governing operations at the site. Restrictions governing the use of irrigation canals for cage culture are in force in several countries, including Egypt and Indonesia (SEAFDEC/IDRC, 1979). In West Java cage fish farmers using irrigation canals must undertake to maintain them and keep them clear of detritus (SEAFDEC/IDRC, *ibid*). They must also site cages at least 3m from any bridge, and are not allowed to use canals where the slope is less than 2.5%. Although no licence is required

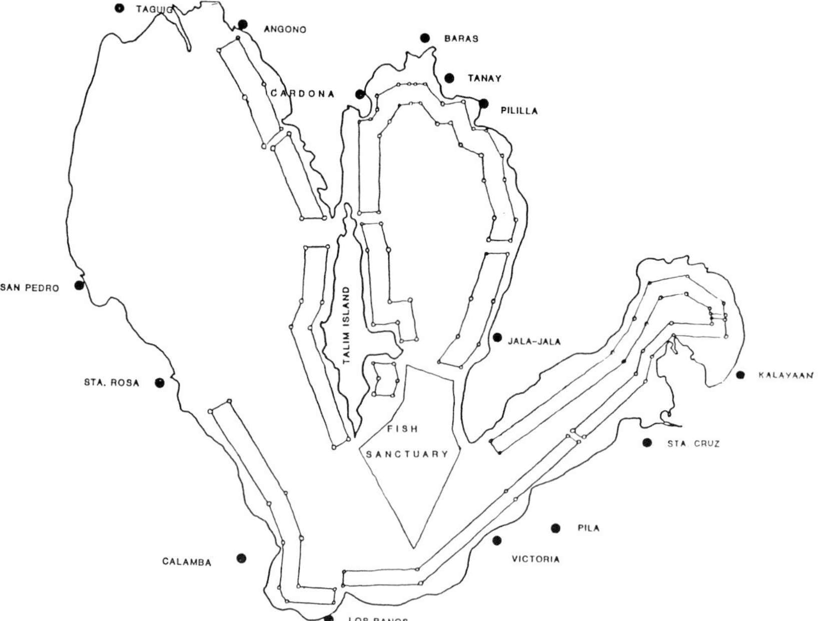


*Fig 4.21* Map of Laguna de Bay, Philippines, showing legal fishpen and fish cage belt, and fish sanctuary (from Beveridge, 1984b).

in the UK, the prospective farmer must agree to manage the site in a responsible manner, or have the lease terminated.

Size restrictions operate in many parts of Scandinavia. In Finland, for example, even small units producing less than 3 tonnes per annum must notify the authorities, whilst farms producing more than 40 tonnes require a government concession (Sumari, 1982). Farms bigger than 350 tonnes are discouraged. For many years size restriction operated at Norwegian salmon farms, cage fish farms being restricted to 8,000m$^3$ capacity and this had the unfortunate consequence of forcing operators to try to squeeze as much production as possible out of their farms, resulting in many disease problems. This law was finally changed in 1983.

Because cage fish farms are often located in coastal areas or lakes frequented by shipping, local navigation regulations may have to be complied with. At present in the UK permission to moor cages at a site must be obtained from the Ministry of Transport, since cages are covered by Part 2 of the 1949 Coast Protection Act, which is concerned with maintaining common law rights to free navigation in British coastal waters [Section 34(1)]. Detailed charts and diagrams of the proposed farm, specifying the size of cages and deployment of moorings must be submitted to the Department of Fisheries, Royal Yachting Association, Coastguards and the National Light House Board and are all then consulted and if objections are raised which are countered by the fish farmer, then a public enquiry may ensue. Some applications, however, fail outright (~10%). The whole process at present normally takes 2-3 months. Recommendations are usually given concerning the installation of buoys or navigation lights.

In some countries the procedures that must be followed in order to gain permission to set up a cage fish farm are straightforward. In Thailand, all that is required is permission from the provincial fisheries office, and the payment of a nominal licence fee (SEAFDEC/IDRC, 1979). In Britain the procedure is somewhat more involved in that in addition to obtaining permission from the Crown Estate Commissioners and from the Ministry of Transport, planning permission from the local authorities may be required for any buildings or access roads that the farmer may wish to build. Nevertheless, few farmers would complain about the process being unduly complicated.

Because of the way aquaculture legislation has been introduced in most countries, laws and regulations can change rapidly. The procedure for gaining permission to rear salmon in cages in British Columbia, Canada, for example, used to involve only five licences and permits (Tillapaugh and Edwards, 1980) but by 1983, one farmer was complaining of having to obtain 23 separate licences in order to farm salmon (Heal, 1983)! However, in an effort to stimulate aquaculture business ventures in the province, the system has recently been overhauled and greatly simplified.

### 4.4.2 Situation, services and shore facilities

For large and intensive cage fish culture developments the availability of sufficient land to construct an office, feed store, laboratory, manager's house, *etc*, close to the fish cages may be an important consideration in site selection (Beveridge and Muir, 1982).

However, if the desired land site is in a conservation area or if there are common grazings or rights of way, then there may be problems in obtaining planning permission.

Services are also important. The availability of freshwater, 3-phase electricity, telephone and postal services, scavenging and sewerage, road and rail services and veterinary assistance should all be ascertained prior to purchasing a site. Availability of local labour and housing can also be critical.

Proximity to markets and food supplies may affect production costs and profitability.

### 4.4.3 SECURITY

Security is a problem for cage fish farmers in many parts of the world (Edwards, 1978; Secretan, 1980; Beveridge and Muir, 1982), since cages are often sited in water bodies which are publicly owned or have unrestricted access and are thus vulnerable to poachers and vandals. Cages located near centres of population may be more at risk. Although there are a number of security measures which can be taken to protect installations (see Section 7.8), farmers may prefer to site their cages where they can keep a careful watch on them. A recent survey of cage fish farmers in the Philippines has shown that this is a prime consideration in site selection (Escover and Claveria, 1985) and farmers are unwilling to locate their cages where they cannot be checked at least once per day by a member of their family.

# 5 Carrying capacity

## 5.1 Introduction

A major consideration in the site selection process should be the carrying capacity of the site *ie* the maximum level of production that a site might be expected to sustain. Intensive cage fish culture results in the production of wastes which can stimulate productivity and alter the abiotic and biotic characteristics of the water body, whilst less intensive methods can result in overcropping of algae and a fall in productivity (see Chapter 2 for review). Both these impacts can radically alter the value of the resource not only to other users but also to fish farmers. A serious deterioration in water quality will stress or even cause mortalities amongst stocks and can encourage disease organisms to thrive, whilst overgrazing of algae can result in poorer growth and an increasing reliance on supplementary feedstuffs. Hence profitability or even viability may be seriously affected.

It is therefore extremely important for all concerned with cage fish farming developments — planners, insurers, banks and farmers — that an accurate evaluation of the sustainable levels of production at a site can be made. This chapter is concerned with a brief review of this process.

## 5.2 Intensive culture

### 5.2.1 Inland water sites

The rationale behind using the models described below is explained in detail in Beveridge (1984b) and is based on the assumptions that algal population densities are negatively correlated with water quality in general and growth and survival of fish stocks in particular, and that phosphorus (P) is the limiting nutrient which controls phytoplankton abundance in lakes and reservoirs. The well-established link between water quality — particularly DO levels — and algal densities is a cornerstone of modern water management philosophy (*eg* OECD, 1982), whilst the relationship between water quality and fish health has been frequently drawn to the reader's attention in various parts of this book. The concept of a limiting nutrient arises from the fact that whilst a range of nutrients is required by planktonic algae, if the supply of any one nutrient is less than the demand, then this nutrient will limit growth. In most aquatic ecosystems it is P that is limiting, since it is the rarest element with respect to algal and higher plant demand (Vallentyne, 1974).

Phosphorus is an essential element required by all fish for normal growth and bone development, maintenance of acid-base regulation and lipid and carbohydrate metabolism (Lall, 1979; Takeuchi and Nakazoe, 1981), which they derive principally from dietary sources (Nose and Arai, 1979). Whilst P requirements are species-specific (*Table 5.1*), most diets developed for intensive culture contain P surplus to requirements or in a form which is partially unavailable to the fish (Beveridge, 1984b). Surplus P is excreted, whilst unavailable P is passed out in the faeces.

In visual-feeding fishes such as the salmonids, which have size-specific preferences for food, damaged pellets may not be ingested, and instead contribute to the pool of P wastes. Other sources of P come from food which is washed out of the cages by both natural currents and turbulence caused by the fishes during feeding. The P losses to the environment are summarised in *Fig 5.1*.

A number of methods have been used to quantify these wastes. Using published data on the P content of feeds, typical FCR (Food Conversion Ratio) values, and the P content of fish carcasses, estimations of total-P (*ie* both inorganic and organic and particulate and dissolved) loadings per tonne intensive salmonid, tilapia and carp production have been made (Beveridge *et al*, 1982; Beveridge, 1984b). These results are summarised in *Table 5.2*. Note that the total-P loading is highly variable and depends on the P content and digestibility of the feed used. For trout, the most common FCR values for cage culture are 1.5-2.0:1, and thus total-P loads per tonne fish produced are 17-25kg. For intensive cage tilapia production FCR values in the range 2.0-2.5:1 are more typical, and thus 23-29kg total-P are added to the environment for every tonne of fish produced. The figures computed here for carp are very high and reflect the high P content of the diet. Other compounded carp feeds may well have lower P contents, and thus have lower total-P loadings.

The figures in *Table 5.2* give no information about the quantities of P

*Table 5.1* Dietary phosphorus requirements of fish, expressed as percentage weight of diet (after Beveridge *et al*, 1982).

| *Species* | *Requirement* | *Source* |
|---|---|---|
| *Anguilla japonica* | 0.29% | Arai *et al*, 1975 |
| *Salmo trutta* | 0.71% | McCartney, 1969 |
| *Salmo salar* | 0.30% | Ketola, 1975 |
| *Salmo gairdneri* | 0.70–0.80% | Ogino and Takeda, 1978 |
| *Oncorhynchus keta* | 0.50–0.60% | Watanabe *et al*, 1980a |
| *Cyprinus carpio* | 0.60–0.80% | Ogino and Takeda, 1976 |
| *Ictalurus punctatus* | 0.45–0.80% | Andrews *et al*, 1973; Lovell, 1978 |
| *Chrysophrys major* | 0.68% | Sakomoto and Yone, 1980 |
| *Oreochromis niloticus* | 0.90% | Watanabe *et al*, 1980b |

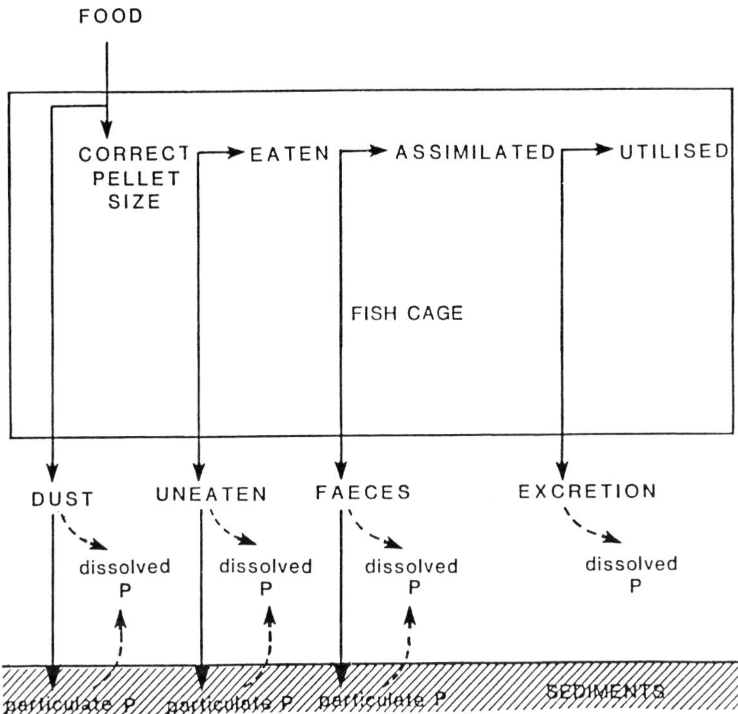


*Fig 5.1* Summary of principal P losses to the environment associated with intensive cage culture (from Beveridge, 1984b).

derived from uneaten food or excretory or faecal sources, and this is an essential step prior to modelling. For intensive trout culture total fed losses (dust + uneaten food) are estimated to be around 20% based on manufacturers' figures for dust (2% of feed) and estimates from various studies of 10-30% uneaten food (Collins, 1971; Hoelzl and Vens Cappell, 1980; Penczak *et al*, 1982; Enell and Lof, 1984). When FCR values for pond and cage culture are compared, those for cage culture are usually at least 20% greater, thus supporting the argument that feed losses from cages are comparatively high. An estimated 20-27% of the P ingested by trout is retained in the carcass (from *Table 5.2*, assuming FCR = 1.5-2.0:1), the remainder being excreted or egested. If a faecal production rate of 260g dry weight faeces per kg food (Butz and Vens Cappell, 1982) and a faecal P content of 1.59% dry weight (Penczak *et al*, 1982) are assumed, then more than 50% of the egested and excreted P is accounted for by excretion (*Fig 5.2*). Thus, around 40% of the total-P wastes from cage trout farming are dissolved, the rest being in the form of faeces and uneaten food.

A number of models have been developed to predict the response of aquatic ecosystems to increases in P loadings. Most are empirical and have been calibrated and tested, verified and modified using a number of data

*Table 5.2* Calculations of total–P losses to the environment during intensive cage fish culture.

(a) *Rainbow trout*
P content of commercial trout pellets — 1.50%[a]
Therefore 1 tonne feed contains — 15.0kg
FCR = 1.0 : 1 — therefore $P_{food}$ = 15.0kg
FCR = 1.5 : 1 — $P_{food}$ = 22.5kg
FCR = 2.0 : 1 — $P_{food}$ = 30.0kg
FCR = 2.5 : 1 — $P_{food}$ = 37.5kg
P content of trout = 0.48% wet weight of fish[b] = 4.8kg tonne fish$^{-1}$
Therefore P losses to environment:
1.0 : 1 FCR = 15.0−4.8 = 10.2kg tonne fish produced$^{-1}$
1.5 : 1 FCR = 22.5−4.8 = 17.7kg tonne fish produced$^{-1}$
2.0 : 1 FCR = 30.0−4.8 = 25.2kg tonne fish produced$^{-1}$
2.5 : 1 FCR = 37.5−4.8 = 32.7kg tonne fish produced$^{-1}$

(b) *Tilapia*
P content of compounded feeds — 1.30%[c]
Therefore 1 tonne feed contains — 13.0kg
FCR = 2.0 : 1 — therefore $P_{food}$ = 26.0kg
FCR = 2.5 : 1 — $P_{food}$ = 32.5kg
FCR = 3.0 : 1 — $P_{food}$ = 39.0kg
FCR = 3.5 : 1 — $P_{food}$ = 45.5kg
P content of tilapia = 0.34% wet weight of fish[d] = 3.4kg tonne fish$^{-1}$
Therefore P losses to environment:
2.0 : 1 FCR = 26.0−3.4 = 22.6kg tonne fish produced$^{-1}$
2.5 : 1 FCR = 32.5−3.4 = 29.1kg tonne fish produced$^{-1}$
3.0 : 1 FCR = 39.0−3.4 = 35.6kg tonne fish produced$^{-1}$
3.5 : 1 FCR = 45.5−3.4 = 42.1kg tonne fish produced$^{-1}$

(c) *Carp*
P content of compounded feeds — 3.09%[e]
Therefore 1 tonne feed contains — 30.9kg
FCR = 1.5 : 1 — therefore $P_{food}$ = 46.4kg
FCR = 2.0 : 1 — $P_{food}$ = 61.8kg
FCR = 2.5 : 1 — $P_{food}$ = 77.3kg
FCR = 3.0 : 1 — $P_{food}$ = 92.7kg
P content of carp = 0.61% wet weight of fish[f] = 6.1kg tonne fish$^{-1}$
Therefore P losses to environment:
1.5 : 1 FCR = 46.4−6.1 = 40.3kg tonne fish produced$^{-1}$
2.0 : 1 FCR = 61.8−6.1 = 55.7kg tonne fish produced$^{-1}$
2.5 : 1 FCR = 77.3−6.1 = 71.2kg tonne fish produced$^{-1}$
3.0 : 1 FCR = 92.7−6.1 = 86.6kg tonne fish produced$^{-1}$

a = average P content of commercial grower feeds used in Europe. Data from Tacon and De Silva (1983).
b = data from Penczak *et al* (1982).
c = estimated from diet formulations of Coche (1982) and Jauncey and Ross (1982), and P content of feedstuff ingredients given in NRC (1983) and Santiago (1983).
d = P content of tilapia, estimated from Meske and Manthey (1983), assuming dry weight = 25% wet carcass weight.
e = estimated from diet formulation for intensive carp feed and P content of feedstuff ingredients given in NRC (1983).
f = estimated from Ogino and Takeda (1976), assuming high P content in diet, and dry weight = 25% wet carcass weight.

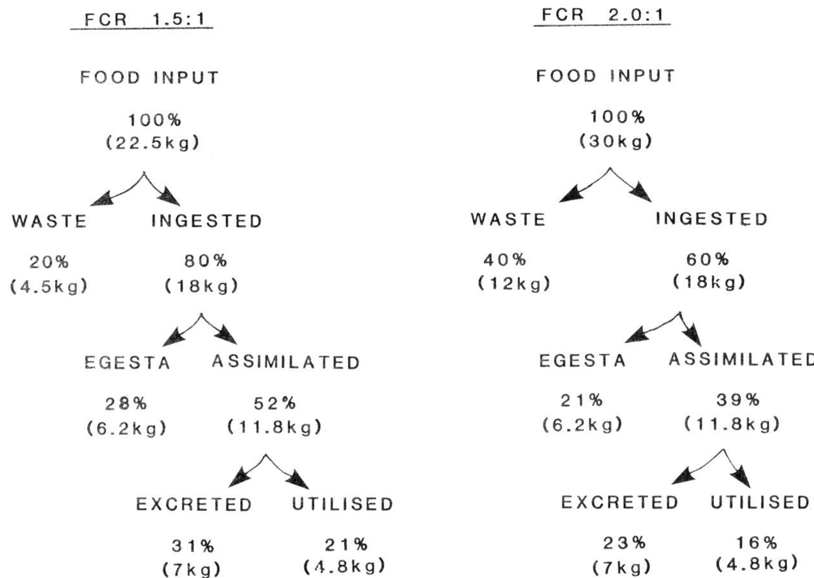


*Fig 5.2* Quantitative estimates of the fate of P in salmonid cage culture, assuming two FCR values.

bases. The two most widely used and tested models are those of Dillon and Rigler (1974) and OECD (1982). The former is a modification of Vollenweider's original model (Vollenweider, 1968), and states that the concentration of total-P in a water body, [P], is determined by the P loading, the size of the lake (area, mean depth), the flushing rate (*ie* the fraction of the water column lost annually through the outflow) and the fraction of P lost permanently to the sediments.
At steady state:-

$$[P] = \frac{L(1-R)}{\bar{z}\rho}$$

where [P] is in g $m^{-3}$ total-P, L is the total-P loading in g $m^{-2}$ $y^{-1}$; $\bar{z}$ is the mean depth in m; R is the fraction of total-P retained by the sediments; and $\rho$ is the flushing rate in volumes per year.

The OECD model states that the total-P concentration in the lake is a function of the concentration of total-P in the inflows [P] and the residence time, T(w):-

$$[P] = [\bar{P}]_j/(1 + T(w))$$

where $[\bar{P}]_j$ is in mg $m^{-3}$; and T(w) in years. However, although this model is widely applicable in most situations, it cannot be used here, since it takes no account of the special nature of cage fish farm wastes (see later). Thus the

Dillon and Rigler model is the more appropriate of the two.

Assessment of carrying capacity is best described in a series of steps.

*Step 1*: Measure the steady-state total-P concentration. In temperate waters this is best determined at the time of spring overturn, when the waters are well mixed. For tropical lakes and reservoirs [P] should be taken as the annual mean total-P concentration of surface waters, and thus should be based on a number of samples taken during the year.

*Step 2*: The development capacity of a lake or reservoir for intensive cage culture is the difference between the productivity of the water body prior to exploitation and the final desired/acceptable level of productivity. Although [P] is used to assess productivity it is, of course, the corresponding levels of algal biomass that we are interested in, and since fish are usually cultured throughout the year it is the *peak* level of algal biomass, as measured by chlorophyll a content, $[\overset{max}{\text{chl}}]$, that is of concern. The desired/ acceptable $[\overset{max}{\text{chl}}]$ is dependent upon the species being cultured and whether or not the water body is multipurpose.

Phosphorus concentrations corresponding to permissible $[\overset{max}{\text{chl}}]$ for different species and water bodies with more than one use are given in *Table 5.3* and *Fig 5.3*. These [P] values can be related to algal biomass using the regression equations in *Table 5.4*. However, since there are no $[\overset{max}{\text{chl}}]$ *vs* [P] relationships established for tropical water bodies, [P] must instead be related to annual mean chlorophyll a levels $[\overline{\text{chl}}]$, using the equation:-

$[\overline{\text{chl}}] = 0.416\,[\bar{P}]^{0.675}$ $r = 0.84$; $n = 16$ (Walmsley and Thornton, 1984)

*Step 3*: The capacity of the water body for intensive cage fish culture is the difference, $\triangle[P]$, between [P] prior to exploitation, $[P]_i$, and the desired/ acceptable [P] once fish culture is established, $[P]_f$.

*ie* $\triangle[P] = [P]_f - [P]_i$

$\triangle P$ is related to P loadings from the fish cages, $L_{fish}$, the size of the lake, A, its flushing rate and the ability of the water body to handle the loadings (*ie* the fraction of $L_{fish}$ retained by the sediments).

$$\triangle P = L_{fish}\,(1 - R_{fish})/\bar{z}\rho$$

$$L_{fish} = \triangle[P]\,\bar{z}\rho/(1 - R_{fish})$$

*Table 5.3* Tentative[1] values for maximum acceptable $[\bar{P}]$ in lentic inland water bodies used for enclosure culture of fish (from Beveridge, 1984b).

| *Water body category* | *Species cultured* | *Tentative maximum acceptable $[\bar{P}]$* |
|---|---|---|
| Temperate | Salmonid | 60 |
| | Carp | 150 |
| Tropical | Carp and tilapia | 250 |

[1] see text (5.2.1).

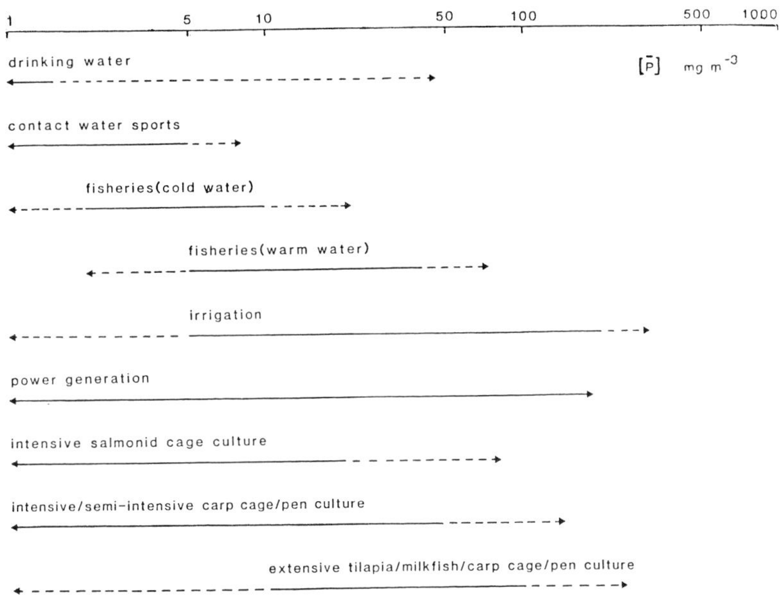


*Fig 5.3* Suggested acceptable (dotted line) and ideal (solid line) P concentrations associated with freshwater bodies used for different purposes (from Beveridge, 1984b).

The acceptable/desirable change in [P], $\triangle$[P] (mg $m^{-3}$), is determined as described in Step 2 above, and $\bar{z}$ can be calculated from hydrographic data obtained either from the literature or from survey work:-

$$\bar{z} = V/A$$

where V = volume of water body ($m^3$) and A = surface area ($m^2$). The flushing rate, $\rho(y^{-1})$, is equal to $Q_o/V$, where $Q_o$ is the average total volume ($m^3$) flowing out of the lake/reservoir each year. $Q_o$ can be calculated by direct measurement of outflows, or in some circumstances can be determined from published data on total long-term average inflows from catchment area surface runoff (Ad.r), precipitation (Pr) and evaporation (Ev), such that:-

$Q_o$ = Ad.r + A(Pr − Ev) (see Dillon and Rigler, 1975, for further details)

$R_{fish}$ is the most difficult parameter to estimate. Using the arguments proposed by Phillips *et al* (1985c) at least 45-55% of the total-P wastes from cage rainbow trout are likely to be permanently lost to the sediments as a result of solids (faeces and food) deposition, and thus only 45-55% of the total-P loadings are in the form of dissolved P. In the absence of any other data, these values will also have to be used for cage tilapia and carp calculations. A fraction of the dissolved total-P component will also be lost to the sediments, and it is suggested that the most appropriate formula in *Table 5.5* is used to calculate this. $R_{fish}$ values are therefore much greater than R for conventional P loadings, and can be summarised as:-

$R_{fish} = x + [(1 - x)R]$

where x = the net proportion of total-P lost permanently to the sediments as a result of solids deposition (*ie* 0.45-0.55), and R = proportion of dissolved total-P lost to the sediments calculated from *Table 5.5*.

The response time of a water body to increases in P loading is a non-linear function of the water residence time (Tw) and mean depth ($\bar{z}$). The expected 95% response time, $t(M)_{95}$, which is used as an approximation to the full response time can be calculated from *Fig 5.4*.

*Step 4*: Once the permissible/acceptable total-P loading, $L_{fish}$, has been calculated, then the intensive cage fish production (tonnes $y^{-1}$) can be estimated by dividing $L_{fish}$ by the average total-P wastes per tonne fish production (*Table 5.2*).

A worked example is given in Appendix 5.1.

### 5.2.2 Marine sites

A similar type of modelling process may be of use in calculating the carrying capacity of marine sites, except that N which is generally regarded as being the growth limiting nutrient in the sea (Dugdale, 1967), would be substituted for P. However, since marine coastal sites generally have much higher flushing rates — days or weeks rather than months or years — than inland water sites, the effect of dissolved N on phytoplankton production may be limited and the impacts of intensive marine cage fish farming on the benthos may be more important. However, this is highly speculative. At present, a number of research programmes funded by the Highlands and Islands Development Board, the Nature Conservancy, the Crown Estate Commissioners and the Scottish Countryside Commission are being undertaken by researchers at the Scottish Marine Biological Association laboratory at Dunstaffnage, Oban, and at Stirling University, to clarify the issues involved, and to develop guidelines for the exploitation of marine sites (see Gowen *et al*, 1985; Phillips *et al*, 1985, for reviews). In the interim, farmers are advised to choose sites carefully (see Chapter 4) and to monitor water quality — particularly $H_2S$ and DO — and the build-up of wastes under the cages.

## 5.3 Extensive culture

The principles of extensive cage culture in inland waters are discussed in Chapter 2. In most cases, fish production and thus carrying capacity are almost entirely dependent upon phytoplankton production. On the basis of studies carried out with tilapias in inorganically fertilised ponds, Beveridge (1984b) has suggested that yields from extensive cage culture are between 1% and 3% of primary production, depending on the level of primary production. Determination of site carrying capacity can be carried out as follows:-

*Step 1*: Determine the annual gross primary production, $\Sigma PP(g\ C\ m^{-2}\ y^{-1})$, of

*Table 5.4* Regression equations[1] relating to annual mean chlorophyll levels $[\overline{chl}]$ and peak chlorophyll levels $\overset{max}{[chl]}$ to each other and to mean in-lake total phosphorus concentrations $[\overline{P}]$, for temperate water bodies.

| | | | |
|---|---|---|---|
| $[\overline{chl}] = 0.61\ [\overline{P}]^{0.69}$ | n = 99; | r = 0.75; | S.E. = 0.335 |
| $\overset{max}{[chl]} = 1.77\ [\overline{P}]^{0.67}$ | n = 65; | r = 0.70; | S.E. = 0.375 |
| $\overset{max}{[chl]} = 2.86\ [\overline{chl}]^{1.03}$ | n = 73; | r = 0.93; | S.E. = 0.199 |

[1] = Regressions based on unscreened data, *ie* no corrections for light or N limitation included. data derived from OECD (1982).

the site. Since many tropical inland water bodies exhibit seasonality in the pattern of primary production (Melack, 1979), regular measurements may have to be made.

*Step 2*: Convert ΣPP to potential annual fish yields, Fy, using *Table 5.6* to convert planktonic carbon to fish carbon, and assuming fresh fish carbon content = 10% net weight of fish (Gulland, 1970).

*Step 3*: The organisation of production depends on a number of variables. The number of crops per year and the size of the fish at harvest should be decided. If, for example, tilapia are being farmed, then two crops of 160g fish (6 fish $kg^{-1}$) may be desirable. However, seasonality of primary production may mean that one crop takes longer to grow. In order to reach target harvest size, the sum of primary production during the crop 1 growth period, ΣPPc1, should approximate that of crop 2, ΣPPc2, although this ignores possible changes in the cropping efficiency of the fish at different algal densities, and may have to be adjusted in practice.

A worked example is given in Appendix 5.2.

## 5.4 Semi-intensive culture

The principle of semi-intensive cage culture is that low quality feeds are given to the fish to supplement their intake of natural food (see Chapter 2). The carrying capacity of inland waters for semi-intensive culture depends on (i) the productivity of the water body and the amount of natural food available, and (ii) the quantity and quality of supplementary food used. Computation can be carried out as follows:-

*Step 1*: Determine the primary production, ΣPP, of the site being considered, as described in Appendix 5.2.

*Step 2*: Calculate the annual fish yield, Fy, from the site using the conversion figures in *Table 5.6*, and a fresh fish carbon content = 10% wet weight.

*Step 3*: Calculate the average annual amount of the various feedstuffs available, ΣFood, and estimate the FCR from the literature (*eg* see *Table 30*, Beveridge, 1984b) in order to determine the fish yield attributable to the supplementary food.

*Table 5.5* Empirical models for calculating the sedimentation rate, $\sigma$, retention coefficient, R ($^1/\sigma$), and the sedimentation coefficient, V, of phosphorus, for both general and specific categories of temperate water bodies (from Beveridge, 1984b).

| *Model type* | *Size of data base* | *Model* | | *Correlation coefficient* | *Source* |
|---|---|---|---|---|---|
| (a) General | 704 | $\sigma = 0.129\ (L/\overline{Z})^{0.549}$ | | 0.81 | Canfield and Bachmann, 1981 |
| US EPA data base and | | $R = 1/(1 + 0.614\rho^{0.491})$ | ★ | 0.79 | Larsen and Mercier, 1976 |
| several European | | $\sigma = 0.94$ | ★ | 0.79 | Jones and Bachmann, 1976 |
| lakes and reservoirs | | $V = 2.99 + 1.7q_s$ | ★ | 0.73 | Reckhow, 1979 |
| | | $V = 5.3$ | ★ | 0.71 | Chapra, 1975 |
| | 73 | $R = 1/1 + \rho^{0.5}$ | | 0.79 | Larsen and Mercier, 1975 |
| | | $\sigma = 0.65$ | | 0.79 | Jones and Bachmann, 1976 |
| | | $R = 0.426\ \exp(-0.271q_s)+0.574\exp(-0.00949q_s)$ | | 0.71 | Kirchner and Dillon, 1975 |
| | | $V = 11.6 + 1.2q_s$ | | 0.68 | Reckhow, 1979 |
| | | $\sigma = 10/\overline{Z}$ | | 0.68 | Vollenweider, 1975 |
| | | $V = 12.4$ | | 0.66 | Chapra, 1975 |
| (b) Reservoirs | 210 | $\sigma = 0.114\ (L/\overline{Z})^{0.589}$ | | 0.83 | Canfield and Bachman, 1981 |
| North American | | $R = 1/(1 + 0.515\rho^{0.551})$ | ★ | 0.80 | Larsen and Mercier, 1976 |
| (c) Natural lakes | 151 | $\sigma = 0.162\ (L/\overline{Z})^{0.458}$ | | 0.83 | Canfield and Bachman, 1981 |
| | | $R = 1/(1 + 0.747\rho^{0.507})$ | ★ | 0.80 | Larsen and Mercier, 1976 |
| (d) Lakes with low flushing rates ($q_s < 10m$) | 53 | $R = 0.201\ \exp(-0.0425q_s)+0.574\exp(-0.00949q_s)$ | | — | Ostrofsky, 1978 |

$q_s$ = areal water loading ($mg^{-1}$).
$\rho$ = flushing rate (volumes per year).
★ = coefficients recalculated by Canfield and Bachmann (1981) using their data base.

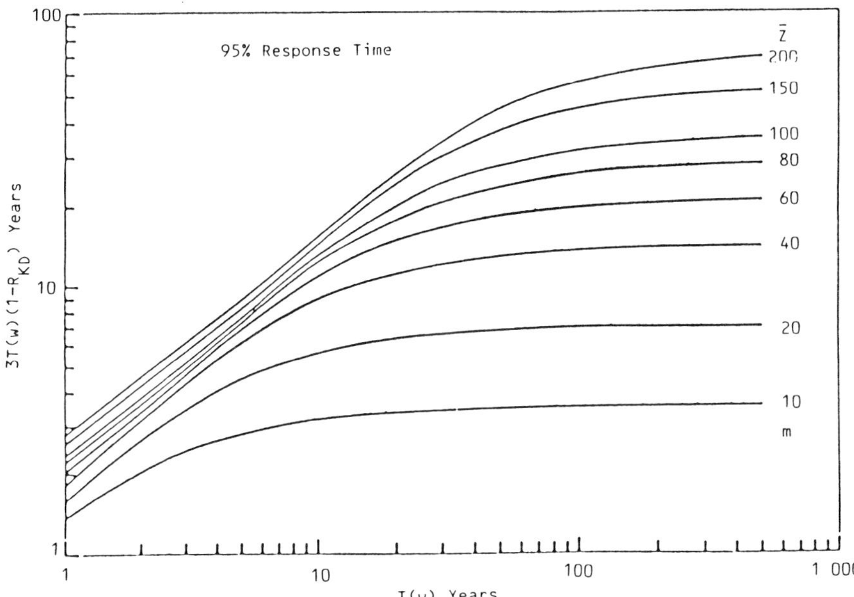


*Fig 5.4* The relationship between response time and water residence time, $T_w$, for water bodies with different mean depths $\bar{z}$ (from OECD, 1982).

*Step 4*: Calculate the total-P loadings associated with the use of supplementary feedstuffs, $L_{fish}$ and using the model described in Section 5.2 above calculate the increase in dissolved total-P. The increase in total-P can then be used to calculate increases in primary production, $\Sigma PP_{fish}$, attributable to fish culture.

*Step 5*: Estimate the fish yields due to $\Sigma PP_{fish}$, using the conversion efficiencies given in *Table 5.6* (see Step 2 above), and calculate the total fish yields from semi-intensive culture $\Sigma Fy$, as:

$\Sigma Fy = (a\Sigma PP) + (\Sigma Food \times FCR) + (b\Sigma PP_{fish})$

where a and b are the expected conversion efficiencies of primary production to fish biomass obtained from *Table 5.6*.

A worked example is given in Appendix 5.3.

## 5.5 Discussion and conclusions

The models proposed here for predicting the carrying capacity of different types of cage fish culture are in the initial stages of development and should be used as guides rather than hard and fast rules to site development. The main problems with each model are outlined below.

For intensive culture, the setting of desirable/acceptable water quality criteria presents a major difficulty. Although various agencies (*eg* OECD, USEPA) have set tentative water quality criteria, these have been concerned with minimising nuisance blooms in multi-use water bodies. However, the establishment of management objectives for lakes and reservoirs where fish culture is the primary or sole objective is made extremely difficult by our poor understanding of the relationships between algae and water quality and water quality and stress, growth, disease and mortality. Other variables which are impossible to quantify include the species being cultured, quality of stock, time of stocking, management methods and the composition of the algal community. Thus the setting of acceptable water quality objectives is a highly contentious area and in view of this the values given in *Table 5.3* must be used with caution and amended through experience and in the light of information collected from environmental monitoring.

Estimates of P-loadings from cage fish culture are likely to be revised, particularly as data on solubility and losses to the sediments are collected.

The model, too, suffers from a number of shortcomings. It is restricted in use to P-limited water bodies. However, some shallow lakes and reservoirs are light-limited due to the presence of high levels of inorganic suspended solids caused by wind-induced disturbance of bottom sediments (Canfield and Bachmann, 1981; Walker, 1982; Hoyer and Jones, 1983), whilst a number of other water bodies are nitrogen-limited (White *et al*, 1982; Wurtsbaugh *et al*, 1985). These problems are discussed more fully in Beveridge (1984b). The model is also most applicable to small, well-mixed water bodies or to sites where the cages are widely dispersed. Cages situated near a lake or reservoir outflow may have much less impact on the water body than predicted by the model.

Beveridge *(ibid)* estimated the overall predictive error associated with the model to be ~55-65%, although in view of recent information on the errors associated in calculating R (Phillips *et al*, 1985c), the total error may be greater. While the magnitude of error involved seems large, predictions should still be good enough to act as a management guide to permissible levels of production and can be adjusted in the light of water quality data collected when the farm is in operation. The importance of instigating a water quality monitoring scheme, particularly on large farms, cannot be stressed too highly.

It is possible to reduce the impact of intensive culture by a number of means (*Table 5.7*), and although apparently expensive, in practice many may be cost effective.

The model for extensive culture is based on limited data for phytoplankton grazing by tilapias in ponds and has yet to be applied to caged fish. Although this type of model may also be applied to other phytoplankton feeders, such as silver carp, it cannot be used to predict appropriate production levels for extensively reared fishes which primarily feed on zooplankton (*eg* bighead carp).

Because it is a hybrid of both the extensive and intensive models, the model suggested for semi-intensive cage culture involves the errors associated with both. Moreover, it is difficult to find data on the P content and conversion efficiencies of materials used for supplementary feedstuffs, and even more difficult to assess how ingestion in combination with natural food (algae, detritus, zooplankton) affects these values.

*Table* 5.6 Conversion efficiencies of ΣPP to areal fish yields for water bodies of different productivites, based on Beveridge (1984[b]).

| ΣPP ($g\ C\ m^{-2}\ y^{-1}$) | *% conversion to areal fish yields* ($g$ fish $C\ m^{-1} y^{-1}$) |
|---|---|
| <1,000 | 1–1.2 |
| 1,000–1,500 | 1.2–1.5 |
| 1,500–2,000 | 1.5–2.1 |
| 2,000–2,500 | 2.1–3.2 |
| 2,500–3,000 | 3.2–2.1 |
| 3,000–3,500 | 2.1–1.5 |
| 3,500–4,000 | 1.5–1.2 |
| 4,000–4,500 | 1.2–1.0 |
| >4,500 | ~1.0 |

*Table* 5.7 Estimated potential for reduction in total-P wastes associated with intensive fish culture through various feed manufacturing and management options. Costs estimated as ranging from * (inexpensive) to *** (expensive) (from Beveridge, 1984b).

| *Option* | *Method* | *Cost* | *Reduction* |
|---|---|---|---|
| Reduction of dust added to water body | — Improved manufacturing (*eg* use of steam conditioning, increased mash transit time in steam conditioner, *etc*[1]) | ** | 2%+ |
| | — Sieving of feeds by farm staff prior to use | * | |
| Reduction of pellet losses to the environment | — Improved feeder design | ** | |
| | — Careful siting of cages | * | |
| | — Careful adjustment of feeding regime to prevailing environmental conditions | * | 10%+ |
| Reduction of total-P load in wastes | — Reduced P content in feeds | ** | 30%+ |
| | — Use of high digestibility diets | * | ~30% |
| Removal of surplus P added to lake or reservoir during culture | — Pumping and removal of wastes from under cages | *** | ? |
| | — Removal of mortalities to site on shore | * | 5%[2] |
| | — Trapping and removal of escaped fish | * | 1.5%[2] |
| | — Utilisation of wastes through combination with extensive culture | ? | ? |

[1] see ADCP (1983).
[2] these figures depend very much on extent of mortalities and number of escaped fish.

In summary, the brief review given above serves to illustrate how little is known about the effects of different types of cage fish farming on the environment — particularly in marine and tropical conditions — but suggests which lines of enquiry are likely to be most profitable. However, until reasonably accurate and simple to use, predictive models are developed, a cloud hangs over the future of this method of culturing fish since it carries an element of risk which could eventually disenchant investors and insurers and bring it into conflict with environmentalists.

*Appendix 5.1* Example of intensive cage rainbow trout production assessment for a hypothetical temperate natural lake (see Section 5.2.1) (modified from Beveridge, 1984b).

*Site*:
Surface area of lake, A, = 100ha (calculated from map).
Mean depth, $\bar{z}$, = 10m (from hydrographical survey).
Flushing coefficient, $\rho$ = 1 $y^{-1}$ (determined from sampling outflows).

*Method:*

*Step 1*: Determine steady state [P], $[P]_i$, prior to development. 15mg $m^{-3}$ as determined from monitoring programme.

*Step 2*: Set maximum acceptable [P], $[P]_f$, following the introduction of fish cages. Assuming no other developments or criteria take precedence, then 60mg $m^{-3}$ is chosen as target $[P]_f$

*Step 3*: Determine $\triangle P$

$\triangle P = [P]_f - [P]_i = 45\text{mg m}^{-3}$

Since $P = L_{fish}(1-R_{fish})/\bar{z}\rho$

$L_{fish} = \triangle P\bar{z}\rho/1-R_{fish}$

$R_{fish} = x + [(1-x)R]$; where R is calculated from *Table* 5.5, and x is assumed to be 0.5.

Thus, $L_{fish} = 45 \times 10 \times 1/0.23 = 1957\text{mg m}^{-2}\text{y}^{-1}$

$= 1.957\text{g m}^{-2}\text{y}^{-1}$

*Step 4*: Since the lake has a surface area of $10^6 m^2$, the total acceptable loading $= 1.957 \times 10^6 \text{g y}^{-1}$

∴ The tonnage of fish that can be produced, assuming a P loading of 17.7kg tonne $^{-1}$ (see *Table 5.2.*)

$= 1.957 \times 10^6\text{g}/17{,}700\text{g} = 111$ tonnes per annum

This value should be used as a pre-development guide to the carrying capacity of the lake. However, a monitoring programme must be implemented and actual production levels adjusted in the light of information collected on water quality — principally algal biomass and DO levels.

*Appendix 5.2*. Example of extensive cage tilapia production for a hypothetical tropical reservoir (see Section 5.3) (from Beveridge, 1984b).

*Site*:
Surface area = 100ha.

*Method*:

*Step 1*: Calculate the annual gross primary production, Σ PP. 1200g C $m^{-2}y^{-1}$, as determined by regular measurement.

*Step 2*: Convert to annual fish yields, using *Table 5.6*.
*ie* = 1.3% Σ PP → fish
= 15.6g fish C $m^{-2}y^{-1}$
= 156g fish $m^{-2}y^{-1}$
= 156 tonnes annual fish production for whole lake.

*Step 3*: Assuming 2 crops per year, determine culture periods.
ΣPPc1≡ΣPPc2, in order for fish to reach target market size.
Σ PP (Nov — May) = 570g C $m^{-2}$
Σ PP (June — Oct) = 630g C $m^{-2}$
One seven month, and one five month cycle are chosen.
Assume 25g fish stocked
Assume 8pcs. per kilo target market size (*ie* 125g each)
∴ each fish grows 100g during culture period.
∴ stocking requirements = 156 tonnes/100g = $1.56 \times 10^6$ fingerlings

*Appendix 5.3*. Example of semi-intensive cage tilapia production assessment for a hypothetical tropical lake (see Section 5.4.) (modified from Beveridge, 1984b).

*Site*:
Surface area = 100ha
Mean depth, $\bar{z}$, = 10m
Flushing coefficient, $\rho$ = $1y^{-1}$

*Method:*

*Step 1*: Calculate the annual gross primary production, ΣPP.
1000g C $m^{-2}y^{-1}$, as determined by regular measurement.

*Step 2*: Convert to annual fish yields, using *Table 5.6*.
*ie* ~ 1.3%Σ PP→Fish
= 156 tonnes annual production for whole lake.

*Step 3*: Assume 100 tonnes of cotton seed meal and 20 tonnes of soya meal are available for feed each year. Using FCR values of 2.69:1 and 3.04:1 respectively (from Beveridge, 1984b, *Table 30*):-
37.2 tonnes can be grown from cotton seed meal, and 6.6 tonnes can be grown from soya meal.

*Step 4*: Total-P loadings from fish grown on supplementary food are:-
$(37.2 \times 23.77) + (6.6 \times 16.97) = 996.24$kg (P-content of feeds from Beveridge, 1984b).
The resultant increase in [P] can be calculated from:-

$[P] = L(1-R_{fish})/\bar{z}\rho$
where L is the areal loading from the fish cages (996.24kg $10^6$m$^{-2}$ = 996.24mg m$^{-2}$); $R_{fish} = x + [(1-x)\ R]$, where $x = 0.50$ and $R = 0.54$ (from *Table 5.5*) = 0.77;
$[P] = 996.24(1-0.77/10 \times 1) = 22.9$mg m$^{-3}$
Using the formula:
$\Sigma PP_{fish} = 31.1\ [\bar{P}]^{0.54}$ (OECD, 1982) to relate increase in [P] to primary production,
$\Sigma PP_{fish} = 31.1 \times 22.9^{0.54} = 34.7$g C m$^{-2}$y$^{-1}$ increase.

*Step 5*: Fish yield due to $\Sigma PP_{fish}$ can be calculated using the conversion efficiences in *Table 5.6*:-
$\Sigma PP_{fish} \rightarrow$Fish = 0.3g fish C m$^{-2}$y$^{-1}$
= 3g fish m$^{-2}$y$^{-1}$
= 3 tonnes fish production for whole lake.
$\Sigma$ Fy; The total fish yield can now be calculated from:-
$\Sigma\ Fy = (0.013 \times 1200 \times 10) + [(100/2.69) + (20/3.04)] + (0.01 \times 34.7 \times 10) = 203$ tonnes fish per annum.

# 6 Management

## 6.1 Introduction

It is the manager's job to be responsible for the fish on the farm from the day they arrive until the day they leave, and the onus is on him or her to get the best production possible out of the system, given the restrictions imposed by the site, stock or type of feed used. In order to achieve this, the caged stock must be kept in conditions which minimise losses and promote good growth, and this means:-

— stocking the fish at densities appropriate to the site, species and methods of rearing;
— feeding the fish in the most cost-effective manner;
— ensuring the best possible water quality within the cages;
— maintaining cages, moorings, anchors and ancillary gear;
— regular checking of stocks for signs of disease, removal of mortalities, and treatment of infected fish.

Good records of water quality conditions, growth, and mortalities should be kept so that management procedures can be properly evaluated and modified if necessary. The manager's job also involves taking an interest in his staff and ensuring that their working conditions are a safe and pleasant as possible. The manager also cannot be indifferent about the quality and transport of fry or fish to the farm, or about the end product of the farm's endeavours. It follows that the choice of management and staff are crucial to any successful cage fish farming venture.

The following sections deal with the general routine of cage fish farming.

## 6.2 Seed supply and stocking

Although it is possible to produce fry of mouth-brooding tilapia species in net cages, juveniles of other cultured species must either be produced in land-based hatcheries or captured from the wild and transferred to net cages at the appropriate time. Similarly, marine salmonid culture relies on smolts or large trout being introduced into cages following a period of freshwater culture which may or may not have involved some time spent in cages. Unless cage farmers also operate a hatchery or smolt production business, they must rely on external supplies of stock and follow the old maxim of 'buyer beware' when it comes to choosing a source of supply. Problems of fry quality bedevil certain industries and are discussed more fully in the relevant sections of Chapter 8.

Reviews of fish transport have been made by Hattingh *et al* (1975), Taylor and Solomon (1979), Okoye (1981), and Solomon and Hawkins (1981). Salmonids and carp being transferred from hatcheries or other culture systems should be starved for at least 24-48h prior to transport, in order to clear their gut of food and to reduce oxygen consumption. For bulk transport,where tens of thousands of these fish are involved, some farmers recommend even longer. *Lates calcarifer* fry, however, require only 1-2h starvation prior to packing (Tattanon and Maneewongsa, 1982). Because of the rigours of the journey, fish should be carefully checked and injured or weak fish removed, and if necessary, treatment for ectoparasites should also be carried out (see Martyshev, 1983, for details). Any veterinary or legal obligations should, of course, be fulfilled prior to dispatch.

The processes of capture, handling, loading and transport are highly stressful to fish, resulting not only in physical damage (*eg* scale removal), but also in changes in blood chemistry, increased oxygen consumption, osmoregulatory problems and increased susceptibility to disease (Solomon and Hawkins, 1981). Some species, such as silver carp, are particularly difficult to transport (Horvath *et al*, 1984). It follows, therefore, that during transport handling should be kept to a minimum. In temperate countries small quantities (a few thousand) of fish may be transported over short distances to cage sites with relatively few problems. Plastic bags can be used which should be one-third filled with water and the remaining space filled with oxygen prior to sealing and double-bagging for safety. Alternatively, an insulated transport box which can hold up to 3,000 *l* and is mounted on a trailer or on the back of a truck can be used (*Fig 6.1*). The tanks should have rounded corners to minimise damage to the fish, and are often connected to aeration or oxygenation equipment. For larger consignments, tankers with a capacity of 10,000 *l* or more are available from specialist transport companies and these are equipped with refrigeration and oxygenation/aeration equipment (see *Fig 6.3*). In the USSR, specially modified rail wagons are also used (Martyshev, 1983). Recommended fish densities for transport are given in *Table 6.1*.

The transportation of fish in the holds of boats is commonly practised in many parts of the world. In Southeast Asia, grouper fry are carried in the water-filled live-bait compartments of traditional pole-and-line fishing vessels from around the Chinese coast to stock the cages of Hong Kong fish farmers (Tseng, 1983). A more sophisticated version of this practice can be seen in Norway. Here 'well-boats', originally developed for carrying live fish such as cod to the important markets of southern Norway, have been used for a number of years to transport smolts to remote salmon cage farms along the Norwegian coast (Edwards, 1978) (*Fig 6.2*). The hold is filled with water, and valves situated below the water line on the sides towards the bow and stern can be opened so as to maintain a flow of water through the hold when the boat is moving. The flow rate is adjusted to the swimming performance of the fish; if too fast, the fish will fall back towards the stern, if too slow, the

*Fig 6.1* Small tank for short-distance road transport of smolts, equipped with a fish pump (from Edwards, 1978).

fish will congregate at the bow. Smolts can be successfully transported over many hundreds of kilometres in this way, although speed has to be kept low during long journeys. Rough weather can also severely affect fish transported by sea and result in high mortalities (Tseng, 1983). Nevertheless, transport by boat is gaining popularity in countries like Scotland where road access to cage fish farms is often difficult.

Transport problems may be aggravated by high temperatures and by salinity. In tropical countries, transportation at night, or packing containers with ice and sawdust (1:1) is often recommended (Tattanon and Maneewongsa, 1982). If fish have to be transported over considerable distances there is also a risk of a build up of toxic metabolic wastes, such as $CO_2$ and ammonia, and increased bacterial numbers. In order to minimise mortalities in these circumstances, a number of modifications to transport practices have been suggested:-

(i) lowering metabolic rate and thus oxygen consumption and waste production through a combination of light sedation and hypothermia (Solomon and Hawkins, 1981; Okoye, 1982; Ross and Ross, 1984)

(ii) absorption of ammonia and $CO_2$ and control of bacterial growth through the addition of natural zeolite, a buffer and an antibiotic to the transport media (Amend *et al*, 1982).

However, these methods are still controversial and their practicality in large-scale commercial fish transport operations has yet to be demonstrated.

Before transferring the fish to cages, care should be taken to ensure that the temperature of the fish is adjusted to approximately that of their new

*Table 6.1* Transport conditions for various fish species. Data from Piper *et al* (1982), Horvath *et al* (1984), and D. Griffith (pers. comm).

| *Species* | *Size* | *Stocking density* ($g\ l^{-1}$) | *Duration* (*h*) | *T°C* |
|---|---|---|---|---|
| Chinook salmon | 40mm | 60–120 | 8–10 max | 5–10 |
| | 60mm | 120–240 | 8–10 max | 5–10 |
| Coho salmon | 100–130mm | 240–360 | 8–10 max | 5–10 |
| Trout (rainbow, brook, brown, *etc*) | 200–280mm | 300–420 | 8–10 max | 5–10 |
| Channel catfish | 100g | 350–600 | 8–16 max | 18 |
| | 10g | 250–400 | 8–16 max | 18 |
| | 4g | 200–350 | 8–16 max | 18 |
| | 2g | 150–200 | 8–16 max | 18 |
| Largemouth bass | 20g (100mm) | 120 | 12 max | 18–30 |
| | 5g (75 mm) | 80 | 12 max | 18–30 |
| | 1g (50 mm) | 60 | 12 max | 18–30 |
| Common and bighead carp | <100g | 280 | — | 5 |
| | | 50 | — | 30 |
| Silver carp | <100g | 90 | — | 5 |
| | | 25 | — | 25 |
| Tilapias | 0.5–200g | 100–200 | 24 max | 8–28 |

*Fig 6.2* Well-boat, transporting Scottish smolts to farms in the remote northwest of Norway (courtesy D A Robertson).

environment. This poses no problem for fish delivered in well boats. However, bags of fish should be placed in the cages and the temperatures allowed to equilibrate prior to release. In warm climates, transfer to cages should be carried out in the late evening or early morning if possible, when temperatures are lower. Again, handling should be kept to a minimum at this stage. Often it is easiest to transport the fish to where the cages are moored, particularly if the fish have been transported by well-boat or in plastic bags. If large transport tanks have been used, then cages may be towed to the nearest suitable landing site, and the truck positioned so that the outflow valve can be opened and the fish + water piped directly into the cage (*Fig 6.3*). Alternatively the transport box may be lifted onto a boat and carried out to the cages. Bagged fish should be transferred by gently tipping them out into the cages, whereas for transport tanks and well-boats, the volume of water is normally reduced prior to the fish being transferred by hand, net or fish pump (*Fig 6.4*) . If nets are used, these should be of fine, knotless mesh to minimise damage. Fish are either counted by eye, or by using a counting board.

The feeding of fish immediately after transfer to cages is generally not recommended, although farmers may not be able to resist the temptation to throw in a handful of food to see if their new stock is in good condition. Some fishes, such as the tilapias, recover fairly quickly from handling, and regular feeding can commence 3-4h after transfer, whereas more stress-sensitive fishes, such as the salmonids, are best left undisturbed for 12-24h before feeding.

Monitoring of blood chemistry following transport suggests that fish take several days to recover from the trauma (Aldrin *et al*, 1979). As mortalities often occur during this period it is best to clarify responsibility for losses at

*Fig 6.3* Large transport truck transferring Atlantic salmon smolts to a fish cage, Scotland. The cage will be towed to the farm site and the smolts redistributed to other cages (courtesy D A Robertson).

the outset of any agreement between supplier and cage fish farmer, and to have this in writing. Smolt producers in Western Europe usually cover any losses that occur during the first week in seawater, or alternatively add 5% extra fish to the order to compensate for any losses. Mortalities incurred during transport are, of course, the responsibility of the transporter, irrespective of whether he is the supplier, purchaser, or an independent agent.

There are two basic stocking strategies: farmers may either decide at the outset to stock the number of fish required to give them the production per unit area/volume they require, taking into account likely mortalities, or more usually, they will stock high numbers of young fish and as they grow, thin them out to other cages. Stocking density can affect growth, incidence of injury and mortality (*eg* Konikoff and Lewis, 1974) although to date this has been poorly researched. Details of stocking practices for various cultured species will be detailed in Chapter 8.

## 6.3 Feeds and feeding

### 6.3.1 INTRODUCTION

Extensive methods of cage fish farming rely solely on natural food, although illumination may be used to attract food organisms towards the captive fish (see Chapter 8). Semi-intensive and intensive operations, however, involve feeding and since feeding costs represent the single largest component in a fish farmers' operating costs, accounting for an estimated 40-60% at intensive fish farms (ADCP, 1983), it follows that the quality of feed and the manner in which it is used will largely determine the yield of a farm and be a significant determinant of profitability.

### 6.3.2 TYPES OF FEED

Fish foods are of two basic types: semi-intensive and intensive. According to Coche (1982), semi-intensive feeds are 'relatively low in protein and made from local materials readily available at low cost'. Fish grown in semi-intensive culture situations still rely heavily on natural feed, which is comparatively high in terms of protein, supplemented by foods high in carbohydrate or fat which will 'spare' the protein from being used as an energy source, and allow it instead to be utilised for growth. This type of aquaculture, of course, is only suited to certain species of fish, such as herbivores, planktivores, detrivores or omnivores, grown in conditions where there is a plentiful supply of natural feed (*ie* productive water bodies). Semi-intensive cage culture is also largely restricted to freshwater species and is most commonly practised in tropical and sub-tropical countries.

The range of feedstuffs used in semi-intensive culture is enormous. Reviews of the materials used and their nutritional values are given in Ling (1967), Jauncey and Ross (1982), Coche (1982), NRC (1983), ADCP (1983) and Pathmasothy (1983). The feedstuffs are fed either singly or in combination, but because of seasonal variations in costs and availability, their use is often sporadic (see Chapter 8 on tilapia culture). Moist feeds are sometimes compounded, using dry ingredients, water, and a suitable binder (*eg* dried green banana powder, cooked plantain, potato *etc*) to form a ball which can be fed fresh or partially dried. Care must be taken with some commonly available materials which contain naturally occurring antinutrients and can adversely affect fish (Jauncey and Ross, 1982; NRC, 1983).

Intensive feeds are used principally in the culture of carnivorous species, although omnivores/herbivores such as the tilapias are sometimes reared in this way where water resources are scarce, if such fish can still be sold at a profit (see Beveridge, 1984b, for discussion). Intensive feeds differ from semi-intensive feeds in that they must supply *all* the cultured species' nutritional requirements, including the right quantities and qualities of proteins, fats, carbohydrates, minerals and vitamins (Halver and Tiews, 1979; NRC, 1981, 1983; Jauncey and Ross, 1982, for reviews). From the early beginnings of intensive aquaculture methods in the latter half of the 19th century, until 35 years ago, the only option available to someone growing trout or any other carnivore was to use fresh or frozen slaughterhouse wastes, augmented perhaps by raw fish. In the mid-1950s the first formulated moist diets were compounded in the United States for salmonids (Phillips, 1956) and this was followed by the development of dry diets in the 1960s. Intensive diets have since developed and improved beyond measure, as our understanding of fish nutritional requirements has grown and fish feed technology improved. Today, there are a number of different types of intensive feedstuffs in use.

*Fig 6.4* Smolt delivery at a Norwegian salmon farm. Fish are netted out of the hold, counted and transferred to the cages (courtesy D A Robertson).

Surprisingly, fresh or frozen minced and chopped trash fish still forms the mainstay feed for a number of important cage fish farming industries such as the culture of yellowtail and seabream in Japan (Kafuku and Ikenoue, 1983), snakehead in Thailand (Wee, 1982), grouper in Hong Kong (Tseng, 1983), and sea bass in Thailand (Sirikul, 1982) and is even still used by some salmon farmers in Norway (Sutterlin and Merrill, 1978). Economic factors and problems with diet formulation, feed storage and distribution, are the principal reasons why this type of feed remains popular in some quarters. At first glance it may seem the ideal type of feed though perhaps not from an environmentalist's point of view. However, there are a number of problems. Many species of trash fish that are readily available, such as members of the sardine and mackerel family, have fat contents which are too high for some cultured species (see Chapter 8), whilst others contain high levels of thiaminase which, if not heat treated, can lead to thiamin deficiency (NRC, 1983). Certain trash fish species may also vary considerably in quality during the year. Capelin, for example, contain 13-14% fat during January-February, prior to spawning, falling to 4-6% after spawning (Edwards, 1978). The availability of some species may vary with season, and certain 'fussy' cultured species, such as the Atlantic salmon, have reportedly refused to switch diet, choosing instead to starve for a period of several weeks (Edwards, *ibid*). Trash fish also has a high moisture content which means that it is expensive to transport and is therefore best suited to farming operations sited close to fish landing or processing centres. A further consideration is the high wastage associated with this type of diet, and its effects on water quality (Sutterlin and Merrill, 1978). Some type of pigmented diet or addition of cooked prawn processing wastes may also be required as a finishing feed to improve flesh colour. Finally, it is known that raw feeds can act as a source of bacterial infection (Ross and Johnson, 1962; Friedman and Shibko, 1972; Shepherd, 1978).

Many salmon farmers in Norway use moist feeds prepared on the farm from minced trash fish or silage (see Section 6.3.2) mixed with a commercially produced binder meal, which typically contains some protein (fish meal), carbohydrate (*eg* wheatmeal), vitamins, a binding agent (*eg* carboxy methyl cellulose) and sometimes a colourant (*eg* carophyll red or shrimp meal). If white fish is used, a fish oil may also be added in order to spare the protein from being utilised as an energy source. The paste is then either fed as a moist dough ball or pelleted to the required size using an appropriately-sized die on a grinder. Details of moist feed manufacture are given in Edwards (1978), Jauncey and Ross (1982) and NRC (1983). Although such feeds are an improvement on trash fish in terms of nutritional qualities and increased stability in water, they also suffer from many of the disadvantages of trash fish diets — seasonal fluctuations in quality, transport problems, and associated high pollution loadings. For these reasons and because of storage problems (see later), wet diets are no longer widely used in North America, and their use in European inland waters is now discouraged (Alabaster, 1982). Outside Norway, where their popularity is due to the ready availability of cheap trash fish, their use is restricted to species which find dry diets relatively unpalatable (*eg* yellowtail, red sea bream) (Nose, 1979), or to areas where dry diets are as yet unavailable.

Dry diets have a number of advantages over moist or trash fish diets: they are less polluting, since they are more stable in water and are more readily eaten by most species of cultured fishes. Moreover, they can be more digestible, and commercially-prepared versions also have fewer anti-nutrients due to the methods used in processing. Dry diets are, however, a little more expensive, the cost of processing adding 1-3% to the cost of the feed (ADCP, 1983). Formulae for commercially prepared dry diets are jealously guarded, although there are many that have been developed by research institutes and FAO scientists which are widely available (Jauncey and Ross, 1982; NRC, 1983; ADCP, 1983). There are a number of different ways to manufacture dry diets. The simplest involves using large commercial food mixers which produce a wet-extruded diet that is subsequently dried, and this is often sufficient for small operations where commercially-prepared feeds are unavailable. Commercial-scale operations, however, involve more sophisticated processing which generally results in better and more uniform quality feeds. A typical production scheme is illustrated in *Fig 6.5*. The dietary ingredients are first pretreated to improve handling and pelleting, to increase the nutritional value, and to help destroy any anti-nutrients that may be present. The materials are then ground and thoroughly mixed before pelleting. Most sinking diets are extruded using low-pressure steam. However, high-pressure steam conditioning prior to pelleting results in an expanded or extruded diet which is comparatively low in density and which consequently floats. Floating pellets were first developed for catfish in 1960 (Avault, 1981) and became popular because the fish could be observed feeding and farmers could thus see if their fish were healthy or not. Floating pellets also have superior water stability properties (Stickney, 1979).

Whilst it is obvious that sinking pellets should be used for species which rely largely on tactile means to locate their food, such as sturgeon, or which remain on the cage bottom to feed, such as turbot, the question as to whether it is better to use sinking or floating feeds in cage fish farming remains largely unresolved. Coche (1979, 1982) believes that the loss of floating pellets in

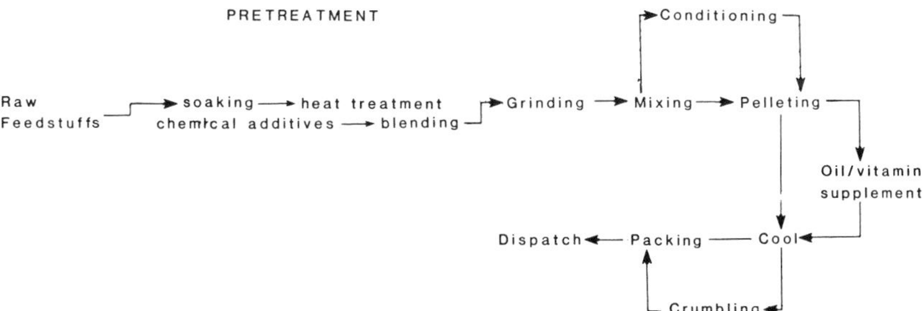


*Fig 6.5* Flow diagram to illustrate industrial manufacture of dry, pelleted feeds (modified from Jauncey and Ross, 1982).

cages is more easily controlled than the loss of sinking pellets, and that in small cages at least, floating feeds may be better for tilapias. However, another authority has stated that sinking feeds are better for caged trout production in that there is less size variation among stock since less dominant sub-surface fish are getting food (see Discussion, p. 105-106, Proc. IFM Symp., Reading. Janssen Services, London, 1980). Moreover, there is evidence that the availability of the carbohydrate fraction of steam-pelleted floating feeds is increased to such an extent that in rainbow trout at least, liver function may be impaired (Hilton *et al*, 1981).

Konikoff and Lewis (1974) claim that as far as caged channel catfish are concerned, floating pellets seem to give a more uniform size-weight frequency distribution at harvest than sinking pellets. However, there is evidence that the density of fish stocked in that particular experiment may have influenced results. Moreover, Newton (1980) states that catfish grown in cages in clear water do not come to the surface to feed.

The above studies suggest that both types of pellet may be suitable, depending on site, species, cage size and stocking density.

The relative advantages and disadvantages of wet and dry feeds for cage fish farming are summarised in *Table 6.2*. Although trash fish diets appear to be cheapest, there are a number of hidden expenses, such as higher transport, storage and feeding costs which can firmly swing the balance in favour of dry diets. The use of trash fish and moist pellet diets is also highly polluting and has been outlawed in inland cage fish farms in several northern European countries. Most observers would therefore agree that dry diets are likely to be the most popular type of feed in the future. Details of species-specific diet compositions are given in the relevant sections of Chapter 8.

*Table 6.2* A summary of the relative advantages and disadvantages of various wet and dry diets.

| *Type of feed* | *Composition* | *Advantages* | *Disadvantages* |
|---|---|---|---|
| Trash fish | Trash fish species, scraps from fish processing industry. Fresh/frozen, whole/ minced/chopped | Cheap to purchase. Palatable to fish | High moisture content, expensive to transport and store, variable quality, high pollution loadings. Some anti-nutrients possible. Coloration of farmed fish may be a problem |
| Moist pellet/ dough | Trash fish and binder meal which includes vitamins, minerals and extra protein | Improved nutritional and water stability properties over trash diets | Requires regular fresh/ frozen fish supply, must be used immediately when pelleted. High pollution loading. Limits choice of feeding system |
| Dry diets | | Stable in water, convenient, consistent quality, relatively low transport and storage costs, long shelf life | Expensive. Unpalatable to a few species |

### 6.3.3 Storage of feeds

Since feedstuffs are usually delivered in bulk, most cage fish farming operations require some sort of storage facilities. Many of the larger cage tilapia farms in the Philippines have guard houses built out on the lake beside their cages, where small quantities of dry feedstuffs may be conveniently stored (*Fig 6.6*), whilst in England, one salmon farmer uses a number of old fishing boats as a mooring facility/living quarters/feedstore. There have also been several radical designs for offshore farms which incorporate feed storage facilities. However, most cage fish farms must rely on land-based stores since they have neither the space necessary for storage of dry feeds nor the power supply usually required for storage of trash fish. Water-based storage facilities also suffer from a number of disadvantages (see later). Even leaving a supply of feed bags unguarded on cage walkways is not recommended, as it invites the attention of gulls and other opportunistic birds which, apart from constituting a disease risk, have been observed taking feed from bags, spilling a great deal in the process. Storage facilities must maintain the quality of the feeds: humidity, heat, insects, rodents, fungi, dirt and other contaminants can destroy or greatly damage feedstuffs, rendering them unpalatable, less nutritious and even toxic to the fish (Chow, 1978).

Trash fish may arrive at the farm in either a frozen or unfrozen state and since fish spoils rapidly, it should be checked for freshness before being stored. Smell and appearance should be sufficient indications of quality. If fish are put into cold storage they will keep for many months and remain free from pest damage. However, temperatures should be low enough to prevent the oxidation of fats, and the longer the fish are kept, the colder the temperature must be. Facilities which can reduce the temperature sufficiently for extended storage are expensive to buy, and costly to run. An alternative is to ensile fish and store as silage which can prove cheaper than freezing, and produces a highly palatable moist feed. There are several different methods of producing silage, and these are detailed by Disney *et al* (1978), Raa and Gildberg (1982) and Raa *et al* (1983). The most common method is by using acid. The fish or fish offal is first chopped and a mixture of 1.5% sulphuric and 1.5% formic/propionic acids added to reduce the pH to less than 4. At this stage an antioxidant, such as ethoxyquin may also be added at 250ppm. The silage can then be used almost immediately, or stored in plastic bins or silage storage tanks where it can be kept for several months. During storage, the essential amino acid tryptophan may be depleted (Kompiang *et al*, 1980), although this problem may be overcome by using a high tryptophan binder. The silage is mixed 60:40 or 50:50 with a commercially produced binder meal containing extra protein, vitamins and a binding agent to form a water-stable moist pellet with a shelf-life of up to three days, depending on storage conditions (see Anon, 1983a).

Dry feedstuffs are delivered to farm sites either in bulk by road tanker, or more commonly, packed in plastic-lined 25kg or 50kg sacks stacked on pallets. Bulk feeds can be stored in hoppers which may be conveniently sited at the water's edge so that the outlet chutes can be swung over the water to

*Fig 6.6* Guard house at a lake-based tilapia hatchery and fishpen/fish cage operation, Laguna de Bay, Philippines.

discharge directly into a boat (*Fig 6.7*). However, the dust content of feeds stored in this manner is reckoned to be a little higher (Edwards, 1978), and there is also a risk that feeds stored near water for a period of time may accumulate moisture and clump, causing problems of pellet release (Webber and Hugenin, 1979). Moist feed retained in such hoppers is often subject to microbial attack. For these reasons it is recommended that bulk dry feed is not stored for long.

Bagged dry compounded feeds and other feedstuffs should be kept in clean, dry, cool facilities used solely for that purpose in order to avoid contamination with other materials often stored on farms, such as insecticides and pesticides, pharmaceuticals, petroleum products or heavy metal-containing compounds (*eg* anti-fouling paint). Both temperature and humidity have a great influence on the rates at which chemical changes take place and on the growth of fungi and insects. High humidity is conducive to the destruction of vitamin C, whilst the combination of high humidity and high temperatures increases peroxidation with the consequent destruction of vitamin E and other fat-soluble vitamins (NRC, 1981). Insects and rodents can not only eat their way through considerable quantities of food, they can also spoil much more than they consume through spillage and contamination with faeces and urine. Faecal contamination of feeds has been shown to be a source of *Salmonella* infection in farmed fish (Friedman and Shibko, 1972). A number of fungal species which invade feedstuffs produce mycotoxins and, unfortunately, some of those feeds, such as cottonseed meal, contain compounds which exacerbate mycotoxicosis in fish (NRC, 1981). Aflatoxins, which are secreted by *Aspergillus* spp., have been shown to produce hepatomas in salmonids and cause a haemorrhagic syndrome in tilapias (NRC, *ibid*; Roberts and Sommerville, 1982). Propionic acid is sometimes added to diets (0.25%) to inhibit mould growth.

Deterioration of stored dry feeds and feedstuffs can be minimised if the following points are followed (adapted from Jauncey and Ross, 1982):-

(i) Feeds should be checked before being put into the store — any infested materials should be dumped or fumigated first to avoid contamination.
(ii) Feeds should be stored in bags and raised off the ground — even pallets will do.
(iii) Ventilation in the storage facility should be maximised and temperatures kept as low as possible. Galvanised iron sheeting is not

*Fig 6.7* A hopper for bulk storage of dry pelleted foods (from Edwards, 1978).

recommended as a building material in hot climates.

(iv) All spilled material should be swept up.

(v) Rodents and insects must be kept under control. Traps and not poisons should be used. Double-skinned buildings in the tropics may become infested and should be regularly inspected.

(vi) Feeds should be checked frequently, and any change in colour and texture (lumpiness indicates fungal attack), friability, or smell (staleness, rancidity) taken as evidence of spoilage. Moisture content should also be assessed. The moisture content of dry feeds is around 10% and if this increases to 13-16%, then it becomes much more susceptible to spoilage. The presence of moulds, insects, and rodents must be carefully watched for, and and any infested materials should be isolated as soon as possible and fumigated.

The shelf lives of both moist and dry feeds are summarised in *Table 6.3*. Fish that are fed contaminated feeds may exhibit abnormal behaviour, loss of appetite, poor growth and loss of condition.

### 6.3.4 Feeding

At most intensive and semi-intensive cage farms, fish are fed throughout the year, except during periods of adverse weather. Extremes of cold and heat mean that fish will not feed, or should not be fed. In the USSR, cages of common carp are over-wintered below the ice surface (*Fig 7.27*) and are not fed until the spring, when the ice melts and the cages are lifted (Martyshev, 1983), whilst in Western Europe, feeding of Atlantic salmon may be temporarily halted if the temperature rises above 18°C (see Chapter 8). In Japan, cages of yellowtail are sometimes lowered below the sea surface during typhoons, and whilst some cages are adapted so that feeding may continue during periods of submersion (*Fig 7.23*), others are not.

*Table 6.3* Summary of storage and shelf-life of various feeds. Specialist publications (*eg* NRC, 1977, 1983) should also be consulted.
*NB* Many commercially-manufactured feeds have expiry dates printed on the sacks.

| *Feed type* | *Storage and duration* |
|---|---|
| Dry supplementary feedstuffs (*eg* rice, bran, wheat middlings) | Providing the moisture content is <10% and materials are stored in cool, dry and pest-free environment, they can remain in storage for several (tropical) — many (temperate) months. |
| Trash fish (frozen) | High fat content — up to 3 months at −20°C.<br>Low fat content — 1 year+ at −20°C. |
| Silage | 6–8 months, providing it has a low fat content and contains sufficient antioxidants. |
| Pelleted, commercial, intensive feeds | In temperate countries, usually contain extremely high levels of antioxidants and vitamins surplus to requirements. Providing that they are stored in clean, dry conditions, feeds will keep for 9 months or more. In the tropics, feeds tend not to have the same levels of antioxidants or vitamins present. Hence they should only be kept for 2–3 months. |

Most cage farms using supplementary feeds, trash fish or moist feeds, feed their fish by hand, since it is cheaper and easier to do so, at least on a small scale. Frozen trash fish is usually thawed first, either at ambient room temperature or by pumping seawater over the blocks, chopped and minced if necessary, carried to the cages in plastic bins and simply broadcast over the surface using a shovel or scoop. However, research in Japan suggests that there may be a number of advantages to feeding frozen trash fish to caged stock, in that bacterial infections and water pollution are reduced (Taniguchi, 1983). Some rainbow trout farmers in Scotland feed their fish on frozen blocks of minced trash fish and shrimp offal which are floated on the water surface. Each block takes several hours to thaw and be consumed.

Dry supplementary feeds and moist pelleted feeds compounded from trash fish/fish meal + binder, are usually presented to the fish in a similar manner to that described above for trash fish. Moist balls of feed which are sometimes used in semi-intensive cage culture of tilapias may be conveniently fed to the fish by placing them on the top net of the cage and gently lowering the net into the water (Coche, 1982).

The principal advantage often quoted for hand feeding is that farmers can assess how hungry their fish are and consequently can adjust the amount of food they use. In this way they can also readily monitor the health status of their stock, since sick or stressed fish will usually stop feeding. For cage farms rearing fish semi-intensively, hand feeding may indeed be the best method since operations are usually small or family based and the quantities and qualities of the materials used may be highly variable. However, hand feeding is labour intensive and may not always be the best way to deliver dry food at large, intensive farms, or if the farmer rears fish on a part-time basis. Hand feeding is also based upon the assumption that the best time to stop feeding is when the fish decide that they have had enough and this can result in overfeeding which will adversely affect FCRs, waste loadings and profitability (see later).

Mechanical feeders have largely superseded hand feeding at most large intensive cage farms in recent years, for reasons of economy of labour. There are two basic types — demand feeders and automatic feeders — and the designs used on cages are virtually identical to those used at land-based sites. Demand feeders typically consist of a feed hopper fitted with a plate on the bottom which is connected to a pendulum rod that protrudes down into the water (*Fig 6.8*). The feeders are usually mounted on a gantry directly over the middle of the cage, and when the pendulum is touched by the fish, the plate moves, releasing small quantities of food. Demand feeders are cheaper and make food available to the fish round the clock so that they can feed whenever they wish. The relative merits of using demand feeders have been reviewed by a number of authors (Avault, 1981; Boydston and Patterson, 1982) and proponents claim less size variability in stock at harvest, better FCRs, higher production, improved water quality and less disease problems.

Many of these claims, however, have not been rigorously tested, or apply to particular species, circumstances and methods of culture. Demand feeders seem to work fairly well with channel catfish, bass and rainbow trout, but

*Fig 6.8* Simple demand feeder in operation at Lake Kossou, Ivory Coast, in a $6m^3$ net cage (from Coche, 1979).

have not yet been widely tested with other species.

Like hand feeding, demand feeders also work on the principle that when a fish is no longer hungry it will stop feeding and that the fish knows best when to stop. Thus demand feeders should always give good results. Carnivorous species, such as catfishes, grouper and salmonids have well-developed stomachs which can store food, and experiments with rainbow trout under constant lighting conditions have shown that at 18°C there are well-defined peaks of feeding activity every 6h (Grove *et al*, 1978) but that at 15°C this rhythm changes to peaks every 8h (Adron *et al*, 1973). Tilapias and carps, on the other hand, which have poorly-developed stomachs or no stomachs at all, require smaller, more frequent meals (Ross and Jauncey, 1981). Appetite is largely controlled by the hypothalamus which responds to stretch receptors in the stomach wall or fore-end of the gut and possibly to blood sugar levels (see Fange and Grove, 1978, for discussion). Thus, appetite and feeding are suppressed as the stomach or fore-end of the gut fills with food, and returns as food passes into the intestine or hind-end of the gut. However, these mechanisms have evolved to deal with food encountered in the wild, and if fish are fed to satiation with artificial food which is much higher in terms of protein and lipid content, then incomplete digestion will occur, resulting in poor FCRs and increased waste production (Goddard and Scott, 1980).

Evidence from the USA with rainbow trout confirms that this can indeed happen at low (<5°C) and high (>16°C) temperatures when fish ingest food from demand feeders in excess of their abilities to process it (see Aquaculture Digest *8* (1), 1983, p. 16, for discussion). This problem, however, may be rectified by adjusting the amount delivered when the feeder is triggered or by filling the feeder with a calculated ration sufficient to supply the caged stock for one day (see later).

Demand feeders are also more commonly used in land-based than water-based culture systems. Results from cage culture suggest that at exposed sites, water currents or waves may be sufficient to trigger the pendulum frequently, releasing more food than the fish can cope with (Pitt *et al*, 1977). However Coche (1977) and Balarin and Haller (1982) report that in several parts of the world this apparent setback has been turned to an advantage in tilapia cage culture, by loading hoppers with small quantities of food only, which when triggered by waves, release the ration continually over a period of several hours.

Automatic feeders differ from demand feeders in that a measured ration is delivered at fixed intervals which are preset by the farmer using some sort of timing device. There are a number of different types, designed to handle either dry or moist pellets. Dry pellet automatic feeders are perhaps the most common type installed at cage fish farms, and typically consist of a plastic feed hopper surmounting a triggering device which releases or fires a quantity of pellets each time it is activated, and are usually operated either by rechargeable batteries or by compressed air. The latter type is illustrated in *Fig 6.13*.

Electrically powered systems either consist of a centralised power supply with control unit linked to a number of feeders, or have feeders which operate independently and which contain their own power source and control device. The former system typically uses a 12v car battery which can operate up to ten units, and which lasts 1-3 months depending on frequency of operation. The control unit determines the frequency and length of the impulses sent to the feeders, so that the number and duration of feeds can be programmed. A photocell is often installed to ensure that the feeders only operate during daylight. The system is commonly installed on a group of cages, cabling being run from the battery to the feeders along the walkways and cage superstructure. The feeders are suspended above the middle of the cages on aluminium poles, and work in a very simple manner: electrical impulses from the battery drive a vibrating plate fitted to the base of the hopper, thus releasing feed into the water. When the hoppers empty the feeders can be pulled to the sides of the cage and refilled.

The system described above suffers from one potential disadvantage: because all feeders are linked to the same power source and timing device, all fire simultaneously, and for the same duration. Feeders which operate independently do not, of course, share this disadvantage, but they can cost considerably more. They are, however, ideal for small numbers of cages or at farms where the cages are not grouped together.

The compressed air system is less popular at cage sites, but can be used where the cages are connected to the land by a walkway and where a 3-phase, high voltage power supply, required to drive the compressor, is available. The compressor part of the system is normally housed inside a building, and pressurises an air chamber from which bursts of air can be released and fed via high-pressure pipelines run along the walkways to the feeders on the cages. The release of air is controlled by a timing device fitted alongside the compressor. The feeders themselves are attached to the sides of the cages and typically consist of a hopper from which pellets fall by gravity into a horizontal delivery pipe which is connected to the compressed air line, sometimes via a subsidiary air chamber attached to the feeder. When air is released from the air chamber, any pellets in the delivery pipe are blown out into the cages. The amount of feed delivered to the fishes is determined not only by the frequency with which air is released from the chamber, but also by the size of the aperture through which the feed falls into the delivery pipe. Edwards (1978) estimates that a compressor with a capacity of 200 $l$ min$^{-1}$ would require a 380v supply, but would be sufficient to operate 20-25 feeders, and fire them four times per hour. Such systems, however, suffer from a number of disadvantages: they are noisy and in sub-zero temperatures water vapour trapped in the air supply sometimes condenses and freezes on feeder valves, thus stopping or reducing efficiency (Edwards, *ibid*).

Some compressed air feeders may be operated from a boat and may be invaluable in distributing feed at sites which use giant cages.

One other type of automated dry pelleted feed system which has been tried at cage fish farm sites involves the use of pressurised water to carry food from a centralised unit to individual cages along pipelines. However, there have been a number of problems with this system, which will be discussed below.

The simplest type of automated moist pellet feeding system used at cage farm sites consists of little more than transporting pelleting equipment on a boat, and pelleting the food directly into the cages. The only prerequisites for adopting such a system are that there is a suitable boat and power supply, and that a modified pelletising unit and delivery tube are attached to the outlet from the feed mixer. Such systems are widely marketed and not uncommon at Norwegian salmon farms, although they can be costly in terms of man-hours and fuel.

A more sophisticated system is employed at some Atlantic salmon farms. Moist feeds are transferred, using pumped water, from the mixing unit to a central distributor unit, and then via pipelines to the cages which can be up to 600m away. A number of refinements may be added to this basic system. The feed distribution operation may be integrated with storage silos and linked to a control room where a microcomputer can record temperature data from each of the cages and adjust the feed rate accordingly (*Fig 6.9*). In this way different feeds — including dry pellets — may be sent to different cages, and detailed records kept of feeds used, FCRs, and daily temperature regimes (*Fig 6.10*). Systems can be installed which handle up to thirty-two cages and which can deliver up to 1500kg of feed per hour.

The great advantage of such systems is that they are responsive to changes in the fishes' environment, and this is a considerable improvement on relying

*Fig 6.9* Computer-controlled feeding operation, showing silos for storing feed ingredients, mixers, and feed distribution system (courtesy Bergen Bartz A/S).

solely on feed tables or on the fishes' response to feeding. Greatly improved FCR values and reduced solid loadings have been widely reported. Similar environmental monitoring systems have also been developed for automatic dry pellet feeders and some of the more sophisticated designs are able to monitor waves and currents and to cease feeding during adverse conditions. However, computerised feed systems can be expensive to install. Moreover, a number of farms using automated pumped water feed delivery systems have reported poor growth or mortalities among stock, which have been attributed to loss of vitamin C from the pellets. Research has shown that up to 67% of vitamin C may be lost from fish pellets after only 10 seconds in water (Slinger *et al*, 1979). Thus farms where the cages are sited a considerable distance from the distributer unit may be at risk unless extra vitamin C is added to the feed, or until better pelleting technology is introduced.

One criticism concerning the feeding of caged fish is that to date there has been little study of feeding behaviour (Phillips, 1985) and there have been few mechanical feeders designed specifically with caged fish in mind. Goddard and Scott (1980) speculated that due to the relatively small surface area to volume ratio of cages, compared with ponds or raceways, caged surface-feeding fishes should be fed over longer periods of time at each feed. Also, little attention has been paid to the distribution of feed — either in land-based or water-based systems. Most types of mechanised feeder deposit the feed over a very small area. However, a new type of feeder, designed in Sweden, has an efficient spreader unit fitted to the base of an otherwise conventional hopper. The unit can be adjusted to spread the pelleted feed over an area up to 13m in diameter. Although this design has not been rigorously tested, those who have used it with cages report excellent results (see Fish Farming International, 9 (5), p.12). On the other hand, there is a great deal of evidence that feed losses from cages are higher than from other systems (see Chapter 2), and this has been attributed to feed being rapidly carried out of the cages by water currents. Thus, it may be better to confine feed dispersal within the central region of the cage.

A number of studies have been carried out into the use of feeding rings, trays or net curtains to reduce feed losses. Plastic feeding rings, 25-150cms in diameter, are normally floated on the surface in the centre of the cage, the feed thus being prevented from floating out through the cage sides. Alternatively, the cage walls near the surface may be fabricated from fine mesh materials. Both are used in conjunction with floating or slow sinking pellets, and whilst there is evidence that they may be effective under certain circumstances (*eg* in fast flowing waters) they have not been thoroughly evaluated. The solid-bottomed cages used in marine flatfish culture not only

FORKON-D
OPERATOR PRINTOUT

FORKON AUTOMATIC FEEDER DATE: 01/05/85

| CAGE NUMBER | CAGE TABLE | | | |
|---|---|---|---|---|
| | 1 | 2 | 3 | 4 |
| STOCKING DATE | 01/07/84 | 01/07/84 | 01/07/84 | 01/07/84 |
| NO OF FISH | 39650 | 23600 | 33400 | 23750 |
| MEAN WEIGHT (Kg) | 0.3 | 0.5 | 0.3 | 0.6 |
| TOTAL WEIGHT (Kg) | 11895 | 11800 | 10020 | 14250 |
| TOTAL FEED (Tonnes) | 0.04842 | 0.11705 | 0.06161 | 0.09393 |
| FEED TYPE | 2 | 1 | 2 | 1 |
| % BODY WEIGHT/DAY | 2.5 | 3.0 | 2.0 | 2.5 |
| ACTIVITY FACTOR | 1 | 1 | 1 | 1 |
| FEED/DAY (Kg) | 12 | 98 | 37 | 63 |
| NO OF FEEDS/DAY | 16 | 16 | 16 | 16 |
| FEED WT/FEEDING (Kg) | 0.8 | 6.1 | 2.3 | 3.9 |
| FEED INTENSITY | 1 | 1 | 1 | 1 |

FEEDING TABLE:

| CAGE NUMBER | 1 | 2 | 3 | 4 |
|---|---|---|---|---|
| LAST RECORDED DATE | 01/05/85 | 01/05/85 | 01/05/85 | 01/05/85 |
| LATEST FEED TODAY | 10.18 | 10.39 | 10.41 | 10.42 |
| TOTAL FEED TODAY (Kg) | 2.97 | 20.38 | 7.74 | 11.77 |
| LAST FEED TODAY (Kg) | 1.5 | 7.1 | 2.5 | 4.0 |
| FEED COMPLETED | 2 | 3 | 3 | 3 |
| CLEANING TIME (min) | 1 | 1 | 1 | 1 |

*Fig 6.10* Typical print-out from computer-controlled operation (courtesy Bergen Bartz A/S).

facilitate the feeding of these fishes, but also prevent loss of feed, and feeding trays, which are normally suspended 20-100cms below the surface, operate in a similar manner. However, like feeding rings, they also seem to have limitations. Coche (1979), for example, has expressed concern that small trays which only partially cover the cage area will benefit only the more aggressive fishes, whilst Ibrahim *et al* (1976) comment that trays provide an unwanted spawning substrate for tilapias.

Loud noises during feeding should also be avoided, as they frighten fish and exacerbate feed losses.

The relative advantages and disadvantages of various types of feeder are summarised in *Table 6.4*. Hand feeding and demand feeders work on the principle that the fish know best when they have had enough, and whilst there is a great deal of truth in this, fish can also overfeed, and it is therefore best to also take cognisance of feed manufacturer's guidelines. Most automatic feeders, on the other hand, administer daily rations derived from feeding tables calculated on the basis of temperature and fish size. However, it is not always practicable to keep track of environmental conditions and to adjust feeding rates accordingly and thus fish may be frequently over or underfed. Attempts to surmount this problem with computerised systems are a great improvement, but these are comparatively expensive, and cannot take account of fish health, an important determinant of appetite. Thus some feeding by hand is always advised.

Specific details of feeding rates for different species will be discussed in Chapter 8.

## 6.4 Routine management

### 6.4.1 MONITORING WATER QUALITY

Water quality parameters relevant to cultured fish in general, and to certain of the commercially more important cage culture species are discussed in Chapters 4 and 8. The aims of instigating a water quality monitoring routine are:-

(i) to avoid losses caused by lethal changes in water quality;
(ii) to evaluate siting and configuration of cages within a lake/bay;
(iii) to maintain optimum stocking and feeding rates;
(iv) to help evaluate how stressed caged stocks are, so that jobs which might further stress the fish, such as grading, can be avoided;
(v) to gain information of long term changes in water quality at a site, so that any proposed changes in production may be properly evaluated (see Chapter 5).

The most valuable data that should be collected is on dissolved oxygen and temperature. Ideally, measurements should be made daily, preferably at those times when they are likely to be highest and lowest (*ie* at dawn and mid-day and at slack tide), and readings both inside and outside cages and at cage surface and cage bottom should be made. However, such a routine

*Table 6.4* A summary of feeding methods at cage fish farms.

| *Type of feeding* | *Type of feed* | *Usage* | *Advantages* | *Disadvantages* |
|---|---|---|---|---|
| Hand-feeding | All types of supplementary and intensive feeds | Used in semi-intensive culture and preferred method at many intensively managed farms, including most yellowtail farms in Japan | No capital expenditure on feeders<br>Enables monitoring of fish health<br>Fish can be fed according to appetite<br>Can distribute feed widely | Can be expensive in man-hours at large farms<br>Ties farmer to farm<br>Under some circumstances may lead to over feeding<br>Poor FCR values, and high waste loadings |
| Demand feeder | Pelleted dry feeds | Used for semi-intensive and intensive culture of a range of species | Cheaper than automatic feeders and less labour-intensive than hand-feeding<br>Feeding rates can be adjusted to suit conditions | Care must be taken to avoid over-feeding<br>Small fish (<10g) can't use most designs<br>Waves can trigger some feeders at exposed sites<br>May not be suitable for some species<br>Poor feed distribution pattern |
| Automatic feeders<br>(a) Air-operated | Pelleted dry feeds | Used for intensive culture of salmonids<br>Not commonly used on cages | Farmer can set feeding rate and duration<br>Relatively good feed distribution<br>Less labour-intensive than hand-feeding | Noisy, can be problems in sub-zero conditions where icing of valves can occur<br>Not suited to all cage sites<br>No feedback from fish to farmer, hence risk of over-feeding |
| (b) Electrically operated | Pelleted dry feed | Widely used for intensive fish culture | Farmer can set feeding rate and duration<br>Less labour intensive than hand-feeding | No feedback from fish to farmer, hence risk of over-feeding |
| (c) Water powered | Pelleted dry or wet feeds | Used mainly in Scandinavia with wet feed for salmon culture | Computerised system responsive to environmental changes<br>Improved FCR values and reduced solids loadings<br>Flexibility in range of feed types used<br>Computerised comprehensive records | May be problems with vitamin losses<br>Very expensive to install |

would be considered far too time consuming by most fish farmers, and so it is suggested instead that farmers take periodic measurements throughout the year, but that they increase the frequency of observations during warm spells. A single temperature and oxygen determination from the middle of a cage within a group will suffice to give an indication of levels affecting the fish stocks.

Data on nitrogen (ammonia, nitrate, nitrite) and dissolved phosphorus levels, pH, Secchi disc and chlorophyll levels are also of importance in that they give the fish farmer a more complete picture of what is happening in the cage farm environment, and in particular help alert him or her to dangerous levels of toxins (ammonia, nitrite), and the effect of the farming operation on algal populations (chlorophyll levels, Secchi disc). Again, these tests should be carried out at regular intervals, but are of particular importance during periods of calm, hot weather. Whilst pH and Secchi disc readings can be readily taken using simple equipment, the other tests are usually carried out by chemical methods. Although this may sound daunting to the uninitiated, such tests are fairly straightforward, requiring minimum skills and equipment. Details of sampling methods, storage and sample treatment are given in Mackereth *et al* (1978), Boyd (1979), APHA (1980), Parsons *et al* (1984) and Stirling (1985). Several portable test kits, suitable for use by fish farmers, are now available, and many of these have been reviewed by Boyd (1980).

Monitoring of water quality is obviously of greatest importance to those cage fish farms which are intensively managed, and which have a high production with respect to the size and nature of the site.

### 6.4.2 Fish husbandry and management

Samples of fish should be taken at regular intervals and weighed, so that the growth of stocks can be monitored. This information, together with records of mortalities (see later), is necessary for making a number of managerial decisions such as the determination of stocking and feeding policies and timing of harvesting.

Nets should be lifted prior to sampling, as not only are the fish easier to catch when concentrated in a small volume of water, but also the sample taken is more likely to be representative of the caged stock, since the fish congregating at the surface may well be larger and more dominant. Fish are captured using a dip net, counted, transferred to a bucket and weighed. The average weight is then easily computed.

As the fish grow, the biomass and thus the stocking density in the cage increases. Unless the fish have been stocked at a level which takes account of the average weight at harvesting, the stock in a cage inevitably must be divided from time to time so as to maintain optimum growth conditions and minimise disease risks. The frequency with which this operation has to be performed depends on how the farmer decides to utilise the cages he has available, and on the relative costs and benefits of disturbing his stocks, since moving fish is a stressful operation which can result in a disruption of feeding, a cessation of growth or even heavy losses, particularly during adverse weather conditions. Fish are usually graded at this stage since, in intensive operations at least, the farmer is attempting to produce fish of a standard size. Also, since feeding rates are usually computed on a percent body weight per day basis, it is best if all the fish in a cage are approximately the same size at any particular time. Some farmers consider that grading should be carried out fairly frequently, particularly when fish are small, since with some species the larger fish in the cage are able to secure more food thus accelerating growth and increasing size variation among the stock. Grading is also carried out at specific times during the growth cycle of salmon, to separate $S_1$ from $S_2$ smolts, and grilse from salmon. Specific details of when particular species are usually graded and moved are given in Chapter 8.

There are several methods for grading fish. The most basic is by eye, which is sometimes practised at less intensive operations and at farms growing large fish. However, at most intensive cage fish farms, and particularly at the larger ones, much of the routine grading is carried out by machine. Hand graders may be used to sort small fish (*Fig 6.11*), especially when fish are stocked initially. However, automatic graders are used at most ongrowing sites and are transported from one group of cages to another by boat, the grading operation being carried out either on board boat whilst moored alongside the cages or, if there is sufficient room, on the cage raft itself (*Fig 6.12*). A typical machine consists of a hopper, into which the fish are loaded, surmounting the top end of a sloping ramp composed of a set of bars which are adjusted so that the gaps between adjacent bars increase from top to bottom. Fish fall from the hopper, slide down the ramp by gravity, and fall through the ramp

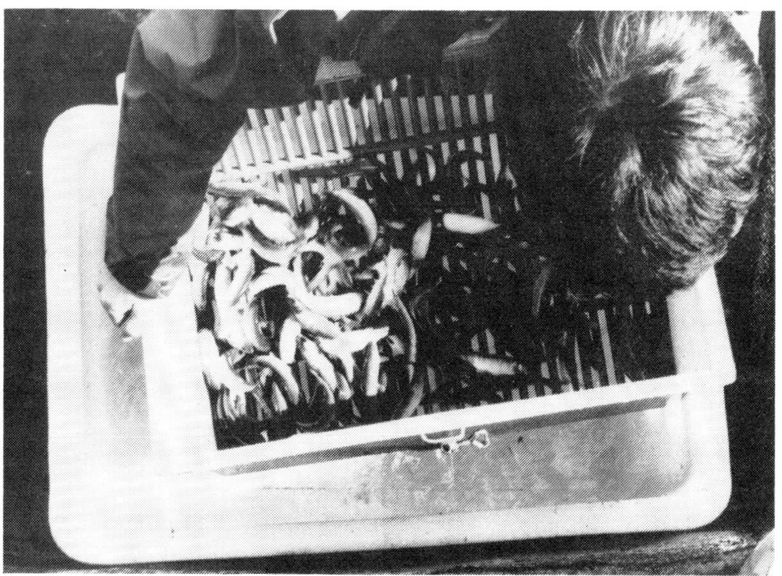

*Fig 6.11* Grading small fish with a hand grader (from Edwards, 1978).

*Fig 6.12* Fish grading machine sitting on cage raft.

whenever the gap between adjacent bars is wide enough. Most machines are designed to handle fish of between 50g and 500g, and can be adjusted to sort fish into four or five different sizes. Fish are then caught in a series of tanks placed below the machine, and subsequently transferred to the appropriate cages. The whole operation may be greatly speeded up by using a live fish pump to transfer the fish from the cage to the grader, and by transferring the graded fish via pipelines to their new homes.

Regular monitoring of fish stocks from a disease point of view is also essential. However, as Roberts and Shepherd (1986) state, 'Before a fish farmer can hope to diagnose and control disease outbreaks, he must be able to appreciate when his stock is healthy and thriving.' This is more an art than a science, and depends a great deal on experience. Often the first signs that something is wrong can be surmised from changes in behaviour. Farmers should therefore be used to observing their fish without unduly disturbing them, and form a general picture of how they are distributed and behave under the normal cycle of environmental conditions which occur at the site, *ie* dawn/mid-day/dusk; slack/flood tide; feeding/non-feeding. A close inspection of the fish during feeding is also recommended, as the fish are normally easier to see, and changes in feeding behaviour are also often a good indication of poor health. If something wrong is suspected, then some fish should be removed from the cage and examined further, changes in general appearance (deformed spine), skin (colour, presence of lesions, rashes, spots or lumps, excessive mucus), eyes (bulging eyes, cloudy lens), fin and tail (erosion) are all signs that something is wrong. Details of disease diagnosis and appropriate treatment methods are given in Chapter 7.

Disease outbreaks apart, some unexplained deaths always occur at a fish farm. Dead fish should be removed from cages as they may be a source of further infection and attract predators. This involves not only scooping out any fish floating on the surface, but also periodic (once per week) lifting of nets and removal of dead fish which are lying on the cage bottom, bearing in mind that lifting nets is both time-consuming and stressful — even injurious — to the fish. Recording of mortalities is essential, as a change in the incidence of mortalities can help warn of the onset of a disease outbreak, and gives the farmer valuable information on the progress of different stocks or management strategies (stocking densities, feeding rates, *etc*). Such records are also essential for any insurance claim.

Dead fish should be removed from the site and preferably disposed of ashore. Some farmers recommend burial in a pit, and covering with lime — particularly if a disease outbreak is suspected. However, the removal of dead fish is not only a precaution against the spread of disease (Arizono, 1979; Needham, 1980), it can also help reduce phosphorus and nitrogen waste loadings (Beveridge, 1984b). All equipment used to transfer dead or diseased fish should be disinfected.

Harvesting of fish is either done continually or in batches, depending on how the production cycle is geared (see Chapter 8). Prior to the harvesting of a batch, the fish should be starved for a day or two, partly in order to firm the flesh and partly to give the gut time to evacuate, since a gut full of food and bacteria can accelerate deterioration and increase the risk of carcass contamination during the gutting process. Fish can be harvested *in situ*, or the cages towed to a quay where the netting operation may be carried out more smoothly, with the aid of mechanised lifting gear, trucks, *etc*. In some parts of the world the cages may actually be towed to market (*eg* in Indochina, Pantulu, 1979). The process of harvesting is simple: in most cage designs the net is pulled up until the fish are concentrated in a small volume of water and netted out using dip nets (*Fig 6.13*). Power assisted hoists, which can be transported to the cages and operated from a floating platform or boat, are sometimes used to operate large dip nets (*Fig 6.14*). Harvesting may become a little laborious if the cages do not have walkways and the fish must consequently be netted out by hand from a boat. Harvesting of the rotating cage shown in *Fig 2.2(b)* is carried out by stretching a net across a section of the cage and then rotating it (Grave, 1975). For most rigid designs, fish must be crowded at one end of the cage and then scooped out (Hugenin *et al*, 1981). However, in the design shown in *Fig 6.15*, the cage is simply rotated and lifted out of the canal, the fish being concentrated in one corner.

Fish should be handled gently at this stage, as undue stress can cause a build-up of ATP in the muscle and a decrease in shelf-life. Fish may also be easily damaged, causing not only ugly skin blemishes, but also bruising of the flesh which will hasten deterioration in quality (Connell, 1980; Neusinger, 1983).

In a number of Southeast Asian countries, species of grouper, snapper, sea bream and even tilapia, are cultured for the live fish market, where they fetch considerably higher prices. At harvest, fish are transferred directly from the cages to boats equipped with holding tanks. In Hong Kong, the fish are either transported directly to a designated landing point by the grower and sold to the wholesaler, or, alternatively, the fish may be collected from the farm by restaurant owners and transported live to their restaurants (Wong, 1982).

At most farms, however, fish are killed prior to dispatch. Dumping fish

*Fig 6.13* Harvesting fish (courtesy J A Stewart).

*Fig 6.14* Floating platform with power hoist, fish harvesting bins and grading machine, at an Atlantic salmon farm, Ireland.

*Fig 6.15* Rotating rigid mesh cages sitting in heated water effluent canal at a power station (courtesy I H MacRae).

into containers and leaving them to asphyxiate may appear distasteful to some; it also increases the lactic acid content of the muscles, causing an increase in the degenerative processes. Although there are a number of electrical fish killers available, some of which can handle up to 500kg at a time, these are not readily portable, and consequently most growers would argue that the only sensible way to deal with small fish species, such as trout or tilapia, produced at a cage farm, is to leave them to asphyxiate.

Larger fish, such as yellowtail and salmon, are more valuable, and are usually killed individually, either by knocking them on the head with a stick, or by bleeding. Bleeding of large fish is still a controversial practice. In Scotland, for example, only a few salmon farmers bleed their fish, whilst in Scandinavia it is universally done. According to advocates, bleeding improves the shelf life, appearance and even flavour of the fish (Edwards, 1978), whilst those that oppose it claim that 2% body weight is lost, and that it is a slow and somewhat inhumane method. The procedure involves piercing the fish behind the gills with a knife in order to sever the blood vessels, and then placing them in tanks of iced or cold running water, leaving them to bleed for a few minutes. Some fish are anaesthetised prior to bleeding by clubbing on the head, or by keeping them in tanks of $CO_2$-enriched water for a short period.

After harvesting the fish are usually dispatched as rapidly as possible, to ensure the freshness of the product to the consumer. Sometimes the fish are simply packed whole between layers of ice in insulated boxes, and dispatched to the buyer or to market. Sometimes the fish are gutted and cleaned before leaving the farm, whilst at other farms fish may be smoked or even dried or frozen prior to dispatch. Although fresh fish always fetches a higher price than frozen, freezing is sometimes a necessary evil — particularly if farmers have been forced to harvest their fish at a bad time, or if the market has taken an unexpected downturn. Where and how cage farmed fish are sold will be discussed further in Chapter 8.

### 6.4.3 MAINTENANCE OF CAGES AND GEAR

Irrespective of the damage that can be caused by storms, predators, drifting objects, shipping, poachers and vandals, all materials used in the construction of cages have a finite lifespan and will eventually wear out (see Chapter 3). Cages, nets and moorings therefore must be checked at intervals for signs of damage and wear and tear and repaired or replaced if necessary, as not only can cages and stock be put at risk, through neglect, but human lives may also be endangered.

Cage nets may be checked for damage during cleaning. However, since cleaning may be infrequent (see later), inspection of the nets *in situ* should also be carried out. In clear water this is often possible from a boat or from the cage walkway, whilst in turbid conditions divers may have to go down and check the nets every week or so, providing the weather is suitable. Rotating cages, of course, may be readily checked without recourse to using divers, whilst predator nets may simply be lifted for inspection.

Although small tears may be able to be repaired by lifting the damaged side of the net and hooking it on to the cage superstructure, for more extensive damage the net will have to be changed (see later) and repaired on shore. There are a number of manuals, giving comprehensive details of net mending methods, which are recommended (*eg* Libert and Maucorps, 1973), although, as has been stated above, it is a tremendous advantage to have someone on the farm who knows about the care and repair of nets.

At most sites, particularly in the marine environment, nets become fouled and unless a rotating design is used (see later), have to be changed and cleaned. In any case, nets of one particular mesh size are often exchanged for ones of a larger size as the fish grow. Care must be taken over mesh size: if too small then exchange is restricted and if too large, some fish species will chew the bars of the net. Some guidelines to mesh sizes are given in Chapter 8. The frequency of net changing varies from once per week to once per year, depending upon site location, materials used, season and management and design of cage. In northern Norway one change per year is sufficient, providing farmers wait until after the period of maximum larval tunicate and mussel settlement in July (Sutterlin and Merrill, 1978). The frequency of net changing may also be reduced if the top 2 metres — the part of the net where fouling is heaviest — of any one side is pulled out of the water and left to dry.

Normally two people are involved in the net changing process, and although there are variations depending on design, the basic principle is the same for most cages. The rigging is released at two adjacent corners, thus freeing one side of the bag from the cage collar. The freed side is then drawn towards the opposite side and gathered up so that the fish are concentrated in a small section of the bag near the surface, the rigging on the other two sides being released from the collar as necessary. One side of the new bag is then attached to the two free corners and drawn under the old bag. The fish are then gently tipped out of the old bag into the new one, prior to the fouled net being removed for cleaning. There are a number of variations on this method (*eg* Yu *et al*, 1979).

The whole net changing process normally takes from 30 min to 2h, depending on the degree of fouling (*ie* how heavy the net is), size and design of cage, and weather. However, as a large farm can have as many as 50 or 60 cages, routine net changing alone can fully occupy two members of staff for much of the year. Several of the more revolutionary designs of industrial cages available have tried to circumvent this by the partial mechanisation of net changing.

There are two basic methods used to clean fouled nets: chemical and mechanical. For rotating designs the cage is turned so that fouled panels are lifted out of the water and exposed to the sea air. The organisms are then usually left to dehydrate and die, a process which typically takes a week or so in temperate regions (Blair *et al*, 1982), although some farmers help remove encrustations with a stiff brush. Comparisons of maintenance time required to keep biofouling to a minimum on rotating and conventionally-designed cages have been made by Blair *et al* (*ibid*) using both types of cage over the same time period at the same site. They estimated that the work involved with the rotating design was reduced to around 5% of that required to maintain the conventional cage.

Some fish farmers stack fouled nets in a heap, cover them with black plastic sheeting and leave the organisms to decay prior to cleaning, whilst others recommend that marine nets are submerged to allow starfish and other predators to remove mussel encrustations (Sutterlin and Merrill, 1978). At most cage fish farms however the usual practice is to hang or lay the nets out to dry for a few days (*Fig 6.16*), thus making them easier to clean. Hard bristle brushes, sticks, or high pressure hoses are used to dislodge adhering material, the latter method being more effective at removing some of the more stubborn fouling organisms such as ascidians. If high pressure hoses are used, then cleaning sould be carried out in an area with a sloping concrete floor (*Fig 6.17*), so that the debris and water can drain away.

In the past, a number of chemical cleaning methods, such as soaking nets for 2-3 days in a 3% formic acid plus 9% copper sulphate solution, or for 3 hours or so in a sodium hypochlorite solution have been recommended (Sutterlin and Merrill, 1978; Moller, 1979). This apparently dissolved mussel byssal threads making subsequent cleaning with a high pressure hose much easier. However, chemical methods are not widely used today largely for economic reasons, and because the farmers fear that the chemicals used may have ill effects on the fish.

A word of caution should be said here about the care of modern synthetic nets. They are costly and easily damaged, and are particularly vulnerable during onshore cleaning operations. Secretan (1979) has estimated that 4-5% of the losses reported by cage fish farmers concern escapes of fish through

*Fig 6.16* Cage bag laid out to dry.

*Fig 6.17* Cleaning cage bags with a power hose.

holes in cage nets. A significant proportion of this he believes to be due to fish farmers' treatment of nets: 'I have lost count of the number of times that I have seen nets, laid out on the ground to dry being trampled on, even driven over by vehicles and trolleys. Such treatment explains away many of the small holes that develop in nets. . . .' Nets not in use should be carefully stored in clean, dry conditions.

As many commercially manufactured cage nets are treated with antifouling

*Fig 6.18(a)* Cu-Ni cage after four years' immersion in temperate, marine conditions (courtesy J E Hugenin).

*Fig 6.18(b)* Pressure cleaning of Cu-Ni cage (courtesy J E Hugenin).

compounds prior to dispatch, fish farmers should check whether or not the nets they have purchased have been treated, and if so, they should ascertain what compounds have been used, and if the nets require to be soaked before stocking with fish. The use of antifouling compounds is discussed in detail in Section 7.6. Most compounds currently in use require that the nets be soaked for at least 24h in clean flowing water, prior to use. If the farmers decide to re-treat older nets with antifouling, they must follow the manufacturer's recommendations as to the necessary precautions they should take before using them again.

The rate of biofouling on polypropylene or metal alloy rigid mesh cages is much slower than on net cages, and hence cleaning needs to be carried out far less frequently (Ansuini and Hugenin, 1978; Milne, 1979; Hugenin *et al*, 1981). Cages are usually cleaned *in situ* using a hard-bristle brush both above and below the waterline to dislodge weed and accumulated debris. Ansuini and Hugenin (1978) estimated from trials conducted over a one-year period in Maine, USA, that cleaning need only be carried out four times per year, and take a total of eight man hours, a saving of more than 90% on the cleaning of conventional cages in that area. Periodically, however, rigid mesh cages will have to be lifted, checked for damage and repaired if necessary. Whilst undergoing maintenance they should be thoroughly cleaned, preferably using pressure hoses (*Fig 6.18*).

Moorings must be checked regularly by divers, particularly after storms. Mooring lines should be kept relatively free from fouling and worn shackles replaced.

# 7 Problems

## 7.1 Introduction

Site selection and site evaluation are largely about choosing the right environment to farm the appropriate quantities of fish so as to minimise mortalities, maximise growth and production, and thus ensure that the venture be as profitable as possible (see Chapters 4 and 5). Unfortunately it is not often possible to choose the ideal site. First of all farmers may be faced with trying to rear a species in a sub-optimal climate (*eg* trout in the tropics; tilapia in Europe). Often, too, many of the best sites in a particular area are taken, forcing farmers to choose what they know are sites with some risks attached to them. Neither is it also possible to foresee all of the problems which might occur, or to examine all risks as thoroughly as one might like. Damage caused by drifting objects or pollution owes much to chance, whilst unless one has a great deal of time and money to spare, it is impractical to investigate properly factors such as fouling.

All sites are to some extent a compromise solution; the price for good water exchange may be swift currents, which can distort nets and make mooring expensive and difficult to install, whilst the trade-off for unpolluted water is often the abundance of wildlife, many of which may have piscivorous tendencies! In the following section, some of the problems associated with growing fish in particular climates or types of site are discussed, and wherever possible advice given to mitigate adverse effects.

This is also an appropriate place to suggest that farmers should investigate the costs of insurance, which can cover all risks involved in cage fish farming from transport of fry to the farm, to shipping the finished product out. Insurance can be conveniently categorised into:-

(i) what is needed;
(ii) what is desirable;
(iii) what is affordable.

In many countries, a certain amount of insurance cover is mandatory; insurance of employees often falls into this category. Category (ii) may include third party insurance, which covers damages caused by a fish farmer's operation to someone else (*eg* drifting cages colliding with yachts), products liability, loss of buildings, cages and stock, and key-man insurance (*ie* medical insurance for manager/owner), whilst category (iii) can cover employees' benefits. In general, insurance cover costs somewhere between 1% and 10% of the value of the insured stock, depending on the extent of cover, and the risks involved. Charges are based on the detailed proposal form completed by the client, and following the occurrence of a loss the insurance company will send pathologists or independent assessors to evaluate the loss and its cause.

There are several companies which specialise in fish farming insurance, and advertise regularly in the aquaculture press.

## 7.2 Acidified water bodies

The pH of a water body is determined by the interaction of the rate of deposition of acids from the atmosphere and the rate of neutralisation in the catchment. Acid deposition, principally in the form of rain or snow, is a natural phenomenon, the pH of pure water in equilibrium with atmospheric $CO_2$ being 5.6 (Overrein *et al*, 1981). Catchments which are composed of hard, slow-weathering rocks such as granite, gneiss or quartz sandstone are poor at neutralising acid precipitation and also give receiving waters their characteristic low ionic and high humic acid content. Lakes and rivers of this type occur naturally in many parts of the world, such as the highlands of Scotland, the Canadian Shield, the Amazon and Zaire river basins, southern Malaysia and Western Australia (Serruya and Pollingher, 1983).

Because many fish species are intolerant of acid conditions, fish diversity in low pH waters is usually low, particularly in temperate climates. Growth rates are also often poor, due to the low productivity and hence the reduced availability of food. It is thus important that cage fish farmers consider these implications for growth and survival of stock when selecting sites and method of culture. However, in recent years it has become apparent that precipitation in certain parts of Europe and North America has become more acidic and that areas which previously had little or no experience of acid rain episodes were becoming increasingly troubled by the problem (Wright and Henriksen, 1979; Likens *et al*, 1979; Overrein *et al*, 1980). Although the major sources are in industrial centres, under certain conditions rain-bearing winds can deposit acidic precipitation in areas far from their origin. Several remote cage fish farms in Scandinavia and Scotland have been affected by acidification. In Lake Liervatn in southwestern Norway (pH 4.9-5.4) which is used for commercial rearing of Atlantic salmon smolts, high mortalities occur (30-40%), especially during smoltification and after grading or disinfection (Rosseland and Skogheim, 1984).

In lakes subjected to the early stages of acidification, the acid is neutralised by any bicarbonate ions present, and only once these are exhausted do changes in pH occur. A number of simple methods to detect acidification, even in these early stages, have been proposed (*eg* Henricksen, 1979). If acidification is taking place, the farmer has two choices; to attempt to control the acidification process or to change to a more acid-tolerant species. The former involves some form of chemical dosing which will either directly change the pH or which will stimulate productivity, the theory being that this will improve the lake's buffering capacity (*ie* indirectly increase alkalinity, and thus the lake's capacity to neutralise further acidic inputs). Over the past 50 years or so a wide range of neutralising substances has been used, including limestone ($CaCO_3$), lime [$CaO$ and $Ca(OH)_2$], soda ash ($Na_2CO_3$), olivine [$(Mg\ Fe)_2SiO_4$], fly ash and industrial slag, and sodium hydroxide ($NaOH$) (Baker, 1984). However, as far as the fish farmer is concerned, limestone is probably the best: it is safest, cheapest and overall the most effective on a wide range of lake types. Once again, a number of different application methods have been tried, although from the cage fish farmer's point of view, direct dosing of the lake as opposed to the inflows or the watershed is likely to be most cost effective. Finely crushed limestone can be added to the lake as a slurry, spread with the aid of a device such as is shown in *Fig 7.1*. Water is pumped from the lake into a tank holding the crushed limestone (mean diameter ~35 μm), and the resultant mixture circulated through the same pump. Part of the slurry is pumped directly into the lake via a short length of hose whilst the remainder is diverted back into the tank, with the object of keeping the mixture agitated (White *et al*, 1984). An alternative is to spread crushed limestone (0.25-0.75mm diameter) into shallow areas where wave action and currents help keep it 'activated' (Hasselrot and Hultberg, 1984; Krester and Colquhoun, 1984).

From the Scandinavian and North American work it seems that dose rates of between 10 and 30g $m^{-3}$ lakewater are used, the exact amount depending upon the severity of the problem, lake retention time, lake type and method used (White *et al*, 1984; Krester and Colquhoun, 1984; Rosseland and Skogheim, 1984; Hasselrot and Hultberg, 1984). In most of the remedial

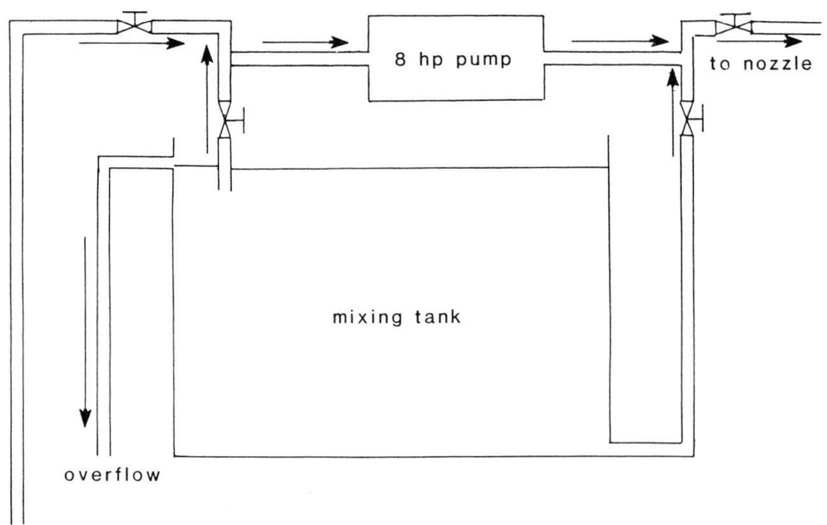


*Fig 7.1* Diagram of crushed lime spreading system (redrawn from White *et al*, 1984).

work that has been carried out, pH is being elevated from around 4-5 to pH 6 and above, both in order to reduce the direct effects of low pH on fish (impaired body salt regulation leading ultimately to death) and to minimise the toxic effects of aluminium. Aluminium is leached in high amounts from soils under acidified conditions. Although the exact mode by which aluminium exerts its toxic effect is unknown, toxicity is highly dependent upon pH, and is most lethal between pH 4.4 and 5.4, with a peak around 5.0-5.2 (see Odonnell *et al* 1984, for review). Experience has shown that lakes with fast flushing rates ($< 1y^{-1}$) are difficult to treat successfully, a single dose being effective for a few months only, particularly if slurry is used. On the other hand, lakes with slow flushing rates have maintained good pH values for 5 years or more following a single dose with crushed limestone (Hasselrot and Hultberg, 1984; White *et al*, 1984).

The alternative to dosing is to try and farm more resilient fish, an option that few cage fish farmers may be willing to consider. Within the salmonids, it is generally reckoned that char ( *Salvelinus alpinus* and *S. fontinalis*) or brown trout are the most hardy, followed by sea trout (*S. trutta*), Atlantic salmon and rainbow trout in decreasing order of toughness (Overein *et al* , 1980).

At the present time dosing seems to be the best option available to cage fish farmers, although it is expensive. At $45 per tonne for 0.5mm crushed limestone, the cost of treating a 50ha lake works out at around $3-4000, excluding delivery and labour. However the outcome is by no means certain. Farmers considering carrying out such an operation are strongly advised to seek advice from government fisheries staff or university scientists, and must also instigate some form of monitoring to observe pH, calcium and aluminium levels following dosing. (Monitoring of all parameters can now be carried out using commercially-available kits). In marginally-affected lakes with short residence times, dosing could be carried out to afford protection over the high-risk periods. Even so, dosing is not 100% foolproof, since acid snowmelt has been known to flow across lake surfaces without mixing. Hopefully current research may lead to improved stocks of acid-resistant fish in the future. As a consolation to the fish farmer, there is also some speculation that cage fish farm wastes may actually be beneficial to sensitive lakes by improving buffering capacity. However, this has yet to be fully evaluated.

## 7.3 Currents

Recommended current conditions for cage fish farming are usually less than 50cm $s^{-1}$; much greater than this, and there can be pronounced deformation of nets, excessive strain on conventional moorings and cage collars, and unacceptable losses of feed (see Section 4.3.3). High flow rates can also affect the metabolic rates of fish, since they are forced to expend energy in swimming against the current. Although ideal flow regimes for different species in culture conditions are poorly documented, according to Russian studies summarised by Privol'nev (1975), the flow rate inside cages should not exceed 20cm $s^{-1}$ for trout, since above this rate energy expenditure on swimming increases markedly and growth and survival are adversely affected. Fast currents, however, are occasionally encountered, particularly in running water sites, and a number of methods have been used to overcome this.

Often, stocking densities are reduced to take account of loss of volume through deformation. An alternative is to use as large a mesh size as possible and to suspend extra weights from the cage walls. However, increasing vertical loading on the netting greatly increases the horizontal forces acting on the net (Tomi *et al*, 1979) and may cause tearing, particularly under adverse weather conditions. This also does nothing to reduce feed losses or other associated problems.

In Indonesia, traditional submerged carp cages used in running water are made from wood (*Fig 2.2*), which not only does not deform, but also greatly reduces current flow. The weight of the wooden cages is sufficient to keep them anchored on the bottom of rivers and canals, even during floods (Vass and Sachlan, 1957). The floating designs used in Indochina (*Fig 1.4*) are also well-suited to the problems of running water. In Russia, 6-10m$^3$ floating cages for use in running water are constructed from 3-4cm wooden slats spaced 0.75-2cm apart nailed to a stout wooden frame and fitted with a galvanised wire mesh lid (*Fig 7.2*), and are held in place by rows of wooden pilings driven into the substrate (Privol'nev, 1975; Martyshev, 1983). When initially stocked with 20-25g fish, capron netting is fitted to prevent the small fish escaping between the bars, and removed a month or so later. Much of the bottom of the cage is solid and acts as feeding platform. Cages of this type are cheap to construct, long lasting (10 years with little maintenance), and reduce currents to an acceptable level within the cage.

Rigid mesh cages have also been used in fast flowing sites. In Bolivia, experimental cage culture of rainbow trout in the outflow of the River Hichukhota involved the use of 1 × 2 × 1m cages constructed from sections of 1cm plastic-coated wire mesh attached to a wooden frame (*Fig 7.3*). However, the flow of water through the cages was still rather high, making feeding difficult. In Italy and Netherlands, where rafts of floating

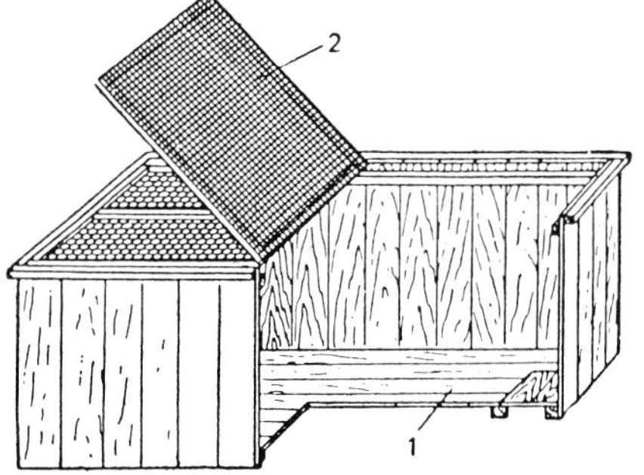


*1*—bottom of cage in the form of a feeding platform;
*2*—lid of cage divided into two halves.

*Fig 7.2* Wooden cage for trout culture in flowing waters. USSR (from Martyshev, 1983).

plastic-coated wire mesh rigid cages are used to culture trout and cyprinids in heated water power station discharges, systems of 'Venetian blinds' have been installed upstream which can be adjusted to deflect some of the water past the cages during periods of high flow (Bronzi and Ghittino, 1980; Hogendoorn, 1984). Metal mesh cages have also been used at marine sites in France and the Netherlands where currents are high (3-6 knots) (Boeuf and Harache, 1980; Anon, 1985).

The floating Cu-Ni expanded mesh rigid cages developed by Hugenin and co-workers (see Ansuini and Hugenin, 1978, *etc*) also have advantages when used at fast-flowing sites, although the computed drag coefficients for the standard 1cm mesh are comparatively low (Gularte and Hugenin, 1984) and thus the reduction in flow through the cages may be less than those produced by other materials.

## 7.4 Diseases

There are now many excellent books available on fish diseases (*eg* Sarig, 1971; Roberts, 1978; Frerichs, 1984; Roberts and Shepherd, 1986). However, with the exception of some of the more common diseases of commercially cultured species which are briefly described in Chapter 8, it is beyond the scope of this book to detail specific diseases, but instead to give a certain amount of basic information on what diseases are, how they are caused and how they may be prevented or treated, with particular reference to the cage fish farming situation.

The types of disease encountered in fish may be divided into six categories (from Kinne, 1980):-

(i) Genetic
(ii) Nutritional
(iii) Environmental, including abiotic factors such as light, temperature, salinity and oxygen, as well as natural and man-made pollutants
(iv) Physical injuries
(v) Parasites and micro-organisms
(vi) A combination of the above

Many of these are non-communicable (*ie* cannot be transmitted from one individual to another) and if diagnosed correctly and the cause ascertained, then changes in husbandry and management practices will often be sufficient to eradicate the problem. For example, genetic disorders, such as spinal defects, are comparatively rare and easily avoided by prudent stock selection, whilst physical injuries can be minimised by handling fish with care, the appropriate choice of netting materials and mesh sizes, the exclusion of predators and the avoidance of over-exposed sites where excessive cage motion can abrade and damage fish (see Chapter 4). Risks of environmental hazards can be greatly reduced by prudent site selection, rational site development and good husbandry. However, farmers can and do have environmental problems and have to turn to some form of remedial action. Mitigative measures which can be used against adverse temperature, light,

*Fig 7.3* Small, experimental, vinyl-coated steel mesh cage, in the outflow of Lake Hichukhota, Bolivia, for trout.

pH and oxygen conditions are described in Sections 7.7, 7.11.2 and 7.11.3.

Nutritional disorders are often difficult to diagnose and treat without specialist knowledge and even where this is available there may still be problems since, for some species at least, research into nutritional requirements is still at an early stage. In the case of intensive fish farming, using commercial feeds which have a proven reputation should minimise the risk of any problems although care must be taken over storage and shelf-life of the feed as well as in the method of feeding (see Chapter 6.3). For semi-intensive operations a varied diet of fresh, uncontaminated feeds is least likely to give rise to problems.

Communicable diseases are caused by parasites, bacteria, fungi and viruses. Parasites spend part or all of their life cycle extracting their nourishment from and at the expense of other living organisms. Fish parasites include members of both the protozoa and metazoa, the latter comprising such diverse groups as acanthacephalans, nematodes, cestodes, trematodes, leeches, crustaceans, bivalve glochidia and even other fishes (*eg* lampreys).

Representatives from more than 20 genera of bacteria have been isolated from fish, of which at least 15 species are recognised as actual or potential pathogens (Frerichs, 1984). Most of these organisms are naturally occurring and widely distributed, living on the dead and decaying organic matter which is universally present in aquatic systems. The change from saprophyte to pathogen is usually precipitated by a stress factor(s) which increases the fish's susceptibility to infection.

Fungi are also saprophytic organisms which are very common in the aquatic environment and under certain circumstances can invade living fish, often through lesions caused by some other agent. Commercially important fungal pathogens of fish include *Saprolegnia* sp., *Ichthyophonus hoferi*, and *Branchiomyces* sp. (Hoffman and Meyer, 1974; Richards, 1978).

Viruses are obligate intracellular parasites which unlike fungi or bacteria are incapable of independent replication, relying instead on their ability to invade cells and direct them to manufacture viral-specific proteins which are subsequently assembled to form new viruses which go on to invade other cells. Cells which become infected often die. There are around a dozen commercially-important fish viruses which have to date been identified, including IHN (infectious haematopoietic necrosis), IPN (infectious pancreatic necrosis), VHS (viral haemorrhagic septicaemia) and SVC (spring viremia of carp).

Wild fish often carry a wide range of parasites and other organisms which rarely prove fatal unless the fish are stressed by some other factor such as pollution, or the balance between host and pathogen is tipped in favour of the latter through the introduction of a novel host or strain of pathogen. However, the host/pathogen relationship on a cage fish farm is very different. First of all, since fish are frequently moved from one site to another and even from one country to another there is always a risk that new or exotic diseases are introduced to the farm. The establishment of a cage fish farm in a water body may disrupt existing parasite life cycles, by increasing the number of available hosts. Thus the risks of parasite transmission from wild to caged stocks can be very high. The cage structure can also harbour parasites. For example, it is suspected that free-living stages of sea lice (*Lepeophtheirus salmonis*) can be harboured by heavily fouled cages, and act as a continual reservoir of infection. Moreover, farmed fish may frequently be stressed due to sub-optimal environmental conditions such as low oxygen, high temperatures or light levels, or through handling or physical damage thus lowering resistance to disease (Wedemeyer, 1970; Snieszko, 1974; Shepherd, 1978). There is also some evidence that the risks of physical damage in cages is greater than in land-based systems. Nets treated with copper-based anti-foulants become hard and can abrade skin, particularly during storms or net-changing operations, possibly pre-disposing fish to vibriosis (Needham, 1980), whilst in the USA, fin erosion in chinook salmon and steelhead rainbow trout have been attributed to cage conditions (Boydstun and Hopelain, 1977; Moring, 1982).

The high densities at which fish are normally stocked, and the frequent close proximity of farms are ideal for the rapid spread of disease.

Disease prevention is always preferable to cure, and there are a number of basic measures which should be taken by all farmers to minimise risks. Some of these have been mentioned elsewhere, but bear repeating:-

(i) Choice of species can be important, since some species or strains of species are more resistant to disease than others. For example, among the Pacific salmon species, coho are reckoned to be more resistant to vibriosis than chum (Needham, 1980), and during cage tilapia trials in the Ivory Coast, a local species *T. heudelotti* had to be substituted for *O. niloticus* because the latter proved highly susceptible to myxosporidian infections (see Pullin and Lowe-McConnell, 1982, p. 262).

(ii) Careful site selection is important (see Section 4.2.3).

(iii) Healthy fish of the appropriate size should be selected for stocking. Some diseases, such as IPN may cause few mortalities in freshwater, but can cause large post-transport mortalities when transferred to sea if fish are already infected. (The movement of IPN-infected fish is statutorily strictly controlled in the UK). Size is an important consideration when moving some fishes such as rainbow trout from fresh to seawater.

(iv) Regular observation of fish should be carried out, and any unusual behaviour or changes in colour or appearance should be taken as a sign that something may be wrong. Early diagnosis of many diseases is essential if mortalities are to be kept to a minimum.

(v) Fish should not be overcrowded.

(vi) Water quality must be monitored and maintained. Site selection and appraisal of carrying capacity are important in this respect.

(vii) Fish should not be overfed.

(viii) Food should be as fresh as possible and free from fungal or other contaminants.

(ix) Mortalities should be removed regularly and destroyed.

(x) Equipment should be regularly disinfected.

(xi) Predators should be discouraged.

(xii) Handling of fish should be done as gently as possible, and kept to a minimum.

Despite precautions, disease outbreaks do occur, but before any action can be taken, a diagnosis must be made. An accurate picture of the pattern of mortality can help here, since different diseases develop in different ways. Viral epidemics can occur very rapidly, whereas bacterial, fungal and parasitic outbreaks may take several days or longer to develop. Although external examination of affected fish may be sufficient to make a diagnosis, in most instances a post-mortem examination will have to be carried out. Tissue samples may also be required for bacteriology and virology. A number of manuals and books give details of these procedures (see above). Whilst experienced farm staff can often make a preliminary diagnosis, one sick fish can look much like another, and so confirmation may require the skills and techniques available at a specialist laboratory. For example, the clinical/pathological features of many of the bacterial diseases of fish include haemorrhagic septicaemia and skin ulcers, and definitive diagnosis is dependent upon culturing the organisms on selective growth media (Frerichs, 1984). Laboratories with suitably trained staff are now widely available throughout Europe, North America and Southeast Asia, although they are still few and far between in most of Africa and South America.

Treatment of diseased stock may eventually have to be faced, and there are three basic methods (Roberts and Shepherd, 1986):-

(i) bath treatment, in which chemicals are added to the water
(ii) direct application of chemicals to the fish
(iii) administration of chemicals in the food.

Soluble chemicals are often used in bath-form to treat fish with certain skin and gill diseases. Whilst this is a relatively simple business in land-based systems, any chemical added to cages would soon be dispersed by currents and the swimming behaviour of the fish. To overcome this problem with conventional floating and fixed cages, the net bag is often lifted to concentrate the fish in a smaller volume of water and enclosed or partially-enclosed by a plastic or canvas bag. In marine situations, treatment is often carried out at slack tide. However, if a large number of cages of fish have to be treated this can be an expensive as well as an extremely time and labour-consuming business. An alternative is to use a small, purpose-built treatment cage, which typically consists of a weighted canvas bag with net liner, suspended from a floating collar. This is then moored alongside the cage of affected fish. In some instances, diseased fish are simply netted out and transferred to the treatment cage containing the appropriate chemicals at the correct concentration, and left for the recommended time before being transferred back to the production cage. Although this may seem to be an economical way to do things, by saving on chemicals, the rather involved procedure of netting and transferring fishes backwards and forwards can prove highly stressful and cause many more mortalities than necessary. A more sophisticated method has been devised by Brandal and Egidius (1979), in which the treatment bag is attached to its collar on only three sides. One side of the production net is then unfastened, passed under the collar to the outside and sewn on to the freed side of the net inside the treatment cage. Fish are transferred by raising the production net bag and depressing the free side of the tarpaulin under the water by approximately 40cm (*Fig 7.4*), so that the fish are encouraged to swim in. Once all the fish are transferred the tarpaulin is raised again. Following treatment, the net bag inside the treatment cage is raised, and the fish gently tipped back inside the production cage. In treating a salmon lice (*L. salmonis*) outbreak, Brandal and Egidius (1979) claim that 8-10 cages of fish, representing 25-40 tonnes, can be treated in a single day (treatment time 20 minutes per cage) by four or five workers, with considerable success.

Many rotating cages can be simply and quickly adapted for bathing fish. In the cages shown in *Fig 7.5*, the cage is set in a half-submerged position, a canvas cover attached to the exposed half, and the cage rotated so that the fish are isolated within the canvas covered section. Chemicals are then carefully

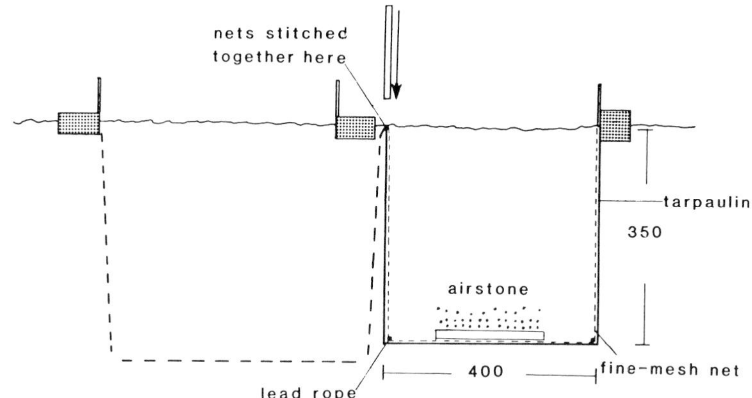


*Fig 7.4* Diagram to show how a treatment cage is attached to the grow-out cage. Fish are transferred by submerging the net walls at the point indicated (modified from Brandal and Egidius, 1979).

added and when the treatment has been completed the cage is turned over and the cover removed.

If the cage is isolated using a tarpaulin, normal water exchange processes are halted, and an oxygen diffuser or alternative aeration device may have to be installed, particularly since oxygen consumption by the diseased fish is likely to increase during treatment. The use of an aerator can also help in the rapid dispersal of the chemical following treatment. A number of other precautions should also be taken (Roberts and Shepherd, 1986):-

(i) Fish should not be fed for 24h prior to treatment.
(ii) Plastic buckets should be used, never galvanised containers, for mixing chemicals. This is especially important if using malachite green.
(iii) The volume of water that the fish will be held in during treatment must be calculated carefully so that the dose concentration is accurate. If possible, the computed dosage should be checked as it is easy to miscalculate by a factor of 10. The use of a dyestuff like Rhodamine-B may be useful in assessing how well the treatment chemical is dispersed in the cage.
(iv) Treatment at high temperatures should be avoided if possible. During summer months, or in tropical climes, treatment should be carried out in the morning, provided that DO levels are satisfactory (*NB* It is particularly important to determine oxygen conditions if there are high densities of planktonic algae).
(v) If it is the first time a treatment has been used at the farm, a trial treatment must be carried out first and 12-24h allowed to pass before the main treatment is carried out, if the trial is successful.
(vi) The fish should be watched continuously during treatment, and the farmer should be prepared to lower the cage net and remove the

tarpaulins rapidly or increase oxygen supply if the fish become distressed.

(vii) Treatment should only be repeated if absolutely necessary, and then at least 36h after the first treatment.

(viii) It is important that courses of treatment are completed, otherwise the disease may return with greater severity, and the cost of the initial treatment will have been wasted (see later). Details of bath treatments used for various diseases and fish species are given in Sarig (1971), Brandal and Egidius (1979), Stevenson (1980), IDRC (1983) and Roberts and Shepherd (1986).

Where valuable fish or broodstock are concerned, individual treatment is sometimes feasible. For example, malachite green is occasionally applied as a paste to fish with fungal skin infections. Injections of individual fish are sometimes necessary (*eg* to induce spawning of eggbound fish), although in many countries this can only be done under veterinary supervision.

Some diseases can only be treated by using drugs incorporated in the feed. Examples include internal parasites, such as acanthocephalans and some of the cestodes, internal protozoan infections (eg *Octomitus*) and bacterial septicaemias caused by vibriosis and *Aeromonas* infections. Unfortunately there are a number of problems with this approach. It can be difficult to persuade manufacturers to incorporate drugs in small quantities of commercially-prepared feed (~1 tonne), leaving the fish farmer with little option but to mix the drugs with the feed himself. When using trash fish or a wet pellet diet, drugs are simply mixed with the feed prior to feeding,

*Fig 7.5* Rotating cage, Scotland (courtesy J F Muir).

whereas dry diets are coated with either a gelatine mixture or corn oil slurry containing the drug and left to dry (Poupard, 1978). These methods should only be carried out immediately prior to feeding, as some antibiotics will degrade within 24h.

As Roberts and Shepherd (1986) state, the most critical calculation in administering drugs orally is the dose rate. They recommend that the feeding rate for medicated diets is reduced by about 0.5% below the normal feeding rate. Thus if fish are feeding at 3% body weight per day when they become diseased, the feeding rate with a medicated diet is 2.5% body weight per day. Details of drug incorporation rates are given in Wood (1974), Hoffman and Meyer (1974), Poupard (1978), Stevenson (1980), IDRC (1982), Roberts and Shepherd (1986). However, all effective dose rates, no matter how carefully computed, can be thrown off course by the fish themselves. Sick fish may well be off their food, making it difficult to persuade them to eat, and the concentrations of drugs required in medicated diets are likely to further decrease appetence. Not only can this render treatment expensive or ineffective, but repeated treatments with sub-therapeutic levels of antibiotics will, in the long term, promote drug resistance.

In many countries the use of antibiotics requires a veterinary prescription and this is of great importance in reducing the risk to animals and man of developing resistant strains of pathogenic bacteria. Finally, it is strongly recommended that fish are not sold for human consumption for at least 4 weeks following treatment, by which time all drugs in their tissues should have been excreted.

It should be borne in mind that most diseases stem from inadequate management or husbandry and that treatment should be very much seen as a last resort. Moreover, many fish diseases cannot be tackled at present by therapeutic means. Control of most viral diseases is best achieved by removal of stressing factors such as low oxygen, crowding *etc* (Liversidge and Munro, 1978). Providing the farmer can obtain an accurate estimate of likely mortalities, the economics of treatment versus permitting the disease to run its course should also be considered. Bath treatments cost little in terms of chemicals, but are very labour-intensive whereas systemic treatment via the diet is usually very costly in terms of drugs.

Treatment of caged fish is a long established practice in some parts of the world. In Kampuchea, for example, cow dung, tree bark and salt, mud and plant mixtures are used to treat parasitic and fungal diseases of caged catfish murrels and gobies, although the efficacy of these measures has not been properly evaluated (Pantulu, 1979). The range of modern drugs and chemicals that fish farmers have at their disposal to treat diseases is small and unlikely to grow. Resistance by fish pathogens to some of the drugs currently in use is growing and fears that antimicrobial resistance might be transmitted to micro-organisms of public health concern may even result in the repeal of some drugs currently approved for use in fish culture (Piper *et al*, 1982). This has stimulated research into the development of vaccines which have the advantages that they do not generate antibiotic resistant micro-organisms and they can control viral as well as bacterial diseases. A number of vaccines licensed for use in fish are now in commercial production in the UK and USA.

## 7.5 Drifting objects

Floating and fixed flexible mesh cages are particularly vulnerable to damage by drifting objects which can tear holes in the netting, restrict water flow, and even completely destroy cages. Logs and other large floating objects are hazards faced by farmers everywhere and cage fish farmers are well advised to keep shorelines adjacent to their sites clear of such objects, since during exceptionally high tides and storms they can be lifted off the beach and carried out to the cages. Jellyfish are commonplace in coastal waters during certain periods of the year, and great numbers can become entangled in cage nets causing net deformation and restricting water flow and so should be removed regularly. Jellyfish have also caused mass mortalities of caged fish in the Baltic (Koops, 1976).

A major problem faced by many cage fish farmers in inland tropical lakes is drifting masses of floating plants, such as water hyacinth (*Eichhornia crassipes*) (*Fig 7.6*). A single square metre of these plants can easily weigh 100kg. Under favourable conditions, the plants can cover substantial portions of a lake's surface and when strong winds blow, floating masses weighing tens and even hundreds of tonnes can bear down on a site destroying moorings, cages and anything else in their path. Fixed cages are particularly vulnerable. In the Philippines, some farmers have abandoned bamboo pole constructions and instead use anahaw palms (*Livingstonia*

*Fig 7.6* Wind-driven accumulation of water hyacinths, Laguna de Bay, Philippines. Water buffalo can be seen in the foreground taking advantage of the cool, shaded conditions.

*rotundifolia*) (*Fig 8.18*). Although they cost 4-8 times as much as bamboos, only a fraction of the number are used, poles being driven into the lake substrate at 2m intervals. They are very strong and because so few are used, offer little resistance to the floating hyacinth mass. They also last considerably longer than bamboos. Top nets are fitted to many of the cages so that during stormy periods, the whole cage may be submerged (see Section 8.4).

An alternative solution is to build a barricade or deflective fence which can divert the weeds and keep the cage area clear. Again, this strategy has been adopted by a number of farmers in the Philippines, where materials such as bamboos are readily available and where skills derived from traditional fish corral construction have been adapted to suit the new cage farming industry.

## 7.6 Fouling

Fouling is something that occurs to a greater or lesser degree at all cage sites, irrespective of cage type. As discussed in Section 4.5.2, marine sites tend to be worse than fresh water in this respect, particularly if the environment is warm and organically rich, and if the prevailing currents, which greatly influence settlement, are low. The degree of fouling is also dependent upon the material used (see Section 3.4.1), mesh size, whether the mesh is knotted or knotless and is, to some extent, influenced by the species being cultured. Biofouling is not a major problem in the culture of herbivores, such as some of the tilapias (Pantastico and Baldia, 1981), and in Japan, yellowtail cages have been found to accumulate fewer fouling organisms than those holding blue fin tuna or red sea bream due to the action of the fish which continually brush past the netting (Kuwa, 1984). Studies of the rates of fouling of net cages and the organisms involved have been detailed for many different parts of the world (Milne, 1970; Moring and Moring, 1975; Cheah and Chua, 1979; Wee, 1979; Kuwa, 1983, 1984, Santhanam *et al*, 1984.

Fouling reduces the actual mesh size, thereby cutting down on the flow of water through the cage and increasing drag (Milne, 1970; Inoue, 1972; Wee, 1979; Wee, 1979; Kuwa, 1983, 1984, Santhanam *et al*, 1984).
Scotland, Milne (1970) found increases in the twine diameter of test panels of various netting materials (*Table 3.6*) and, using the equations given in Section 3.4.1, calculated the increases in current forces (*Fig 3.9*). For 25mm nylon and polyethylene netting, these corresponded to approximately 12-fold increases. The increases in the weight of fouling recorded by Milne (*ibid.*) (*Table 3.6*) not only affect routine handling, but also flotation. Milne (1976), and Hugenin (1978), have stated that a number of net failures, due largely to fouling, have occurred at several large commercial marine cage fish farms.

If there is a severe problem at a site, the farmer has a number of options:-

(i) he or she can decide to accept the problem, and adjust management procedures accordingly
(ii) attempt to reduce the problem with the aid of chemical or biological agents

(iii) opt for a fouling-resistant or rotating design.

The first option is discussed in Section 6.4, which describes the routine business of net changing and cleaning that most farms have to carry out on a regular basis. If fouling is severe, then the farmer can simply choose to carry out this procedure more frequently.

An alternative is to use chemical antifouling agents. According to Lovegrove (1979) most antifouling agents contain a biocide which slowly leaches so that the surface is permanently surrounded by a thin layer of toxic material which prevents the susceptible juvenile planktonic stages of fouling organisms from gaining a foothold. The ideal compound should release the toxin at a controlled rate over an extended period. There are two principle types in use: (i) soluble matrix or soft film antifouling and (ii) contact leaching tough film antifouling. The soluble matrix antifoulants contain both a water-soluble biocide and a part water-soluble resin. When immersed in seawater, the paint gradually becomes porous, thus allowing access to the biocide in the deeper layers. This type of antifoulant has the advantage that fairly low levels of biocide are included in the paint. The latter type, however, relies on water-resistant media which, so that adequate leaching occurs, must contain much higher levels of water-soluble biocide.

Today the most common antifoulants are still either organotin (tributyl tin oxide or fluoride) or copper-based, both being sold in a variety of forms under different brand names. Antifoulants can either be applied at the factory (*Fig* 7.7) or, using less sophisticated equipment, at the farm. Nets are

*Fig 7.7* Copper-based anti-fouling material being applied to cage nets (courtesy W & J Knox, Ltd).

dipped in the compounds for 10-30 minutes or so, and hung up to dry for several hours in a fully open position so that none of the meshes become clogged. This is particularly important if using copper-based antifoulants which tend to form a stiff, relatively inflexible coating on the net. Manufacturers recommend that copper-treated nets are immersed as soon as possible after drying. Organotin-treated nets, on the other hand, are less rigid and can easily be stored for six months prior to use. Once the newly antifouled net is installed, it is advisable to leave it for 24h before stocking.

Antifouled nets provide a measure of protection for many months, although the degree of protection declines rapidly after 6 months or so. In temperate regions, therefore, nets may have to be treated each year and it is important that newly treated nets are installed in April-May to give maximum protection over the summer months when larval settlement is greatest. The cost of treatment is approximately the same for organotin and copper-based products, for although the former is more expensive, less is required. The cost of treatment will add approximately 20-25% on to the cost of a knotless nylon cage net.

Both the copper and organotin compounds used as antifoulants are toxic to shellfish, although the latter appear to be worse (Good *et al*, 1979; Waldock and Thain, 1983). For this reason it has been suggested that copper-based products be used in preference to organotin at sites close to important shellfish beds (Lovegrove, 1979). Tin-containing antifouling paints used to protect cages have also been implicated in deformities of yellowtail in Japan (Anon, 1982b). Research into new antifouling compounds, which are less toxic and give more extended protection, is very active, and one group of compounds that is generating a great deal of interest is low surface energy coatings. These are non-toxic and work by making the surface so slippery that fouling organisms find it impossible to settle.

Whilst the organisms which foul net cages may augment the diets of cultured herbivores or detrivores (see earlier), they are only occasionally ingested by carnivorous species such as the salmonids and play a negligible role in their diet (Moring and Moring, 1975). However, in several countries herbivores have been stocked alongside commercially more important species in attempts to reduce biofouling. In Palau, Hasse (1974) successfully used siganids (*S. canaliculatus* and *S. lineatus*) at densities of approximately 2 fish $m^{-3}$ to reduce fouling in cages holding strings of oysters (*Ostrea nomades*), whilst Chua and Teng (1977) and Aquacop (1975) have found siganids effective in curtailing fouling in cages of grouper and carangids. The striped and spotted knifejaws, *Oplegnathus* spp., have been used in cage mariculture in Japan (Kuwa, 1983, 1984) (*Fig 7.8*), and low densities (1 fish $5m^{-3}$) of rohu (*Labeo rohita*) have proved effective in maintaining the cleanliness of carp cages in Nepal (Sharma, 1979). Marine prawns have also been used in salmon cages, and mullets with pompano and other species (Swingle, 1972; Tatum, 1974; Aquacop, 1975; Rensel and Prentice, 1979). However, there can be problems: siganids have been found to damage cage netting as a result of grazing (Ben Yami, 1972), and knifejaws have been observed attacking the tails and fins of sick yellowtail (Kuwa, 1984).

*Fig 7.8* Spotted knifejaws inside a yellowtail cage, Japan (courtesy M Kuwa).

Alternatively, farmers may consider the use of fouling-resistant materials in the construction of the cages. A number of companies now manufacture polyethylene netting inlaid with copper wire and whilst this material may be readily fabricated into conventional net bag designs, and thus used with existing cage collars, it is much more expensive and costs at least twice as much as a nylon net bag of similar dimensions. Moreover, the copper gives limited protection, and the nets must either be replaced every year or two, or treated with conventional antifouling compounds after the copper wire has corroded.

A number of rigid cages have also been designed using more fouling-resistant materials such as galvanised steel, PVC coated wire and copper or copper-nickel compounds (Milne, 1970; Swingle, 1971; Powell, 1976; Ansuini and Hugenin, 1978; Ojeda and Strawn, 1980; Hugenin *et al*, 1981; Kuwa, 1983, 1984). Of those materials, copper and copper-nickel appear to be the most effective, although galvanised steel is also excellent, providing that some anti-corrosion safeguards are taken (see Section 3.4.1). As might be expected, copper and copper alloy mesh cages are considerably more expensive than conventional nylon bag systems, and cost between 3 and 5 times more on a per unit volume basis (Ojeda and Strawn, 1980; Hugenin *et al*, 1981). Although Hugenin *et al* (*ibid*) claim that the high initial expenditure can be offset during a 10-year period due to reduced operating costs, they base their comparisons on cages which have very high labour costs.

As discussed in Chapter 2, rotating designs seem to be effective in controlling fouling and reduce cleaning to a minimum. However, in countries where UV levels are high, the prolonged exposure of netting to intense sunlight that is necessary to kill the organisms has resulted in accelerated degradation of the nets (Porter, 1981). Most of the commercially available designs are also very expensive and can cost several times as much per unit volume as conventional types of cage.

In summary, the majority of commercial fish farmers still seem to prefer to use nylon or polyethylene net cages and either accept the increased labour involved in cleaning (some managers believe that frequent checking of nets for fouling results in the staff being much more aware of the status of fish health on the farm), or resort to the use of antifouling compounds. The savings in manpower offered by rotating designs or copper alloy cages may not alone justify the much higher initial capital expenditure. However, they have a number of other advantages over conventional net bag systems which should be carefully weighed up by prospective farmers (see Sections 7.7, 7.8, 7.9 and 7.11).

## 7.7 Oxygen

The dissolved oxygen (DO) regime in a fish cage is a balance between input and output. As oxygen input by diffusion across the air/water interface is negligible — Kils (1979) estimated that it provided somewhere in the region of 0.5% of the oxygen respired by caged fish — the major source of DO is water flowing through the cage as a result of external currents and the movements of the caged fish. At most sites the DO content of surface waters approximates saturation levels and as long as the cages are maintained relatively free from fouling there should be few problems (see Section 4.2.1.2). Under exceptional weather conditions, high temperatures can create difficulties by reducing the absolute amount of oxygen that the water can hold in solution and by increasing the oxygen demand of the fish. Providing the farmer carefully monitors DO levels and reduces feeding rates and avoids handling (*ie* keeps oxygen consumption to a minimum), the event usually passes with only a small interruption in production.

DO levels can fall far below saturation in natural water bodies under the following circumstances:-

(i) Where the phytoplankton community is large (*ie* under bloom conditions), night-time respiration can cause a marked decline in DO, reaching a minimum around dawn.
(ii) At the end of a phytoplankton bloom, bacterial decomposition of the algae can lead to anoxia.
(iii) Upwelling of anoxic water from the hypolimnion to the surface can occur at certain sites given the right combination of environmental and climatic conditions.

In many instances these problems can be avoided during the site appraisal stage by rejecting eutrophic or strongly stratified water bodies. Restricting fish production to a level appropriate to the site will further reduce risks at intensive farms at least. Nevertheless this is not always possible or even desirable (*eg* extensive fish farms are best sited in eutrophic waters) and problems have been reported at both marine and freshwater sites in temperate and tropical environments (Inoue, 1972; Koops, 1972; Grove, 1975; Barica, 1976; Kils, 1979; Landless, 1985).

Remedial action involves either increasing the DO content of the water and/or the rate of water flow through the cage. The factors which govern the solubility of oxygen in water have already been discussed in Section 4.2.1.2. The transfer of oxygen to water is dependent upon the difference between the maximum level of oxygen that can be dissolved in the water at specific conditions of temperature, salinity and atmospheric pressure, and the actual level. It is also influenced by the area of contact between the water and air or oxygen supply, such that:-

$$R = K_L A(C^\star - C)$$

where $K_L$ is the mass transfer coefficient (this has to be determined experimentally); A is the interfacial area between the gas and liquid; $C^\star$ is the equilibrium concentration of dissolved oxygen; and C is the dissolved oxygen concentration of the oxygenated water.

In practice, oxygenation is achieved by maximising the area of contact between water and air or oxygen either by introducing fine bubbles of gas into the water or by breaking the water into fine droplets in air. Air can be delivered to the water from an air blower, via coarse bubble (large volumes at low pressure) or fine bubble (small volumes at high pressure) diffusers and airlifts, whilst droplets of water can be generated by various mechanical devices, including surface aerators, surface agitators, venturis and submerged pumps (*Fig 7.9*). Details of how these systems work are given in Colt and Tchobanoglous (1981) and Sowerbutts and Forster (1981).

Oxygen may be delivered from a pressurised oxygen source. The most common source of oxygen used at fish farms is bulk liquid oxygen which is more than 90% pure and is stored under pressure in tanks or cylinders. However pressure swing absorption (PSA) devices which produce 90%-95% pure oxygen by removing nitrogen from air as it is forced through a series of 'molecular sieves' are gaining in popularity since, according to manufacturers, they can be more economical than alternative oxygenation systems for small-medium size production units ($<15 m^3$ oxygen $h^{-1}$).

For fish farmers, a key comparative measure is transfer efficiency, defined as the quantity of oxygen dissolved per unit energy expended (kg $O_2$ kW $h^{-1}$) and this is usually determined under standard conditions (*ie* 0% saturation in clean fresh water at 20°C). Transfer efficiencies for various devices, under standard conditions, are summarised in *Table 7.1*.

Most studies on aeration come from the water treatment industry, and few of the oxygenation or aeration devices have been rigorously tested under fish farm conditions where the driving force ($C^\star - C$) is usually much lower than that used in standard tests. Transfer efficiencies for field conditions are often

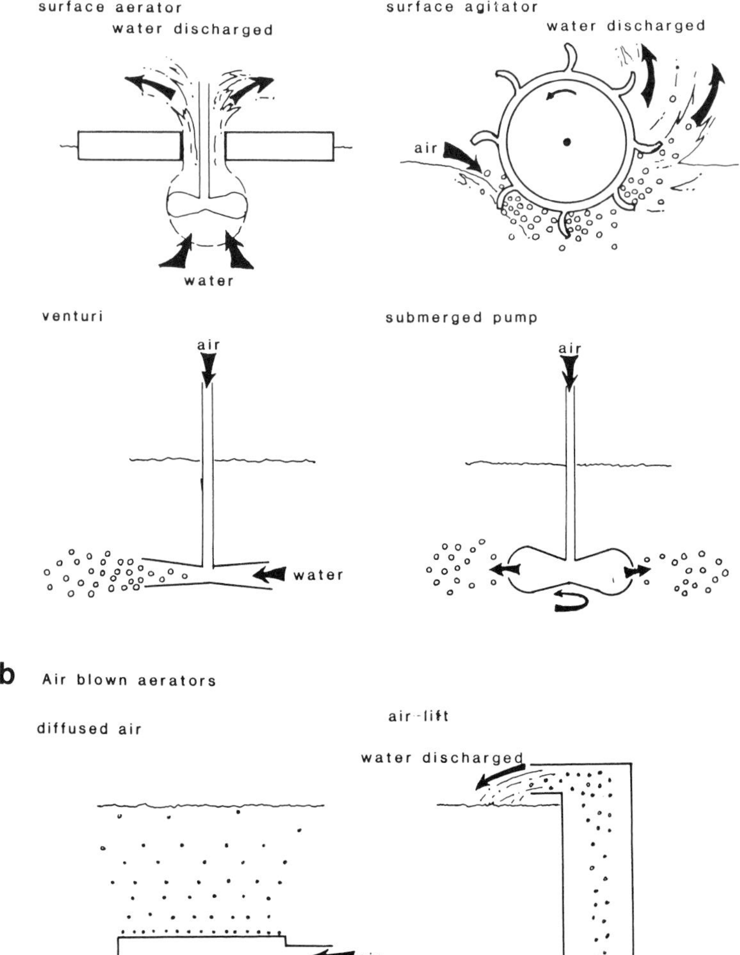


*Fig 7.9* Aerators of various design (redrawn from Sowerbutts and Foster, 1981).

*Table 7.1* Typical oxygen transfer rates of various devices used in fish culture systems (modified from Colt and Tchobanoglous, 1981).

| *Aeration system* | *Transfer rate* $kg\ O_2\ kW\ h^{-1}$ *Standard*[a] | $6\ mg\ l^{-1}\ O_2$[b] |
|---|---|---|
| Diffused air system | | |
| fine bubble | 1.2–2.0 | 0.25–0.42 |
| medium bubble | 1.0–1.6 | 0.21–0.34 |
| coarse bubble | 0.6–1.2 | 0.13–0.25 |
| Surface aerator (low speed) | 1.2–2.4 | 0.25–0.80 |
| Surface aerator (high speed) | 1.2–2.5 | 0.25–0.50 |
| Venturi aerator | 1.2–2.4 | 0.25–0.50 |
| Pure oxygen system | | 1.20–1.80 |
| fine bubble | | |

[a] 20°C, tap water, 0 mg $l^{-1}$ D.O., $\alpha = \beta = 1.0$.
[b] 20°C, $\alpha = 0.85$, $\beta = 0.9$, D.O. = 6 mg $l^{-1}$.

computed from data obtained under standard conditions using the following equation:-

$$N = N_o \left[ \frac{\beta C^{\star} - Cd}{9.17} (1.024)^{T-20} \alpha \right]$$

where N = kg $O_2$ kW $h^{-1}$ transferred under field conditions; $N_o$ = kg $O_2$ kW $h^{-1}$ transferred under standard conditions; $\beta$ = salinity – surface tension correction; $C^{\star}$ = equilibrium dissolved oxygen concentration; Cd = design dissolved oxygen concentration; T = temperature, °C; $\alpha$ = oxygen transfer correction for waste. In column 2, *Table 7.1*, mass transfer rates have been calculated for a freshwater pond or tank system, assuming T = 20°C, $\alpha$ = 0.85, $\beta$ = 0.9 and Cd = 6.0mg $l^{-1}$ (*ie* Cd = 70% saturation). Note that values are around 20% of those derived under standard conditions. However, if one tries to achieve 90% saturation instead of 70% saturation, then the mass transfer rate falls to around 10% of the standard values, *ie* it takes *twice* as much energy to achieve 90% saturation as it does to achieve 70% under those conditions.

Salinity determines surface tension forces, which in turn determines bubble size, and therefore transfer efficiency. Thus, transfer efficiency in seawater is higher than in freshwater (Kils, 1977).

The depth of water in a cage is greater than that in most tanks and ponds. There is also little restriction in exchange between the cage environment and that outside. Thus the cage fish farmer has a problem in trying to oxygenate the entire cage volume without aerating the external environment as well. Surface aerators and agitators are unlikely to be very efficient when used in cages since they aerate only the top few metres: moreover, paddlewheel devices would be impossible to use in such confined spaces without causing a great deal of damage! Submerged devices (pumps, venturis, air-lifts and air stones) would therefore appear to be best.

The formula given above for calculating oxygen transfer efficiencies, and the values shown in column 2, *Table 7.1*, apply to freshwater pond and tank conditions, although in the absence of any data for cages they can be used as a general guide in cage aeration computations. Surprisingly, studies of aeration in deep tanks, conducted by Lister and Boon (1973), have shown that the efficiency of oxygen transfer under a range of conditions is unaffected by the depth at which fine bubble aerators are installed (between 1m and 8m). The greater depth at which aeration devices are installed in cages (3-4m compared with 1-2m in ponds and tanks) results in increased contact time between air and water which may offset the increased energy costs of delivering air at those depths.

In choosing and designing a system the cage fish farmer must first of all establish the time of year problems are likely to occur, and have some idea of how serious conditions could become. Given the present state of knowledge on aquatic oxygen budgets and the effects of cage fish farming, this is best accomplished by 'guesstimate' or through prior experience, bearing in mind that a balance must be struck between a design which can cope with worst possible conditions, and costly over-engineering.

Based on temperature, species and fish size, the highest hourly oxygen consumption of the farmed fish during the crisis period can be computed (see *Table 7.2. NB* More detailed figures for different species under different conditions can be obtained from the literature), whilst the hourly oxygen requirements per cage can be derived from stocking density and cage size data. Using the oxygen mass transfer efficiencies given in *Table 7.1*, the power requirements for providing all or part of the fishes' oxygen requirements can then be calculated. The lower value should be used where

*Table 7.2* Some oxygen consumption values for farmed fish.

| *Species* | *Size (g)* | *Temperature (C°)* | *Feed rate* | *Oxygen consumption ($g\ O_2\ kg\ fish^{-1} h^{-1}$)* | *Source* |
|---|---|---|---|---|---|
| *S. gairdneri* | 100 | 15 | ? | 0.3 | Liao, 1971 |
| | 100 | 15 | production levels | 0.3 | Muller-Fuega *et al*, 1978 |
| *O. nerka* | 28.6 | 15 | unfed | 0.23 | Brett and Zala, 1975 |
| | 28.6 | 15 | 3% body wt day$^{-1}$ | 0.28 | |
| *I. punctatus* | 100 | 30 | unfed | 0.56 | Andrews and Matsuda, 1975 |
| | 100 | 30 | satiation | 0.81 | Andrews and Matsuda, 1975 |
| *C. carpio* | 100 | 10 | fed | 0.17 | Beamish, 1964 |
| | | 20 | fed | 0.48 | Beamish, 1964 |
| | | 25 | fed | 0.70 | Beamish, 1964 |
| *H. molitrix* | 15 | 20 | — | 0.20 | Muhamedova, 1977[1] |
| | 240 | 23 | fed | 0.25 | Vetskanov, 1975[1] |
| *O. niloticus* | 50 | 25 | unfed | 0.16 | Ross and Ross, 1984 |
| | 50 | 30 | unfed | 0.24 | Ross and Ross, 1984 |
| | 50 | 35 | unfed | 0.40 | Ross and Ross, 1984 |

[1]From ADCP 1984.

the DO gradient ($C^\star - C$) is small, and the more optimistic figure used where $C^\star - C$ is large. A worked example is given in *Table 7.3*.

Note that the estimated power needed to supply all of the fishes' oxygen requirements are around 1.2kW tonne $fish^{-1}$, which approximates to the rule-of-thumb value of 1kW $tonne^{-1}$ for rainbow trout held at 15°C, given by Sowerbutts and Forster (1981). For economic reasons cage fish farmers in the UK and Ireland install systems with power outputs much less (5-50%) than this, and rely on aeration merely to top up available supplies when necessary. At some farms, aeration equipment is switched on and left running for the entire risk period, whilst at others, the system includes an oxygen transducer, the signal from which switches on or off the equipment at predetermined levels.

A typical aeration system consists of a diesel or electrically-driven blower or compressor connected to a flexible PVC hose which distributes air to an aerator or system of aerators suspended 3-4m below the cage surface. Compressors are cheaper and are often easier to obtain in tropical countries, although they are less efficient and will give a lower oxygen transfer rate. An oil-free air supply is of prime importance and so an air filter or oil trap may also have to be fitted. Blowers utilise either a single (low pressure) or multiple (high pressure) fan system, the choice depending on the extent and nature of the pipework involved. If the cages are attached to the shore via a walkway, then the blower/compressor and motor can be installed on land and the airline run out across the walkway. If the cages are moored offshore then it must be decided whether the blower/ compressor and motor remain on shore and a floating airline used to deliver air to the cages, or whether the blower/compressor and motor are installed at the cages — either on the cage raft or on a boat moored alongside. In considering these options it must be remembered that the longer and narrower the pipeline, the greater the drop in pressure and hence the lower the overall efficiency of oxygen transfer.

The capital cost of an aeration system is influenced by the size of the engine, blower, *etc*, and whether the system is purchased 'off-the-shelf' or assembled by the farmer. For example, second-hand diesel engines and ex-milking parlour blowers have been used with considerable savings in capital costs.

As indicated above, there has been little study of aeration in cages and thus it is difficult to assess which aeration devices (airstones/venturis/air lift pumps/submerged pumps) are best, or even how effective aeration in cages is. A number of examples are cited by Landless (1985) in which air-lift devices have proved effective in maintaining good DO levels at cage farms subject to blooms. However, according to Kils (1979) submerged airstones can actually make oxygen conditions worse by creating a rising stream of oxygenated air which rapidly flows across the surface and out of the cage, the displacement of cage water resulting in an upwelling of deoxygenated hypolimnetic water which, if the cages are located near the lake or sea bottom, could also carry in cage wastes.

Kils (*ibid.*) designed and tested a modified submerged pump (*Figs 7.10a* and *7.10b*) under field conditions and claimed that it gave excellent results

*Table 7.3* Example of calculation for aeration system design for freshwater rainbow trout culture, assuming air-lift pumps are employed.

Expected problem period: late summer
Projected worst conditions: 20°C, 50% DO saturation
Size of fish = 100g
Estimated respiration rate ~300mg $O_2kg^{-1}$ $h^{-1}$ (from Muller–Fuega *et al*, 1978)
Size of cage = 5 × 6 × 5
Stocking density = 15kg $m^{-3}$
One cage holds 2.25 tonnes
Oxygen requirements = 0.68kg $O_2h^{-1}$ $cage^{-1}$
Assuming aeration must supply all of the fishes oxygen requirements, and a poor mass transfer rate (*ie* 0.25kg $O_2kW$ $h^{-1}$), power requirements will be:

0.68/0.25 = 2.7kW $h^{-1}$ = 3.8 h.p.

If the farm has 10 cages, then total power requirements = 27kW $h^{-1}$ or 38 h.p.

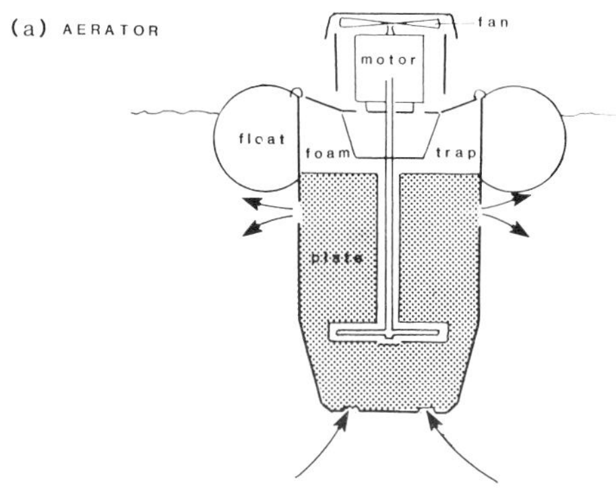


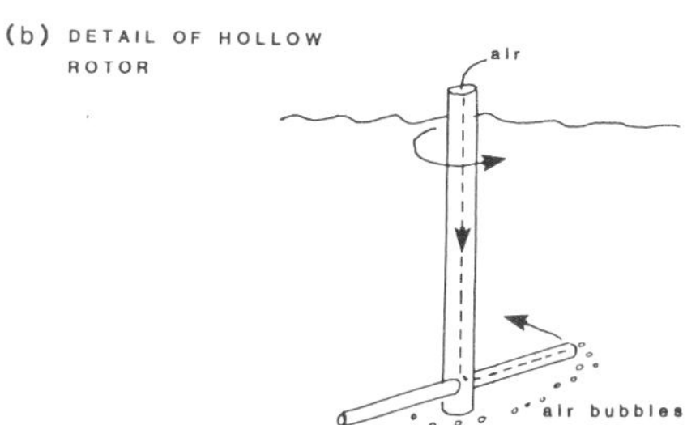


*Fig 7.10(a)* Schematic diagram of cage aerator, showing casing motor with fan, and hollow rotor. Water is sucked in through inlets in the bottom of the casing, and aerated water is discharged via outlets in the side. The transverse plate reduces agitation of the water, whilst the air space at the top acts as a foam trap (redrawn from Kils, 1979). *(b)* Hollow rotor of aerator (redrawn from Kils, 1979).

with a very high oxygen transfer efficiency (2.3-3.2kg $O_2$ kW $h^{-1}$). However this design has yet to be tested in large or more conventional type cages. No details of venturis (which require pumped water) having been used in cages are available, and most practical experience has been with air lift devices (*Fig 7.11*). Either fine or coarse discharge holes are used, the former being more effective in oxygenation. However, capital costs for fine-bubble systems are higher, and the aerators are also more difficult to clean and maintain. Whilst air-lifts appear to be successful, once again there is no good evidence as to how effective they are.

The use of submerged aeration devices in cages can have a number of other effects in addition to oxygenation. Their use in maintaining ice-free conditions is discussed in Section 7.11.2, and they are also used to reduce water temperatures (and hence improve oxygen content) through de-stratification of temperate lakes used for salmonid culture (*eg* Anon, 1980; Landless, 1985). Aeration has been shown to affect fish behaviour in cages (Kils, 1979), although whether for better or worse has yet to be determined.

There are two alternatives to aeration. One is to use oxygen instead of air, the other being to increase the rate of water exchange through the cage using a low speed propeller. For cage farmers, bulk liquid oxygen may, for the moment, be best suited as a back-up to aeration, since figures from tank studies suggest that the cost of oxygen transfer is likely to be at least twice as high as even the least efficient aeration systems (Hopwood, 1982). However, the economics of oxygen (especially PSA oxygen generators) use on cage fish farms merits further study. Low-speed propellors (*Fig 7.12* ), which generate water flow and create vertical as well as horizontal mixing, are also advocated as an alternative to aeration and have been successfully tested in catfish ponds in the USA where they were found to use a fraction of the energy required for aeration (Bush and Goodman, 1981). Similar devices are used at some salmon farms in Norway and have been found to improve DO conditions in cages. However, in freshwater lakes and reservoirs mixers should be used near the surface, as devices installed on or close to the lake bottom may adversely affect water quality by resuspending sedimented wastes which could stimulate algal growth. Again, there have been no published accounts of how effective these mixing devices are.

## 7.8 Poaching and vandalism

In general, cages are more vulnerable to poaching and vandalism than any other type of aquaculture venture. Farms sited some distance from shore-based offices and staff accommodation can easily be visited without observation, particularly at night. They are also difficult to protect since they cannot be fenced and conventional electronic devices such as IR beams are easily triggered by the bobbing motion of the cages. Fish may be simply and quickly removed by hand net, whilst vandals can cause many thousands of pounds worth of damage and loss of valuable stock in minutes using nothing more sophisticated than a sharp knife.

Security is often a major consideration in site selection, and it is

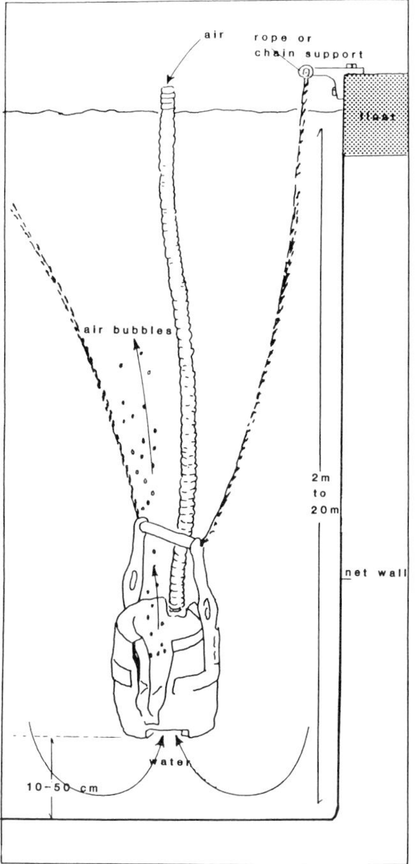


*Fig 7.11* Submerged venturator (redrawn from Landless, 1985).

undoubtedly true that proximity to the owner's house is a deterrent to would-be poachers. In many parts of Southeast Asia guard houses are built on the cage rafts and nightwatchmen with dogs employed to guard stock (IDRC/SEAFDEC, 1979; Tseng, 1983) (*Fig 6.6*). In the UK one large farm has moored an old fishing boat which serves as accommodation for staff on night duty, alongside the farm. Certain types of cage are less vulnerable than others. Rigid mesh cages cannot be vandalised so readily, and it is also more difficult to remove fish without the appropriate equipment. A number of devices are under trial to help protect cages, including a high pressure cable system, first developed for safety applications in the mining industry, which is laid along cage walkways and can be laced through the netting. When twisted, compressed or stretched, it can activate a range of alarm systems, including a klaxon, warning lights, radio paging device and camera (see Anon, 1984a, for details). Radar has also been successfully employed at a number of larger salmon farms in the UK and will warn of anyone entering the site. Lockable lids are also incorporated into some smaller cage designs.

The reasons why poaching and vandalism are commonplace in many parts of the world are discussed in Beveridge (1984b). Sometimes socio-economic problems lie at the root of the problem. For example, conflict of interests over the utilisation of a resource can occur. In Laguna de Bay in the Philippines, the virtually unchecked expansion of the fish pen and fish cage industry which occurred during the late 1970s and early 1980s disrupted the traditional livelihoods such as fishing, duck rearing and even transport of the local people. Many of the fish pen owners were from outside the area and employed guards from Manila to protect their investments, thus leading to a great deal of friction and even violence between the local communities and the people associated with the fish pens. At other lakes in the Philippines, development was controlled and the villagers encouraged to set up their own farm units, aided by government loan schemes. In these areas, problems were largely avoided.

In some parts of the world, the penalties for poaching have been increased. In Alabama, USA, for example, a bill was introduced in the early 1980s to stiffen penalties for anyone convicted of catfish rustling. First time offenders can now be fined up to $1,000, and sentenced to up to one year in jail, whilst second-time offenders can be fined twice that amount, and must serve a mandatory 30 days in jail.

## 7.9 Predators

There has been a great deal of comment in the fish farming press concerning predation at fish farms, but there has been very little detailed study of the actual extent of the problem. There are thus no reliable figures of the numbers or species of predator involved, or the extent of the losses incurred at cage fish farms. A small survey carried out by Mills (1979) in Scotland showed that nearly 90% of fish farms (including cage fish farms) were visited at some time or other by piscivorous birds. However, as the study carried out by Meyer (1980, 1981) into heron predation at trout ponds has demonstrated, it is extremely difficult — even by careful observation — to arrive at a reliable estimate of numbers of fish taken. Whilst it is sometimes convenient for farmers to blame unexplained losses on predators, it must be accepted that many cage fish farms do have problems with predators.

The range of predators reported from cage fish farms is enormous, and includes squid, fish, turtles, lizards, sea snakes, birds and mammals (*Table 7.4*). Many of these are seasonal visitors. For example, cormorants (*Phalacracorax carbo*) in Scotland typically begin to arrive at freshwater cage fish farms in August, increase in numbers over the autumn and winter months, and disappear at the start of the breeding season in April. In a study of one freshwater cage farm on the west coast of Scotland, Ranson (1982) observed that the daily number of visiting cormorants increased to an average of 15 per day at the end of December, although on one occasion more than 70 birds were counted. A study of cormorant distribution by Mills (1965) has

*Fig 7.12* Mixer installed at Atlantic salmon farm to improve water flow through cages (courtesy J F Muir).

shown that the number of birds visiting freshwaters depends on the availability of food at sea, and thus, to some extent, the weather. A marked seasonal pattern in the number of herons (*Ardea cinerea*) arriving at fish farms has also been observed (Meyer, 1980).

Most predators, particularly birds and mammals, exhibit diurnal patterns of activity, preferring to attack their caged prey at dawn and dusk when staff are usually absent from the farm. Clear, calm nights, too, are often used by predators. Although there are great differences between one farm and another, and between different parts of the world, marine farms seem to have more problems, attracting the greatest range of predators. Most predators (though not all; see later) are attracted by the presence of the large concentrations of enclosed fish and try to gain access to the cages either from above or below the waterline.

Predators cause problems by:-

(i) killing or damaging fish
(ii) damaging nets
(iii) stressing fish so that feeding is disrupted and resistance to disease is reduced
(iv) spreading disease

Many of the predators listed in *Table 7.4* will consume fish, providing they can get access to the cage. If the mesh is big enough, squid, sea snakes and predatory fish will enter and consume as much as they can, although they are usually unable to get out again. At the end of one experiment carried out in the Philippines, involving the cage culture of milkfish, more than 12 species

*Table 7.4* Predators at cage fish farms.

| *Country* | *Cultured species* | *Predator* | *Comments* | *References* |
| --- | --- | --- | --- | --- |
| USA (Washington) | Salmon | Spiny dogfish (*Squalus acanthias*) | Particular problems with larger mesh sizes | Mahnken, 1975; Lindbergh, 1979 |
| USA (Maine) | Trout | Mink, otters | Used vinyl-coated metal mesh cages to deal with problem | Anon, 1981 |
| Canada (New Brunswick) | Salmon | Gulls (*L. argentatus* and *L. marinus*), Great blue herons (*A. herodias*), cormorants (*P. auritus*) | Large number of unrecorded losses attributed to birds | Henderson, 1980 |
| UK (Scotland) | Salmon and trout | Seals, otters, cormorants, mergansers, guillemots, gannet, goosander, herring and black gull | List compiled from 38 farms, both marine and freshwater | Mills, 1979 |
| UK (England) | Trout | Gulls, grebes, cormorants | Damage to nets by diving birds | Anon, 1978 |
| Norway | Salmon | Herons, gulls, cormorants, seals | More than 50% losses at some farms attributed to predators. Recommend anti-predator nets | Moller, 1979 |
| Israel | *Sparus* sp | Moray eels, squid | Lots of small fish also found in cages | Pitt *et al*, 1977 |
| Malaysia | Marble goby | Snakeheads, water snakes | Damage to nets and mortalities in freshwaters | Jee, 1978 |
| Malaysia | Grouper sp | Otter (*L. sumatrana*), gulls, puffer fish (*Sphaeroides* sp.) | Nets damaged by puffer fish. Dogs used to scare otters. Stout PE anti-predator nets introduced | Chua and Teng, 1980 |

of fish, including croakers and barracuda were discovered in the cages. These fish were possibly attracted to the cages for the same reasons that fish are attracted to FADs (Fish Aggregation Devices; see Beveridge, 1984b, for discussion), and must have entered the cages whilst small enough to pass through the meshes. Predators are also sometimes inadvertantly introduced when the cages are initially stocked or when nets are changed (Reksalegora, 1979).

Herons can take fish from unprotected cages by fishing from the walkways, whilst mink and otters can easily clamber over cage collars and into the net bags. Other attempts to catch fish, however, are more destructive. Seals, mink and otters will often grasp a fish through the sides of the cage causing considerable damage to nets and on occasion creating tears through which stock can disappear. Predatory fish can also on occasion damage nets (Jee, 1980). Chua and Teng (1980) reported that puffer fish (*Sphaeroides* sp.) damaged the net bags of cages of grouper resulting in 20% loss of stock, and on the Pacific coast of Canada dogfish are apparently a major cause of concern to fish farmers (J F Muir, pers. comm). The success of predators at catching fish in this way is to some extent dependent upon mesh size, larger mesh nets — and hence bigger and more valuable fish being more vulnerable (Mahnken, 1975). Often partly-consumed fish will be found stuck firmly in a mesh, whilst on other occasions, horrific wounds may be seen on the sides of a fish which has had a lucky escape. Ranson and Beveridge (1983) have suggested that diving birds, such as cormorants rarely manage to extract fish from a cage, although they can certainly wound a great many. During a survey of a freshwater rainbow trout cage farm in Scotland it was estimated that up to 6% of all caged fish showed signs of wounding by birds, at certain times of the year (Ranson and Beveridge, *ibid*). Diving birds have also been held responsible for damage to nets.

The attraction of birds and mammals to a cage farm can affect disease risks, since many animals play major roles in the life cycles of commercially important diseases. The predatory heron is an important intermediary host in the life cycle of the cestode *Diphyllobothrium* sp., a parasite which is prevalent at a number of UK cage trout farms, whilst otters (*Lutra* sp.) are known to act as final hosts for the digenean *Haplorchis*, a common parasite of tilapias (Roberts and Sommerville, 1982). Ranson and Beveridge (1983) speculated that the extent of scale loss and wounding on many of the fish attacked by birds (*Fig 7.13*) if not sufficient to kill them directly, have provided a site for invasion by fungi or bacteria.

Needham (1980) has also suggested that even unsuccessful attacks on caged Atlantic salmon can stress fish sufficiently to upset feeding and suppress their immune system thus increasing susceptibility to disease.

Farmers can and do, of course, retaliate. The use of top nets to deter gulls and herons, and antipredator nets to foil underwater attacks is common. Top nets are usually made from stout 5-10cm square mesh netting, treated to retard the effects of weathering. PE and nylon monofilament are probably superior to woven nylon; they are cheaper, and since they do not absorb water do not sag in the middle. It is perhaps stating the obvious, but top nets

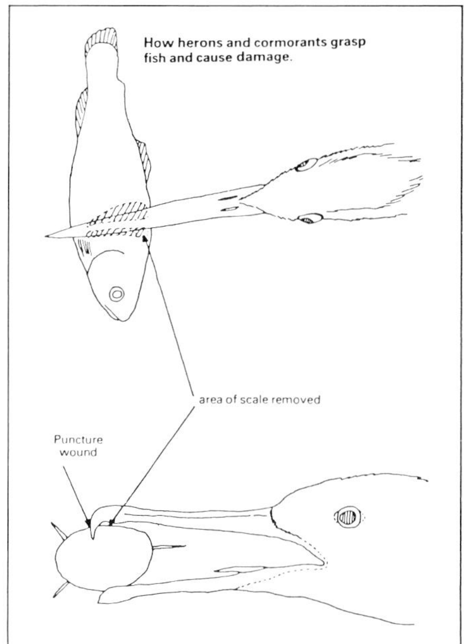


*Fig 7.13* Bird damage to fish (from Ranson and Beveridge, 1983).

must not only be secured firmly, but also should not be permitted to touch the water surface in the cage, as they can gill or damage fish. Groups of birds have also been observed standing on slack top nets thus weighing them down and gaining access to the fish. For large cages, a supporting pole mounted on a float in the middle of the cage is recommended.

Antipredator nets provide a considerable degree of protection against most underwater predators. The material, design, size and mesh size vary a great deal depending on the predatory species involved. In Scotland, most cage fish farmers generally use curtains of 10cm mesh PE nets, 5-10m deep, which can surround two cages and are fixed approximately 1-2m from the cages. However, on one farm nets with an 8cm mesh size and which surround each cage have had to be used, since grey seals (*Halichoerus grypus*) in the area have become an enormous problem and have learned to utilise the narrow spaces between cages to attack fish. Small mesh anti-predator nets are also popular in Shetland, where guillemots (*Uria aalgae*) are common predators. Tangle nets are sometimes used, although there are concerns about diving on farms where they are installed.

Many of the diving birds, such as cormorants, can dive to depths of 20 or 30m, and come up under the cages. Dogfish, too, are reported to scavange for dead fish by attacking the bottom of cages. However, to use nets which extend to these depths or which protect the cage bottom, is prohibitively expensive. Like the other nets in use on a cage fish farm, anti-predator nets require to be serviced regularly and antifouled if necessary.

Other measures commonly employed against predators include the use of dogs in tropical countries to deter otters and rats (IDRC/SEAFDEC, 1979). A number of deterrents, such as scarecrows, flashing lights and devices which emit loud noises or distress calls are marketed to try and frighten off birds and even seals, but to date these methods have had limited success, as most predatory species soon habituate to them (Meyer, 1980; Martin, 1982; Salmon and Conte, 1982). Live traps are sometimes employed against mink. Shooting is also resorted to by a number of farmers. However, there are two points to bear in mind with regard to shooting. First of all, it is illegal in many countries. In the United States, for example, all birds are protected by federal and state laws, and cannot be killed without both a permit issued by the Law Enforcement Division of the Fisheries and Wildlife Service and a State permit. Federal permits are only issued after all non-lethal control methods have been exhausted (see US Fisheries and Wildlife Service leaflet, 'Controlling birds at fish hatcheries'). Secondly, shooting often does not work. For example there are numerous accounts of how shooting cormorants makes no difference to the numbers feeding at a farm, since the visiting birds usually come from several different colonies (Ranson, 1982).

Much of the above discussion assumes that farmers are using flexible mesh cages. Rigid mesh cages are, of course, much superior in terms of resistance to damage and infiltration by predators (Hugenin *et al*, 1981), providing that they are fitted with adequate protection above the waterline. Operators using the more conventional flexible mesh cages, and who are troubled by predators are best advised to use appropriate top nets and anti-predator net curtains, and to keep these well-maintained. Cage nets should be rigged and weighted so that the netting remains taut, as this will reduce the success rate of attacks. Scavenging fish, such as the dogfish (*Squalus* sp.) are often attracted by dead fish on the cage bottoms, and so these should be removed at regular intervals.

## 7.10 Wastes

Results from a number of studies (*Table 7.5*) have demonstrated that the quantity of material sedimenting under fish cages at intensively managed farms is usually at least an order of magnitude higher than that recorded at control sites or in undisturbed water bodies (Kadowaki *et al*, 1978, 1980; Enell and Lof, 1983; Collins, 1983; Merican and Phillips, 1985; Phillips *et al*, 1985c). As most of this material is uneaten food (see Chapter 5), higher sedimentation rates are likely at the more intensively farmed units. However, there is enormous variability in the rate of waste build up due to local site conditions, species, feed type and management. The closer the cage bottom is to the lake or sea bed the greater the proportion of wastes that is likely to sediment directly under the cages and where prevailing currents are low, few of the wastes — especially the larger, denser waste feed particles — will be carried away. Anchoring rafts of cages on single-point moorings ( *Fig 3.31*) will result in the distribution of the wastes over a much greater area and hence greatly reduce the areal rate of waste accumulation (see Section 3.4.4).

Although feed losses associated with trash fish or moist diets are much

*Table 7.5* Waste sedimentation rates at cage fish farms.

| *Species* | *Country* | *Diet* | *Sedimentation rate* (*g dry matter* $m^{-2}d^{-1}$) | *Reference* |
|---|---|---|---|---|
| Rainbow trout | Sweden | dry/moist | range 17–26 | Enell and Lof, 1983 |
| ,, | Scotland | dry | mean 87 | Collins, 1983 |
| ,, | Scotland | dry | range 14–203 mean 16.43 | Merican and Phillips, 1985 |
| Yellowtail and sea bream | Japan | trash anchovies | range 4.1–5.9 | Kadowaki *et al* 1980 |
| ,, | Japan | trash mackerel | range 17.0–21.6 | Kadowaki *et al* 1980 |

higher than those associated with commercial dry diets (Warrer-Hanson, 1982), dry weight is only 20% (*cf* 85% dry weight for commercial dry pellets), thus offsetting waste inputs to the sediments as measured in terms of dry matter. Kadowaki *et al* (1980) and Nishimura (1983), noted that the sedimentation of solids varied with season at both caged yellowtail and red sea bream farms, the quantities of waste increasing with fish size and feeding rate. Merican and Phillips (1985) found that solid waste production was independent of fish size but was highly correlated to feeding rate, although there was some residual variation which the authors suggested was probably due to differences in size and type of pellet used, stocking density and fish health.

The sedimented wastes can adversely affect fish production in a number of ways. From organic pollution studies it is known that a small increase in the rate of supply of organic material to the sediments will stimulate bacterial, fungal, and macroinvertebrate activity thus causing an increase in the oxygen demand of the sediments. Increases in bacterial numbers and macroinvertebrates have been observed at a number of intensively farmed freshwater sites (Eley *et al*, 1972; Kilambi *et al*, 1976; Korzeniewski and Korzeniewski, 1982; Tsutsumi and Kikuch, 1983), whilst Enell (1982) and Enell and Lof (1983), in studies of laboratory-incubated cores, found that the rate of oxygen uptake was more than twice as high in cage sediments than those taken from a control site. Where the rate of waste sedimentation is high, the rate of oxygen supply may be insufficient to meet the respiratory requirements of the macrobenthic and microbial community. Hence the sediments become anoxic, the benthic community changes to low oxygen tolerant and anaerobic species, and the end result of biological and chemical activity is reduced inorganic and organic compounds such as lactate, ammonia, methane, hydrogen sulphide and reduced metal complexes. Evidence for anoxic conditions in sediments under fish cages has been reported in a number of studies (*eg* Arizono and Suiza, 1977; Braaten *et al*, 1983; Nishimura, 1983; Stewart, 1984).

The implications for fish health of a build-up of wastes under fish cages are not yet fully understood. High rates of oxygen consumption by sediments may cause deoxygenation of overlying water, as has been observed during the summer months at a number of Japanese yellowtail farms (Nishimura, 1983; Tsutsumi and Kikuch, 1983), and pose a threat to fish health during periods of low DO, especially if upwelling occurs. The hydrogen sulphide ($H_2S$) produced during anoxia undergoes a number of fates. Since it is highly soluble in water it is present in solution in sediment poor water. However, it is readily precipitated as ferrous sulphide, giving the anoxic sediments their characteristic black colouration. Some of the $H_2S$ may also be utilised by sulphur bacteria present at the sediment/water interface, whilst under certain conditions bubbles of $H_2S$ gas may be released from the sediments. Although $H_2S$ is known to be highly toxic to fish (*eg* Reynolds and Haines, 1980), it is suspected that much of the gas may be oxidised before it reaches the water surface (Lieffrig, 1985). Nevertheless, methane/$H_2S$ bubbles have been implicated in outbreaks of gill disease at Norwegian salmon farms (Braaten *et al*, 1983).

Another theory which remains untested is that there may be a build-up of facultative disease organisms in the sediments under cages, and that disease outbreaks at long-established farms may be more common (Kafuku and Ikenoue, 1983).

In summary, the relationship between conditions in the sediments and fish mortalities has been poorly studied and is very much open to speculation. There is, however, some evidence from Japan of a close link between the two. In a study of yellowtail farms, Arizono and Suizu (1977) demonstrated that major disease outbreaks, where more than 1% of stock were lost, were related to conditions in the sediments, which they measured as EI (Environmental Index), such that:-

$$EI = (TS/DO) \times 100;$$

where TS = concentration of sulphides in the mud (mg $g^{-1}$ dry mud), and DO = dissolved oxygen concentration of the water immediately above the sediments (ml $l^{-1}$) (*Fig 7.14*).

The waste-rich sediments also act as a major potential nutrient store, which under the appropriate conditions releases nitrogen and phosphorus into the overlying water, thus stimulating algal growth (see Chapter 5).

The accumulation of wastes under and around fish cages should therefore be of concern to fish farmers. Apart from careful site selection and prudent management, and the measures that the feed manufacturers may one day be able to implement (see Chapter 5), what can be done to minimise risks? The Norwegians suggest that the following are taken as signs of a growing problem (Braaten *et al* , 1983):-

(i) Periodic smell of sulphur and methane, especially during summer and autumn,
(ii) The appearance of gas bubbles at the water surface,
(iii) The rapid build-up of anoxic (black) sediments under cages, with

pockets of trapped gas. (Divers should periodically examine the muds under the cages).

(iv) Extensive gill damage. Gills may appear pale, sometimes with dark patches. Filaments are swollen, fused, or badly damaged.

If the accumulation of wastes under cages is perceived as being a major threat to fish production, then the wastes must either be removed from the vicinity of the caged fish, or the farm relocated at a new site. Scavenging red (*Chrysophrys major*) and black ( *Mylio macrocephalus*) sea breams have been stocked inside grouper cages in Hong Kong in an effort to reduce waste feed output (Tseng, 1983). The collection of solid wastes as they fall through the cages has been attempted on an experimental basis at a number of cage fish farms. In Poland, large filter-funnel shaped collectors completely enclosing the cage bottom were slung under rainbow trout cages and the sedimented wastes periodically (weekly — monthly) recovered by pump (Tucholski and Wojno, 1980, Tucholski *et al*, 1980a, 1980b). Results showed that 45% of solid wastes could be removed in this way, although the reductions in nitrogen and phosphorus loadings were only in the order of 15-20%.

In Sweden, a computer-controlled system was tested at a 20 tonne commercial rainbow trout cage farm (Enell *et al*, 1984). Large PVC funnels, similar in design to those used in Poland, were attached to the bottom of the fish cages. Immediately following feeding, submerged pumps were switched on for several minutes and the waste feed and faeces pumped to the surface where it was washed and filtered and the uneaten food separated and returned to the cages (*Fig 7.15*). Analysis of one season's results has shown that 71% of the phosphorus added to the system in the form of feed could be recovered,

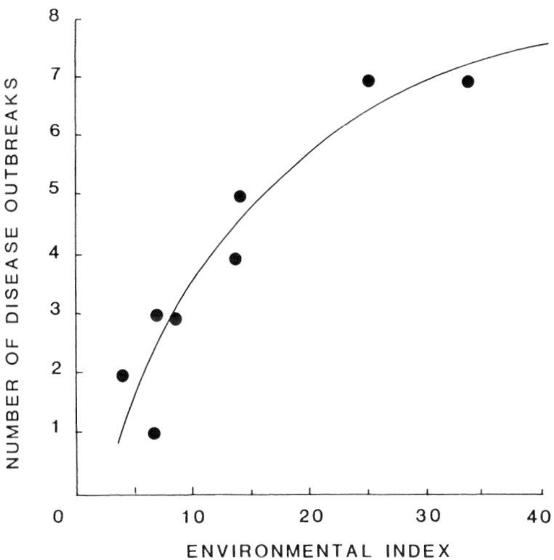


*Fig 7.14* Environmental Index versus number of disease outbreaks at Japanese cage fish farms (redrawn from Arizono *et al*, 1977).

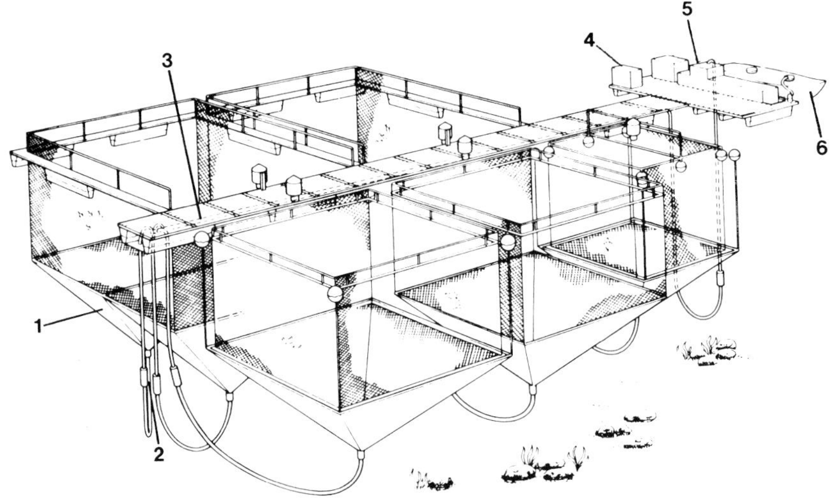


1 PVC CANVAS FUNNEL
2 AIR-LIFT PUMP
3 PIPE FOR TRANSPORTING WASTES/WATER TO FILTRATION UNIT
4 COMPUTER FOR CONTROLLING FEEDING, PUMPING, AND FILTERING OPERATIONS
5 FILTER UNIT
6 WASTE RECEIVING TANK

*Fig 7.15* Pilot computer controlled waste collection and uneaten feed retrieval system, Sweden (from Enell *et al*, 1984).

whilst savings in lost feed resulted in a reduction in feed costs of 25-30%. Cost benefit analysis suggests that the system could pay for itself within 2-3 years, although the assumption that the recovered feed is equal in quality and palatibility to the original product requires verification.

The periodic removal of sedimented wastes, using submersible pumps, has been carried out at a number of large commercial salmon farms in Norway for five or six years, and according to Braaten *et al* (1983) several companies specialising in waste removal have become established. The wastes, being rich in organic minerals and nutrients, may be used as fertilisers. However, although the method is claimed to be effective, it is difficult and not without risk, and it is suggested that all fish be removed from the vicinity prior to pumping.

Dispersal of wastes, using submerged electrically-driven mixers (*Fig 7.16*), has been employed at a number of Scandinavian farms, including the Austevoll Marine Aquaculture Station in Norway. Here Braaten *et al* (*ibid*) found that dispersal was most effective if the mixer was operated from a boat

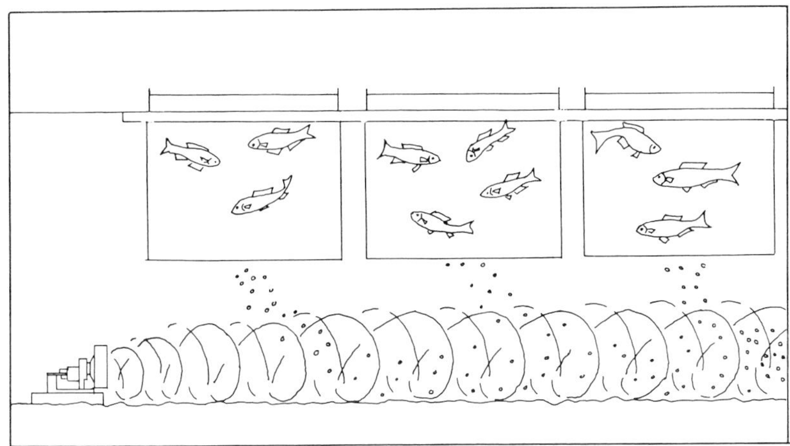

*Fig 7.16* Diagram of low-speed mixer installed on the bottom of a cage site to disperse wastes. The mixer can also be raised, or the angle changed to improve water flow through the cages.

so that it could be moved around the farm, and if the device could be positioned above the sea bed during operation. As a precautionary measure, cages in the immediate vicinity of the mixer were towed to an undisturbed site. In a three month trial period the accumulated 40cm of wastes were reduced by an estimated 60-75%, and the researchers concluded that the annual operational costs of preventing further accumulation at the 35 tonne production site would be around £1,800 (1983 prices).

The removal of fish cages, in order to let sediments recover, has also been advocated as a solution to the build up of wastes. However, this implies that the fish farmer has another site available or that he will be able to find one, if necessary. In most countries existing planning regulations do not take account of the possibility that a farmer may have to move to a new site after a period of time, or that he may have to alternate production between two sites. (Remember, too, that dispersal/removal of wastes may also require a standby site). However, in Norway at least, there are signs that the laws governing site concessions may be changed to take account of these needs.

Few studies of the efficacy or costs of the above measures have been published, and therefore it is difficult to advise which is best. As with most other problems, prevention is preferable to cure and the industry should concentrate efforts on reducing waste production rather than simply avoiding the consequences. Since there are few intensive cage farm sites more than a decade old it is impossible to pinpoint where or when troubles will occur. Some of the more intensively managed sites with poor water circulation in Japan and Europe have developed problems within 5-10 years. Moreover, it is not known if simply leaving culture grounds fallow is sufficient to effect recovery, or if sediment removal methods must also be employed.

The economic benefits of waste entrapment have not been satisfactorily proven, but deserve further study. The use of mixers to disperse wastes is best reserved for marine sites, since resuspension of wastes in lakes and reservoirs would be more likely to aggravate problems through the stimulation of algal growth.

## 7.11 Weather and climate

### 7.11.1 Rough weather conditions

Certain areas suffer more violent storms than others by virtue of their geographical location or local topography (see Section 4.3). During storms, high winds blowing over exposed stretches of water will cause wave heights to increase and wavelengths to decrease. The resulting wave action can not only damage cage structures and moorings, it can also injure fish through the pitching motion of nets, and can make routine jobs such as net changing and harvesting extremely hazardous. In some parts of the world the best marine sites have already been developed, and if more exposed locations are to be utilised, then either the environment must be modified so that the effects of waves are reduced, or cages which are designed to withstand rough conditions or which can take evasive action by submerging during adverse weather, must be used.

Breakwaters are often used to reduce the effects of waves on coastal marine installations such as harbours and marinas, and a number have also been installed to protect coastal aquaculture projects in various parts of the world (Tazaki *et al*, 1975; FAO, 1977; Kogan and Romanycheva, 1979). There are two basic types; fixed bottom-resting and floating. The former are usually substantial concrete or stone-built structures which work principally by reflecting wave energy. They have an extended lifespan and low maintenance costs, but are extremely expensive, costing at least £1,200 $m^{-1}$ (extrapolated from De Young, 1978, to 1985 prices), and suffer from a number of other disadvantages from a cage fish farming point of view:-

- (i) They reduce or interfere with prevailing currents;
- (ii) They are largely restricted to shallow waters, otherwise costs become prohibitive;
- (iii) They are difficult to install where bottom sediments are soft;
- (iv) Once constructed they are difficult to modify to meet changing circumstances;
- (v) They are 'over-engineered' if protection is only required for short periods during the year.

Floating breakwaters, on the other hand, have been suggested by a number of authors as being much more suited to the requirements of cage fish farming; they are relatively inexpensive to build can be moored in deep or shallow water, do not interfere with currents and can easily be modified as the farm expands or cages are moved (Tazaki *et al*, 1975; McGregor, 1978; Kato, 1979; Kerr *et al*, 1980).

Floating breakwaters redistribute wave energy in a number of ways (McGregor, 1978):-

(i) reflection
(ii) dissipation within the structure
(iii) transmission of unsupressed energy
(iv) diffraction (bending) of the wave train passing the end of the breakwater, into the sheltered lee of the structure
(v) generation of waves by the movement of the breakwater itself, which interfere with the incident waves.

The relative importance of each of these effects depends on the design. At one extreme are floating pontoon designs which reflect most of the energy, whilst at the opposite extreme are floating sheets, which absorb and dissipate almost all the energy by friction (Kato *et al*, 1979). Floating pontoons, whilst very effective are difficult to anchor securely, and floating sheets are impractical since in order to obtain a sufficient damping effect it is necessary for the sheet to be more than ten times the length of the incident waves. Floating structures which incorporate the properties of the two designs (*ie* redistribute incident wave energy both by reflection and absorption) without suffering from either's disadvantages would therefore be ideal.

Much research and development has thus gone into floating breakwater design and a number of commercially built structures are now available. In Japan, for example, Bridgestone breakwaters, developed over the past 15 years, are now in use in several areas protecting aquaculture projects. The design consists of floating synthetic rubber-filled fibreglass modules (12 × 7m or 10 × 7m) linked together as shown in *Fig 7.17* on the windward side of the fish farm. Effectiveness depends on incident wavelength and how the modules are deployed.

*Fig 7.17* Bridgestone breakwaters, Japan (courtesy Professor J Katoh).

Details of commercial floating breakwaters and how they should be installed can be obtained from the manufacturers.

One type of floating breakwater which has excited a great deal of interest is the floating tyre breakwater or FTB. FTBs were first developed in the United States in the early 1970s. Although a number of designs which differ in flexibility have been proposed, the most widely used today are flexible mats based on the Kowalski or Goodyear module (see Kowalski, 1974, 1976). In this design, 18 tyres are tied together to form a module which floats with all tyres in the vertical plane (*Fig 7.18a*). Air trapped in the crown of the tyres gives each module sufficient buoyancy to support the weight of a man. Movement of the modules caused by wave action is usually enough to replenish the air pockets and keep the structure afloat, unless prolonged periods of calm occur during which trapped air may slowly dissolve in the sea-water. Lacerations and heavy fouling may also adversely affect buoyancy.

Both truck/tractor and car tyres can be used to construct the modules. Because truck/tractor tyres are considerably larger, fewer modules are required to effect comparable wave attenuation and since wave attenuation is to some extent dependent upon draft, they are probably more effective than car tyres in deeper (>10m) water. However, they are less buoyant since most truck and tractor tyres are tubed and do not have the airtight butyl or chlorobutyl lining which car tyres possess, and the heavy duty steel beading used around the rims makes them much more abrasive and thus they are much more likely to cause problems with tying materials (see later). Moreover they are heavier and consequently more difficult to handle. The final choice may sometimes be indicated from computer simulations of wave characteristics at the site (see later), but more often than not is dictated by availability and economics.

Nylon, polypropylene and dacron ropes, stainless steel wire, closed and open link galvanised chain and conveyer belting have all been used as tying materials. However, a number of trials conducted at cage fish farms on the west coast of Scotland have demonstrated that ropes tend to abrade and wear out very quickly and that either strips of 3-ply synthetic rubber conveyer belting, 35mm wide, or 8mm galvanised chain are best. Chain is expensive, heavy and prone to wear inside the links, whilst the conveyer belting which may be obtained as scrap, is considerably less durable, two of the three brands used by McGregor and Gilbert (1983) to link modules having snapped through fatigue within six months of installation.

De Young (1978) describes a simple to manufacture tyre rack which can be used to aid construction of the modules. However, he estimates that two workers can put together a module in 5 to 10 minutes simply by stacking the tyres on the ground via a 3-2-3-2-3-2-3 configuration, standing on top, and securing the bindings in such a way that the module is rigid without the tyres being distorted. Greased high-tension shackles or closing links work best with chain, whilst black nylon or stainless steel bolts or shackles may be used to secure the conveyor belting (*Fig 7.19*). (*NB* There have been problems with corrosion of the stainless steel washers).

Assembly of the FTB from modules requires that the four outside tyres be swung out as shown in *Fig 7.18c,d* and four individual tyres are then required to link a group of four modules together. Experience suggests that small sections should be put together on land, launched and linked up with other units in shallow water, before being towed to the site and ultimately connected to other completed sections. The size of the sections completed on land is dictated by available machine power. A 4 x 10 module unit can easily be towed by a boat fitted with a 40 hp engine during calm weather.

Mooring the FTB can be achieved in a number of ways. One suggestion is to secure a windward corner of a completed section to an anchor and anchorline, tow it to the site using the anchorline as towrope, and to drop the anchor when the section floats into the correct position. The section can then be secured with as many anchors as necessary. The second section is then towed into position and attached to the first prior to the anchors being dropped. This procedure is then repeated until all the sections are in position.

The actual design of any FTB is site-specific, depending on the wave height, wavelengths and wave periods generated over various fetches at the site, the size and orientation of the farm and the size and type of cages used. It must also take account of any navigational regulations. A number of specialist consultant engineers are able to carry out design trials for a particular site using wave tanks, model tyres and a wave generator which can simulate the wave spectra of the site based on the appropriate meterological and topographical data. FTBs developed in this way have performed well,

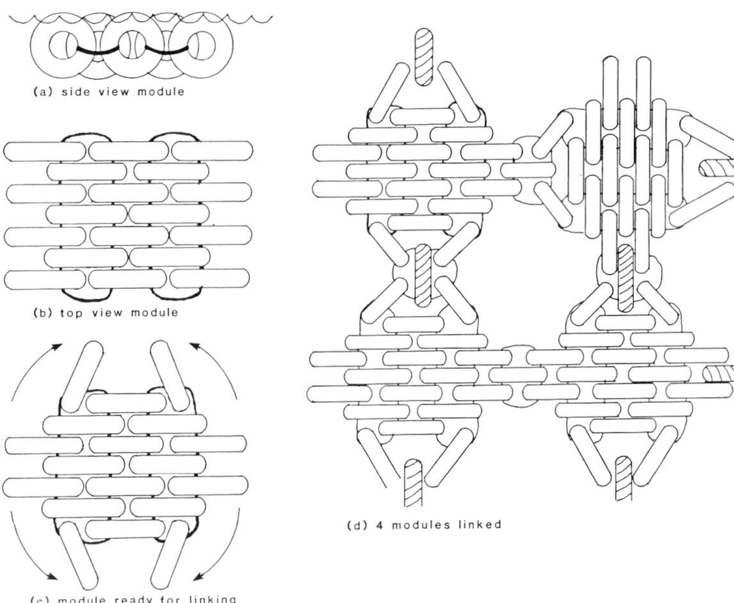


*Fig 7.18* Floating tyre breakwater (redrawn from several sources).

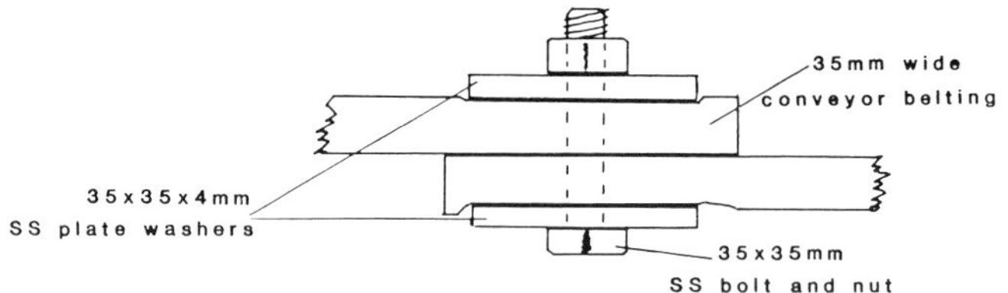


*Fig 7.19* Method of joining scrap conveyor belt ties (modified from WFA design).

and details of the design procedures involved are given in McGregor (1978), and McGregor and Gilbert (1983). However, it is also possible for a farmer to improvise an FTB himself, based on a number of assumptions. Because the breadth of an FTB largely determines its wave dampening characteristics, and since the most destructive wavelengths are between 0.5 and 1.25 times the length of the structure (see earlier), as a general guide Kerr *et al* (1980) suggest that the breadth of the breakwater should be around 1.5 times the width of the structure to be protected. Thus, a group of conventional 4-tonne production cages, exposed to gale force winds over an effective fetch of 3-4km should be 4-5 modules in breadth. The length of the FTB should at least be as long as the cage raft(s) it is protecting, since there will be some refraction around the edges of the breakwater.

The anchoring system is designed to withstand the peak mooring load, and depends upon the breadth and draft of the structure, the depth of water, wave steepness (the ratio of wave height : wavelength) and substrate. Numerous methods of mooring any marine structure can be found in the literature (see Section 4.4), and details relative to FTBs are given in De Young (1978). McGregor (1978) has estimated that average mooring loads are approximately 250N $m^{-1}$, with extreme values around five times this value. As an indication of what this means in practical terms, McGregor and Gilbert (1983) used 250kg block anchors every 9m on the windward side of a 3 module-broad FTB anchored in soft sediments in shallow water, whilst De Young (1978) estimated around twice that holding power was necessary for an 8 module-broad structure installed in a similar depth of water. Anchorage on the leeward side should be around 20-25% that on the windward side (see Section 4.4 for details of anchor types). A buoy is placed at the top of each riser chain for support and to facilitate removal and maintenance of the structure. Each buoy should be connected to at least two modules in order to distribute the mooring load as evenly as possible (see *Fig 7.20*). The FTBs should be well marked and navigation lights installed if necessary.

The best protection is achieved by orienting the breakwater at 90° with respect to the most destructive waves. However, if more than one fetch is involved a compromise must inevitably be chosen. As a general guide, De Young (1978) recommends that the cages are moored on the leeward side within four wavelengths of the breakwater, although the actual position at the FTB relative to the farm is best decided by trial and error. One advantage of the FTB is that it can be moved relatively easily. Assuming that the tyres are scrap, the initial cost of an FTB works out at around 10-20% that of a fixed breakwater, depending on the breadth, the type of anchors and tying material used and excluding the costs of transport. However, maintenance costs can be heavy (up to 30% capital outlay per annum) as the ties inevitably wear out and fail, and so it is recommended that moorings and ties be checked every month or so, and that the whole structure be beached regularly — each year — and serviced. Fouling should be removed, and any badly damaged tyres or ties replaced. Modules which lose their buoyancy and sink can be temporarily refloated using compressed air, and the crowns of leaking tyres filled with high density polyurethane or polyethylene foam (McGregor and Gilbert, 1983). There may also be problems during winter with the formation of ice on the leeward side of the structure, and in severe climates the movement of ice floes has caused a great deal of damage. If desired, maintenance of the FTB can be assigned to a contractor. However, his responsibilities should be clearly defined; he should be immediately available for any unscheduled maintenance or repair work that may be necessary, and he should keep accurate records of all work that is carried out.

A number of manuals on the design and construction of FTBs may be recommended, including De Young (1978) and Bishop (1980). Several cage manufacturing companies also offer breakwater construction and installation services.

Another method for modifying waves is discussed briefly by Milne (1970, 1979) and involves using a pneumatic barrier, produced by pumping compressed air through a submerged perforated pipeline (*Fig 7.21*). Limited trials conducted by Milne demonstrated that both wave height and turbulence could be ameliorated to some extent, although temperature/salinity profiles were also disrupted. To date, there have been no further developments of this system for protection of cages.

The alternatives to using FTBs or bubble curtains are to use tougher cages

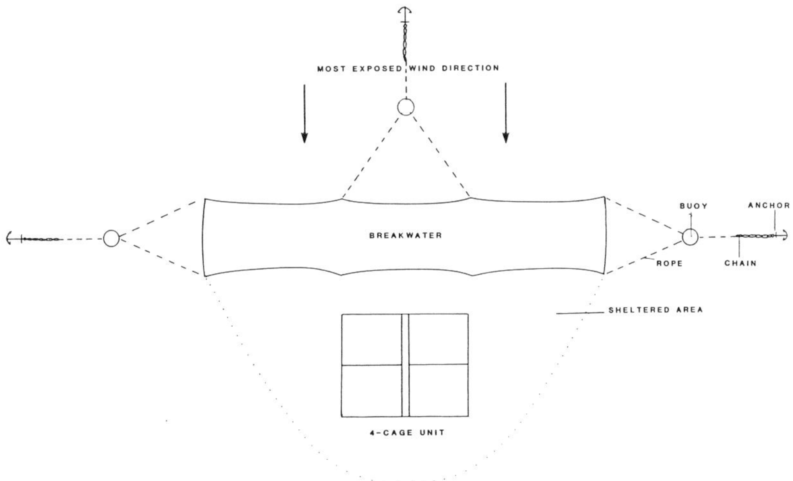


*Fig 7.20* Suggested mooring scheme for FTB.

or cages which can be lowered below the surface during storms. There are two basic solutions to building cages which can withstand rough seas — either flexible collar designs, or rigid collar cages built to high specifications using strong materials. Flexible collar cages of the type shown in *Fig 7.22* have been used for a number of years in Japan, where they were first developed, and more recently in the USSR, western Europe and North America. The commercially-available designs are constructed from 10m,

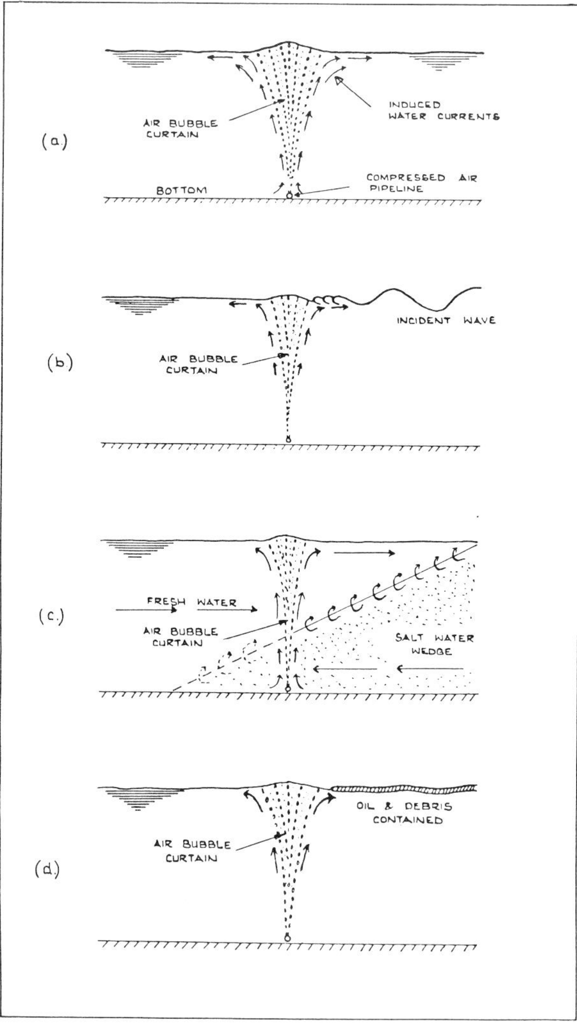


*Fig 7.21* Pneumatic barriers used (a) to induce water currents (b) to reduce wave heights (c) to prevent incursion of salt water, and (d) to contain oil and debris (from Milne, 1972).

14m or 16m lengths of synthetic rubber, 420mm in diameter, which are simply bolted together to form a 6, 8 or 10-sided collar from which a cage net is suspended. The rubber units are hollow and are filled with gas at high pressure (7kg $cm^{-2}$) to aid flotation and increase rigidity. The cage is supplied complete with floats, stanchions, handrails and mooring ropes. In rough weather conditions the huge rubber frames ride out heavy seas by absorbing and dissipating the wave energy through internal friction in much the same way as an FTB functions (see earlier). Anchoring can be a problem. The cage installed by one fish farm in Ireland was eventually secured by 6 $\times$ 10 tonne scrap steel block anchors, one at each corner. Several flexible frame cages in use on the east coast of Japan are reported to have survived 7m waves and winds of 29m $s^{-1}$ which occurred during a typhoon. The surface area of the largest structures which can be built from the sections is over 1,000$m^2$, and according to the manufacturer, performance in adverse weather conditions is independent of size.

A number of northern European companies have designed rigid collar cages fitted with nylon net bags for use at relatively exposed locations. They are best described as industrial units and owe much to technology developed in the ship-building and oil industries. Most are constructed from sections of welded, surface treated steel plate and come fitted with accommodation, storage for feed, a slaughterhouse, and cold store facilities. Some have their own power source, and so can operate independently of shore facilities. Cranes and/or fork lift trucks are used to assist in the routine chores of net changing, harvesting *etc*.

The designs are based on empirical concepts rather than on well-established principles. It is postulated that because of the strength of the

*Fig 7.22* Bridgestone flexible frame cage (courtesy B Whelan).

materials used the structure can withstand the forces exerted during rough weather. However, because they transmit much of the absorbed wave energy to the moorings (more than flexible collar designs will), the anchoring system is extremely important. Few of these units have actually been constructed or have been in use for very long. Those that have, have worked well, albeit at not particularly exposed sites.

Rigid copper alloy mesh cages are also reported to be extremely resilient to storm damage and several are currently in use at some of the more exposed North American locations (Woods Hole Engineering Associates, 1984).

Submersible designs are in commercial use in a number of countries, sometimes as a strategy against stormy conditions, and sometimes to avoid problems with ice (see later). There are both fixed and floating types. Fixed submersible cages are used in several lakes in the Philippines to culture tilapias (*Fig 8.18*). The cages are fitted with top nets, and when a storm warning is received, the rigging is untied, the cages pushed below the surface and secured a metre or so under the water. The posts which anchor the cage to the substrate and project above the surface offer little resistance to waves.

In Japan, floating submersible cages are used to culture yellowtail, red or black seabream at some of the more exposed sites. In a typical design, the headrope of the cage is attached to mooring buoys on the surface, and during normal conditions these are fixed so that the cage remains 2-5m below the surface. Fish are fed via a feeding tube. During storm conditions the feeding tube is tied shut and the ropes are lowered so that the cage is submerged to a depth of 10m or so (*Fig 7.23*). According to Fujiya (1979), some of the submerged units are as large as 100m$^2$, and are congregated together in groups of 10-12.

A more sophisticated type of submersible cage has been developed and tested in Japan by Tomi *et al* (1979) and involves the use of variable buoyancy synthetic rubber floats (*Fig 7.24a*) which can be filled or emptied with compressed air or seawater from the surface. Floats are linked together by high compression rubber hoseline and the cages laid out in a grid formation as shown in *Fig 7.24b*. Cages remain on the surface and are only submerged during periods of rough weather. However, problems of net deformation due to deep water currents were encountered.

Submersible and submerged floating cages have also been used at marine sites in Europe to culture salmonids, sea breams, and sea bass (Boeuf and Harache, 1980; Woods Hole Engineering Associates Inc., 1984; Cook *et al*, 1984). In France two designs appeared during the late 1970s/early 1980s. One type, in use for several years at three rainbow trout farms in Brittany, involved raising the 64m$^3$ volume cages each day to feed the fish, and then lowering them again to a depth of 15-20m. The second design, once used in Belle Isle, Brittany, involved the maintenance of rainbow trout in cages submerged at 20m. Shortly after the trout were introduced to the cages, it was observed that many appeared 'distorted'. The problem was solved by submerging a large air bubble in a bell, presumably enabling fish to replenish the air in their swim bladders (G. Boeuf, pers. comm).

In Martinique, sea bass, sea bream and sar (*Diplodus sargus* ) have been

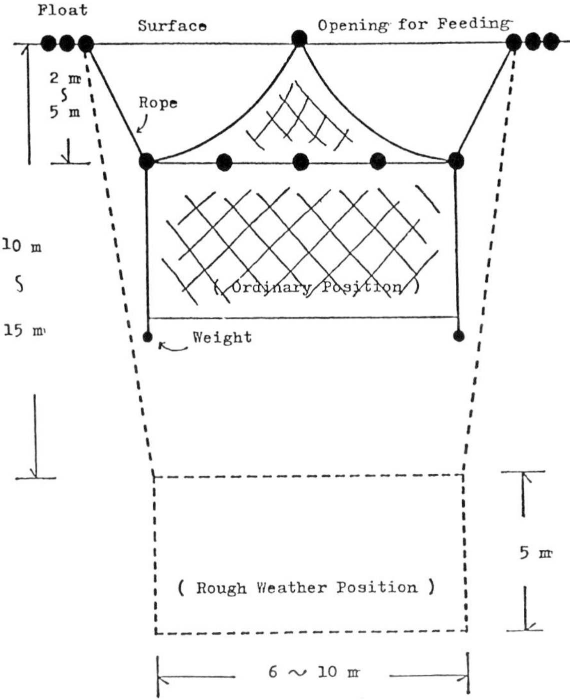


*Fig 7.23* Submersible cage for yellowtail (from Fujiya, 1979).

cultured in rotating cages which, if a typhoon occurs, may be swiftly sunk to the sea bottom (René, 1984).

The reason for using submerged cages is that the water movements produced by waves are largely a surface phenomenon. As a deep water wave moves over the sea, the water particles do not move forward with the wave, but instead rotate in circular orbits (*Fig 7.25*). At the surface, the diameter of the orbit is equal to H, the wave height. However, the orbit of the particles decreases exponentially with depth according to the formula:-

$$D_Z = H \exp(2\ z/L);$$

where $D_Z$ = diameter of the orbit; z = depth; and L = wavelength (Pond and Pickford, 1978). At a depth of around L/9, $D_Z$ is approximately halved, and at L/2, $D_Z$ has decreased to 0.04 H. Thus submerging cages, particularly during storms when the waves increase in steepness (wave height : wave length ratio), can dramatically reduce wave forces and the pitching motion of the cages.

It is extremely difficult to support any one solution to the problems of farming fish in cages in exposed locations, partly because the structures themselves have only been tested in limited conditions, and partly because of economic considerations. Despite various successes some doubts have also been expressed as to the suitability of submerged designs for all fish species. It would therefore seem best that each site be appraised in terms of the frequency and severity of storms and the cost of the various options available. In the end a balance must be struck between choosing a design with an adequate margin of safety and the cost of overengineering for rare and extraordinary conditions.

### 7.11.2 Ice

Ice can be a problem in both marine and freshwater environments where fish are grown in cages throughout the year. Ice which occurs in many maritime temperate lakes and reservoirs for short periods during the winter months rarely presents a problem, since it is relatively thin and can easily be broken up and dispersed. Often, the movement of the fishes in the cages helps keep the environment ice free (Koops, 1976; Paetsch, 1977). Alternatively, an air lift or submerged propellor device may be used to transport warm, deep water up to the surface, and thus keep ice at bay (see Anon, 1984) (*Fig 7.12*). Care must be taken to regulate the flow so as to avoid splashing at the surface which can cause severe icing of overlying antipredator nets (Kiekhafer, 1983). Disruption of sedimented wastes may also be a problem with this type of system. Drifting ice is rarely a problem in these environments.

In many of the continental lakes of central Europe and central North America, however, thick ice cover is present for several months during the winter. Although staff can keep the area immediately around the cages

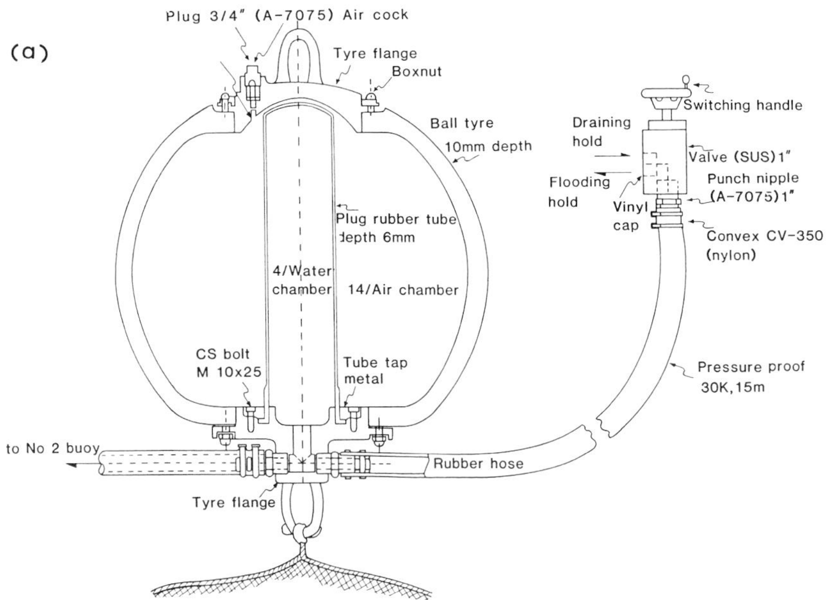


*Fig 7.24(a)* Diagram of submersible float (from Tomi *et al*, 1978).

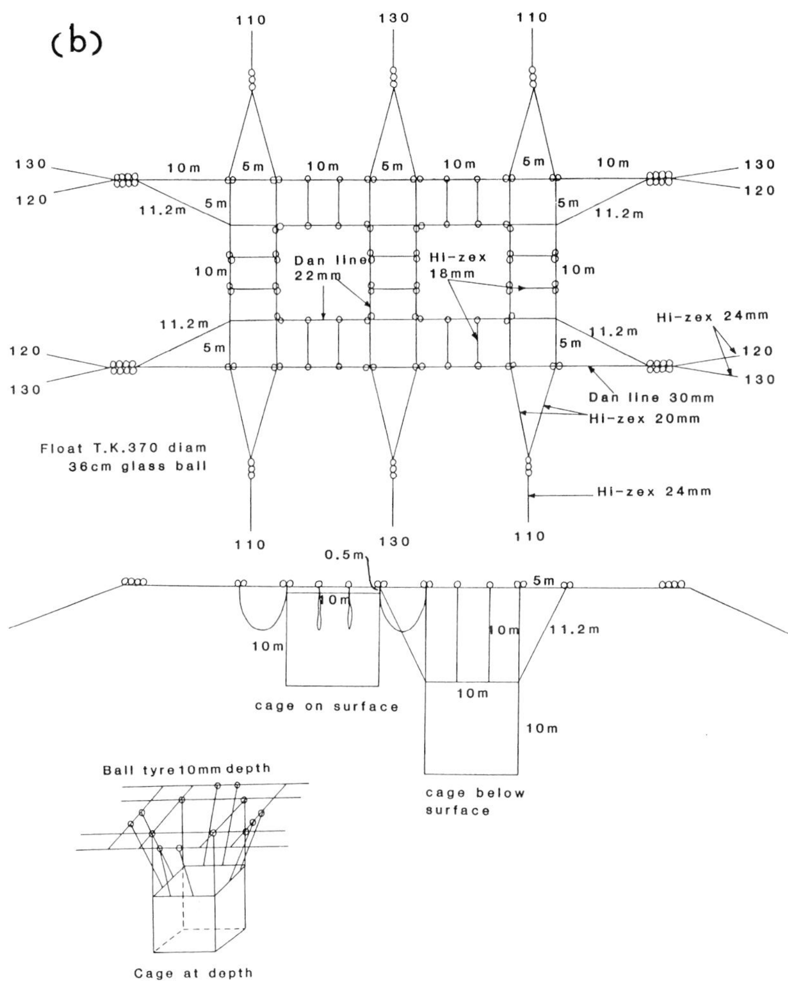


*Fig 7.24(b)* Layout of cages and deployment of submersible floats (from Tomi *et al*, 1978).

ice-free, problems occur in the spring when strong, warm winds cause the rest of the ice to break up and move. A layer of ice, 10cm thick and covering 1ha, will weigh 1,000 tonnes, and can easily crush cages trapped in its path. Sheets of ice frozen around mooring lines can also drag anchors and completely disrupt moorings when the ice begins to break up and drift. One solution is to site the cages on the windward side of the lake so that ice is driven away from the cages by the prevailing winds. An alternative solution is to adopt the system used in the Federal Republic of Germany and several Eastern European countries. Using iron bars and anchors which are deliberately trapped in the ice, large sheets surrounding the cages are

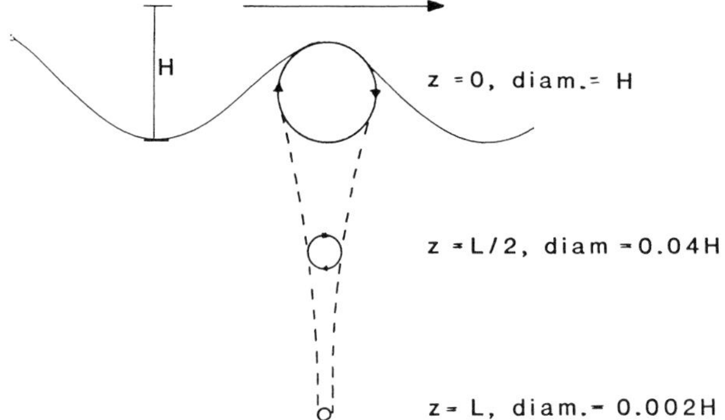


*Fig 7.25* Rotational properties of water particles in a wave (redrawn from Pond and Pickard, 1978).

anchored to trees on the shore and held secure until the ice becomes so thin that it is no longer a problem (Behrendt, 1984). A similar technique is described by Kieckhafer (1983). Temperatures inside cages in lakes with winter ice cover are obviously extremely low, resulting in poor or even negative growth. In the USSR, two types of submersible cage are used to avoid ice problems entirely; one for fishes, such as rainbow trout, which require feeding during the winter months, and another design for fishes which do not (*eg* catfish, carp) (Martyshev, 1983). In the autumn, prior to ice formation, rainbow trout are transferred to cages of the type shown in *Fig 7.26*. The cages measure 4 × 4 × 1m, and are fitted with a 1 × 1 × 1.2m plexiglass funnel which serves not only to support the cage when the ice cover has formed, but also as a window through which light and air can penetrate, and food can be delivered. It is apparently important that the top of the plexiglass funnel be kept free from ice and snow. A cage of the above dimensions can, according to Martyshev (*ibid.*), hold 50,000 10-15g trout (31-47kg $m^{-3}$) throughout the winter. The fish are fed with a vitamin-enriched diet at a rate of 1% body weight, twice per week. The cages are freed from the ice prior to the spring thaw to prevent them being damaged by drifting ice.

The design shown in *Fig 7.27* is used to overwinter carps and catfishes at a depth of 1-1.2m below the surface of the ice.

The actual freezing of marine sites is rarely a problem, unless the farm is located in a fjord or estuary with a substantial freshwater input, where due to differences in density inflowing freshwater can flow over the surface of the underlying saltwater and freeze. This has happened at several sites on the west coast of Scotland, and on one occasion, a fishing boat had to be employed to break up the ice.

Drifting ice floes pose more of a threat (see Chapter 4), and have been responsible for a number of insurance claims (Secretan, 1980). To date, farmers have either accepted the risks, or have moved cages before winter has set in. Some sort of floating boom may be of use in deflecting ice, providing that the ice is sufficiently broken up. There has also been some interest

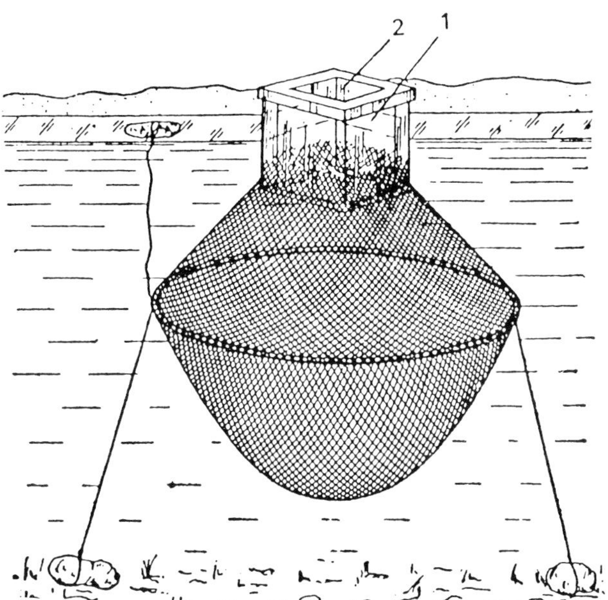


*1*—funnel 1 m × 1 m, height 1.2 m, from which cage suspended; in winter funnel freezes in ice and fish are fed through an opening in it; *2*—abatjour of plexiglass for penetration of light in cage.

*Fig 7.26* Submerged trout cage, USSR (from Martyshev, 1983).

expressed in the use of submersible or submerged cages to overwinter salmon in areas prone to storm and ice damage. However, whilst such cages have proved successful in the culture of yellowtail and sea breams, there are some doubts concerning their suitability for salmon (see Section 7.11.1).

### 7.11.3 Light and Weathering

The prevailing light levels at a farm are something that all fish farmers should be aware of, both because light is an important component in the fish's environment and because of its effects on many of the materials used at the farm.

The light reaching the outer edge of the earth's atmosphere comes mainly from the sun, and has a broad and unequal spectral distribution (*Fig 7.28*). As it passes through the atmosphere, there is a selective scattering and absorption of wavelengths, particularly at either end of the spectrum, so that the light reaching the ground is greatly modified. The intensity and quality of light arriving at any particular point is determined by the angle of incident light, the distance through the atmosphere it must travel, and prevailing atmospheric conditions, resulting in geographic, seasonal, and even local

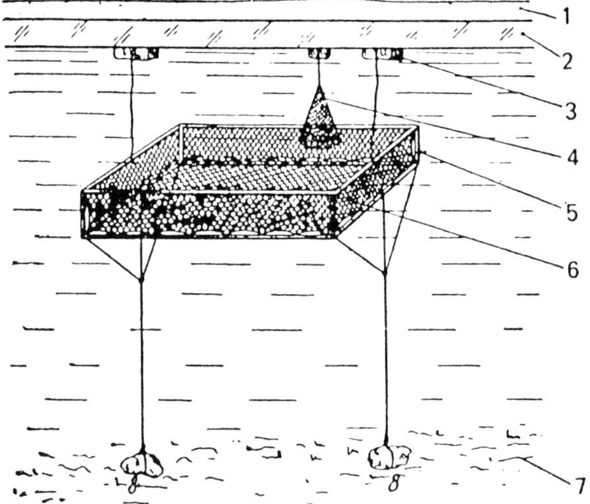


*1*—snow; *2*—ice; *3*—floats; *4*—sleeve; *5*—frame of box;
*6*—net; *7*—bottom; *8*—sinkers.

*Fig 7.27* Submerged carp/catfish cage, USSR (from Martyshev, 1983).

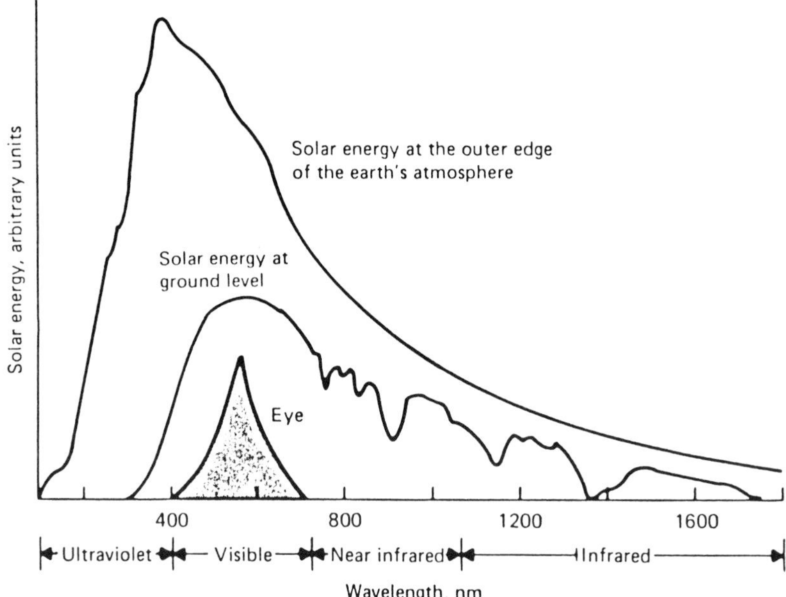


*Fig 7.28* Solar energy as it reaches the earth's atmosphere at ground level. The black area shows the sensitivity of the human eye (400-700nm). Thermal radiation (not shown) is at about 5,000 to 14,000 nm and together with infrared radiation accounts for about half the solar energy at the lake surface (from Goldman and Horne, 1983).

variations. In general, however, ground level light intensity is much higher in the tropics than in the mid or low latitudes (*Fig 7.29*, see Straskraba, 1980, for discussion) and increases in a linear fashion with altitude (see later).

The most damaging wavelengths in terms of the materials used in aquaculture and on fish and other living organisms are at the lower (*ie* ultraviolet) end of the electromagnetic spectrum. Ultraviolet radiation (UVR) is harmful principally because of the comparatively high energy photons that it consists of. Photobiologists usually divide UVR into three sections in order of decreasing wavelengths; UV-A (400 - 320nm), UV-B (320 - 280nm) and UV-C (280 - 200nm), on the basis of their physical properties and biological effect (Parrish *et al*, 1978). Only UV-A and UV-B reach the earth's surface,

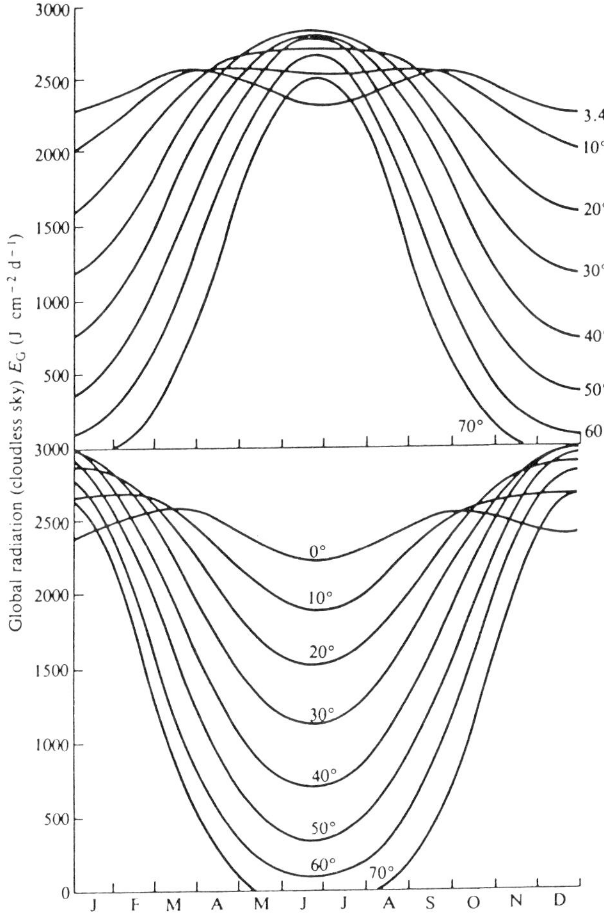


*Fig 7.29* Annual variations of daily integrals of global radiation reaching the ground on clear days. Northern hemisphere — upper panel; southern hemisphere — lower panel (from Straskraba, 1980).

the former in much greater quantities. However it is the group which encompasses the shorter wavelengths (*ie* UV-B) which is by far the more damaging.

Both the cage collar and any netting components which appear above the waterline — top nets, exposed portions of cage bags, *etc* — are prone to weathering, which is a term covering the combined effects of UVR, rain, wind, industrial smokes and gases and microbial attack, although of these factors, UVR is reckoned to have the single greatest deleterious effect (Klust, 1982). The various synthetic materials used differ in their susceptibility. A number of tests designed to measure the deterioration in breaking strength of undyed netting materials, have been carried out in different parts of the world and under different conditions, and in a league table compiled from the results, PVC fibre comes out on top, being extremely resilient, whilst PA (nylon) monofilaments and PVA, PVC and PES fibres show moderate resistance. Normal nylon fibres (continuous filament and staple) have a resistance similar to cotton and other natural fibres, whilst PP filaments, even with the inclusion of antioxidants and radiation absorbers, come out at the bottom of the list (Little and Parsons, 1967; Klust, 1982). Light absorbers, incidentally, are also included in many other synthetic fibres such as PE.

The lifespan of most netting materials can be further extended by the addition of dyestuffs which absorb the light. The dyeing process is carried out by the fibre producers, netmaker or by the fish farmer himself, depending on the materials used. PE and PP are usually dyed by the fibre manufacturer whilst nylon netting can be treated either by the farmer or by the net manufacturer. A group of dyestuffs commonly used to treat nylon netting is coal derivatives, such as bitumen or coal tar, which in addition to retarding fouling is also believed to improve resistance to degradation by light. Unfortunately net stiffness is also increased, thus increasing the risk of abrasive damage to caged stock. In most temperate countries, however, where fouling is a greater cause for concern to fish farmers than the effects of weathering, nets are more commonly treated with antifouling compounds than dyed. New Zealand is an exception where, as a result of the high prevailing UVR, most nylon netting used for cages is dyed black.

Fibre manufacturers often add dyestuffs to PE and PP polymers before extrusion as monofilaments. The pigments used are very fine particles ( 1μm) of synthetic compounds which, once incorporated, cannot be washed out. Not all dyes help retard weathering. In *Fig 7.30*, the breaking strains of various dyed and undyed netting yarns, after exposure to 1500h of sun, is shown. PE samples dyed orange and green were relatively well protected, whilst the orange and blue pigments which were used to treat the PP samples acted not as light absorbers but instead had a photocatalytic effect, thus accelerating deterioration (Klust, 1982). The colour of synthetic fibre ropes is of less concern, since they do not readily succumb to photodegradation, the harmful UVR being absorbed in the outer few millimeters.

Polystyrene is another material commonly used in cage construction which suffers badly from weathering (*Fig 7.31*). Although it is highly resistant to microbial attack, it is prone to photodegradation due to impurities, such as

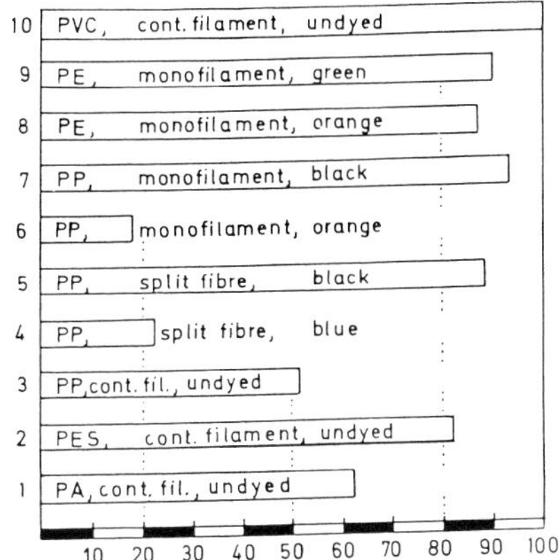


*Fig 7.30* Remaining breaking strength of netting yarns (of approximately 2-3mm diameter) in percentage of the initial breaking strength after exposure to 1,500 sunhours (from Klust, 1982).

*Fig 7.31* Block of polystyrene, following several years immersion at a temperate, marine site.

acetophone groups, which are formed by thermal oxidation during the manufacture of the polymer (Geuskens, 1975). Under intense sunlight, the outer surface becomes yellow and brittle and is easily eroded, exposing the underlying material to UVR. However weathering is also due to physical damage by wind, waves *etc*, and resistance to this form of attack is largely dependent upon how the material was manufactured. Where weathering is a severe problem, it is recommended that the polystyrene be encased in glass reinforced concrete or epoxyresin, plastic drums, oil drums, or tyres.

Only around 5% of light striking the water surface is reflected, the rest being transmitted through the water column. As the UVR faction travels down through the water, it is gradually attenuated, the rate of attenuation being primarily dependent upon turbidity and colour (Calkins, 1982). The humic acid compounds which commonly stain dystrophic highland waters yellow are especially efficient at absorbing UVR. However, it is a commonly-held fallacy that UVR is absorbed within the first few centimeters of water, as calculations by Zaneveld (1975, in Bullock, 1982) have shown that the depth at which surface levels of the more harmful UV-B is reduced to 10% varies from 15.4m in clear oceanic water to 0.66m in the more turbid coastal waters.

Although the synthetic fibre nets gain some protection from the UVR when submerged, levels as low as 0.01% of the typical surface UV-B levels can elicit a response in fish skin (Bullock, 1982). Fish skin is very different in organisation from that of higher vertebrates, the protective pigment layer being located in the upper dermis as opposed to the epidermis, thus leaving the outermost layer of the skin highly vulnerable to UVR damage. This is perhaps of little consequence in the wild, where fish can avoid harmful light levels by seeking shaded areas or deep water. However, as a number of studies have demonstrated, fish constrained in shallow aquaculture facilities such as tanks or cages without adequate cover can experience sunburn especially if the prevailing light levels are high (Hastein *et al*, 1980; Eiras, 1982; Bullock, 1982, 1984; Bullock and Coutts, 1985). Typical gross evidence of UVR damage includes greyish focal thickening in certain dorsal areas of the fish, particularly along the outermost region of the dorsal fin or on the head (*Figs 7.32* and *7.33*). In many cases these regions will spread, and will be followed by necrosis and sloughing — sometimes of the entire dorsal fin — exposing the underlying dorsal musculature. Histopathological examination of the centre of the lesions reveals the characteristic presence of sunburn cells in the outer layer of the epidermis (see Bullock, 1984, for review).

Although there is evidence that fish have repair mechanisms which, given favourable conditions, can deal with moderate levels of radiation damage in the outer epidermis (Bullock and Coutts, 1985), severe lesions are often invaded by opportunistic pathogens which flourish in the fish farm environment. It is also probable that prolonged exposure to high sunlight levels may stress fish sufficiently to catalyse the onset of a disease outbreak (Bullock and Roberts, unpublished data), and in this respect light levels should be considered in a similar manner to temperature and water quality, as an important component in their environment.

Some form of shade should be provided for caged fish whenever possible and this is particularly important where UVR levels are high, and where species known to be susceptible to sunburn are being grown. There is an approximate 4% increase in UVR levels for every 300m increase in elevation, and it is therefore not surprising that trout grown at high altitudes in a number of tropical countries, such as Bolivia, Kenya and Mexico, have been found suffering from UVR-induced damage (Bullock, 1984). Unfortunately, there has been no work carried out on the materials or extent of cover

*Fig 7.32* Characteristic UV damage to head of fish (courtesy A M Bullock).

*Fig 7.33* UV damage to dorsal spine region of fish (courtesy A M Bullock).

*Fig 7.34* Floating trout cage, Lake Titicaca, Bolivia, fitted with canvas area for protection of fish from UV.

necessary to give adequate protection against UVR, and so once again this must be determined through trial and error by the fish farmer. The alternatives are either to use a small mesh top net to cover the entire cage or to restrict shading to certain areas of the cage (see *Fig 7.34*), or to use a completely solid (*eg* wooden) top, such as is commonly used for small fish cages in the USA.

# 8 Cage culture of various species

## 8.1 Introduction

As discussed in the overview (Chapter 2), cage culture has a relatively short history and at present contributes only around 4% in overall terms to world farmed fish, shellfish and crustacean production. However, in some sectors of the industry, cage culture is extremely important and indeed is for the present the most economically feasible method of producing certain species. Commercially viable cage farming has also developed rapidly in certain areas of the world, but the methods and lessons which have been learned have still to percolate through to a wider audience. Lastly, there are certain methods of rearing aquatic organisms in cages that although of little commercial interest are interesting and deserve further consideration and development.

In this section several of the economically more important cage farming industries are examined in detail. However space is also devoted to some of the more innovative methods of cage aquaculture. For information on other species or subjects the interested reader is urged to turn to Coche's useful and comprehensive bibliography of published material on cages (Coche, 1983).

## 8.2 Yellowtail farming in Japan

### 8.2.1 Introduction

The yellowtail, *Seriola quinqueradiata*, is a warm-water fish which occurs in the seas around Japan and its neighbouring islands, between latitudes 25°N and 54°N in the Pacific Ocean. The fish spend the autumn and winter months in the southern part of their range but migrate northwards in the spring to the spawning grounds off the southern coast of Kyushu island. Most spawning takes place here around late April/mid May, although spawning can in fact take place at any time between January and August in areas further to the southwest where suitable temperatures (16-29°C) occur (Fujiya, 1976; Mitani, 1979). A southerly migration takes place in August-October.

The eggs are small (1.15-1.44mm) and pelagic and hatch after 50hrs or so (temp. = 18-25°C). The hatched larvae, just 3mm in length, are carried northwards by the Kuroshiro current towards the large masses of floating *Sargassum* seaweed which occur off the coasts of southern Honshu, Kyushu and Shikoku islands during the summer months (Mitani, *ibid*; Kafuku and

Ikenoue, 1983). The juveniles congregate under the seaweed mat and feed on small zooplankton such as *Caprella* spp. and fish eggs. They grow rapidly and by the time they are 30-40cm long they are feeding on fish larvae — often other yellowtails — and large zooplankton such as copepods (Mitani, 1979; Kuronuma and Fukusho, 1984). The larger fish feed on sardines, mackerel and squid and can attain a size of 13kg (Kafuku and Ikenoue, 1983). Adults are fast-swimming and wide-ranging in their distribution.

Yellowtail is one of the most popular fishes in Japan, and next to the bigeye (*Thunnus obesus*) and yellowfin (*T. albacares*) tunas, commands the highest market prices (Yamaha Fishing Journal, 1980). The fishery dates back almost 500 years, and yellowtail culture is one of the earliest form of marine enclosure fish farming in the world. It began in 1928 at a site in Kagawa Prefecture on the island of Shikoku (Fujiya, 1979; Kafuku and Ikenoue, 1983), although the industry did not really begin to develop until the mid-1960s, when floating net cages were introduced on a large scale and the Japanese government took an active part in promoting fish culture. Production soared during the late 1960s and 1970s, and by 1978 approximately 150,000 tonnes per annum were being produced from some 4,000 farms (Nose, 1985) (*Fig 8.1*). Since then production has stabilised around this figure. Nevertheless since its inception, yellowtail culture has grown dramatically and now accounts for 90% of all marine farmed fish produced in Japan.

Most yellowtail are cultured in the south and southwestern parts of the

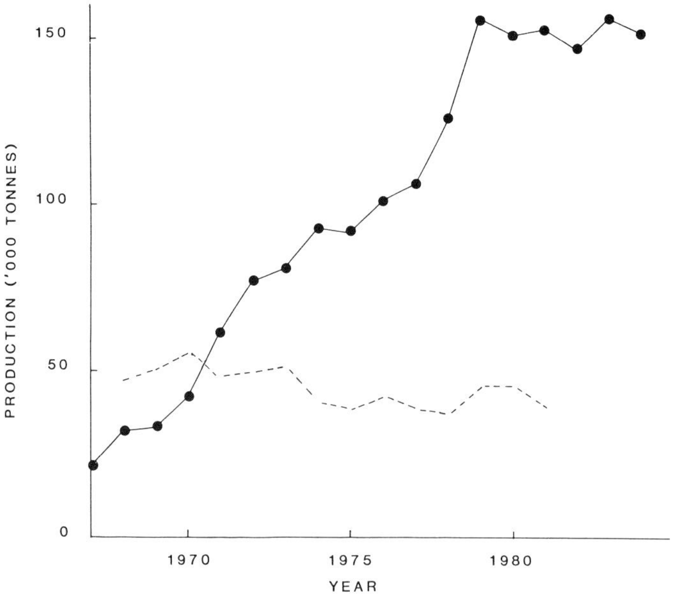


*Fig 8.1* Yellowtail production in Japan. Solid line = farmed fish; dashed line = fishery production (from various sources).

country, near to the fry collecting grounds and to the major markets of Tokyo and Osaka, although in areas avoiding pollution (see later).

In 1974, there were just over 3,000 largely family-owned yellowtail farms in Japan (Brown, 1983). Of those, 20 used pond enclosures formed by stone, concrete or earth embankments with sluices to facilitate tidal water circulation, and a further 80 farms were based in enclosures formed by net barriers. The remaining farms used net cages. Both types of enclosure — net and embankment — are described in detail by Milne (1972, 1979). However, despite farmed yellowtail production having almost doubled in the ten years since then, the use of such enclosures had declined drastically until almost none are in use today and production is virtually restricted to cages (Kafuku and Ikenoue, 1983). The first type of enclosure (embankment) became prohibitively expensive to build (Milne, 1979) and has such poor water exchange that stocking levels were sometimes as low as 1% (1-2 fish $m^{-3}$ initial stocking density) those of cage culture (Fujiya, 1979; Brown, 1983). The net enclosures were usually only stocked at 10% (5-20 fish $m^{-3}$ initial stock density) of the cage levels.

### 8.2.2 Fry rearing and collection

Despite successes in induced spawning of adults and rearing of yellowtail fry under experimental conditions (Fujita, 1977; Giovanardi, 1981), the industry still relies almost entirely on the collection of fry from the wild (Fujiya, 1979; Mitani, 1979; Kafuku and Ikenoue, 1983; Kuronuma and Fukusho, 1984). However, this pattern seems slowly to be changing. In 1982, three hatcheries in western Japan produced almost 3 x $10^6$ fry for sale (Anon, 1983b). Although this is less than 1% of Japan's current fry requirements, it was double the previous year's production. Constraints to development of this sector of the industry are production costs and technical difficulties of rearing larval and early fry stages (Kuronuma and Fukusho, 1984).

The fry, termed mojako, are captured from under the floating *Sargassum* masses in late April — early May. The fry collection and retail business is dominated by a few large operators — usually fishermen — who use dip or haul nets and operate from small 2-7 ton fishing vessels of the type commonly employed in pole or long-line fishing. The fry are graded and stored in large tanks at the rear of the vessels, until transferred to rearing cages (see later).

Most fry are collected in this way, from Mie and Kochi Prefectures (Brown, 1983). To avoid overfishing, a special licence is required to capture and sell the fry, and an annual catch limit is set. Despite these measures, demand for fry is high and increasing, so that each year the catch limit has had to be raised until it is now in excess of 900 million fry per annum. Unfortunately, this has had the effect of decreasing the average size of mojako caught from the ideal of 4cm+ (1g+) at which size they can feed happily on the minced raw fish commonly used in culture, to 25-30mm (Mitani, 1979). The huge demand for fry has also adversely affected the yellowtail fishery (Mitani, *ibid.*).

During May and June the mojako are reared in small to large (2-50$m^2$ in area, 1-3m in depth), fine, synthetic-fibre mesh net cages, which are prone to fouling and must be checked regularly. During this 4-6 week period they are fed on minced available fish (sand eel, horse mackerel, saury, anchovy) and shrimp, although according to Fujiya (1979), mortalities are minimised if low-fat, white-fleshed fish is used. Feeding rates during this period are 50% body weight per day (Brown, 1983). The use of lights at night to attract zooplankton for food is also practised at a few sites (Fujiya, 1979; Kuronuma and Fukusho, 1984). Growth is rapid, and because the fry are cannibalistic, particularly at high stocking densities where 50% mortalities can occur, grading must be carried out regularly. This is achieved using various mesh sizes of net; 5g mojako are retained by 8.5mm netting, 6-9g by 11mm and 10-20g by 15mm (Honma, 1980). The fry are eventually sold by the fishermen for ongrowing when they are 5-10cm long (8-50g), the 10cm size being preferred. The fish farmers usually transport the mojako by boats to their farms and then transfer them to the ongrowing cages.

### 8.2.3 CAGE CULTURE

A range of cage designs is currently in use for Japanese yellowtail culture. Most are either floating or submersible, the latter type having been developed in the late 1960s and early 1970s. A few fixed cages are used in shallow water inshore sites.

Floating cages for yellowtail were originally small and square (2 × 2m) in configuration, with rigid collars constructed from 10-15cm diameter bamboo poles, and held afloat by steel drums. As the industry grew so did the size of the cages, and today most floating units are around 300m$^2$ in area, by 3-6m deep (Nose, 1985). Despite the increase in size, many cages are still designed and constructed in much the same way as they were in the past, although new materials such as 3-5cm galvanised steel pipe and polydrums or polystyrene cylinders covered in polyethylene sheeting have been adopted by many farmers to replace the traditional bamboo, cedar wood and steel drum collars (Fujiya, 1979) (*Fig 8.2*). Although more expensive, they have a much longer life span. Most cages have single net synthetic fibre bags fitted with a top net, the double net type described by Milne (1972) being little used due to the work involved in maintenance. In the 1970s some farms began using galvanised or nylon coated steel mesh bags to minimise fouling and to reduce net changing to once per year (Milne, 1979; Kafuku and Ikenoue, 1983; Kuwa, 1983, 1984). However, because of expense and poor durability these materials have not proved commercially popular. Most farmers treat their nets with anti-fouling compounds.

Large, flexible rubber collar cages of the type shown in *Fig 2.2e* are gaining in popularity and are being used in more exposed but less polluted offshore sites. Submersible cages are also used in some exposed sites in the Japan Sea, Inland Sea and other areas (Fujiya, 1979). A typical design is shown in *Fig 7.23*. The top net often has a net chimney, 1-2m in diameter which is held afloat by buoys and when the fish are not being fed it can be tied and closed off. Cage volume is maintained by a combination of floats on the top rope and

*Fig 8.2* Typical yellowtail farm, Kagashima Prefecture. Cage size is 9 × 9 × 9m (courtesy Professor T Sano).

weights on the toe rope. Sometimes, in sites with strong currents, a rigid frame constructed from galvanised steel or sand- filled plastic pipe is attached to the bottom of the bag. When the weather is calm, the cage sits just below the sea surface. During rough weather the feeding chimney is closed and the cage lowered to a depth of 10-15m. Submersible cages are often very large — 1,200-1,500$m^3$ volumes having been recorded (Honma, 1980).

The siting of yellowtail cages is strictly regulated by the Prefectural governments who charge a site rental fee (Brown, 1983). Conventional cages are usually set out in parallel rows held in place with wires and concrete block anchors, as shown in *Figs 8.3* and *8.4* (Kafuku and Ikenoue, 1983), or less commonly in rafts of 10 or so, anchored at each corner (Milne, 1972). In order to minimise damage caused by wave action in exposed areas, floating breakwaters are sometimes used (see Section 7.11), and in some parts of Mie and Aichi Prefectures cages may seek the shelter of the prolific floating pearl rafts (Honma, 1980).

Stocking density depends on the size of the fish, and the prevailing environmental conditions. Recommended stocking densities, based on 'average' conditions are shown in *Table 8.1*. Stocking density increases from a minimum of 0.9kg $m^{-3}$, when the mojako are first stocked in the grow-out cages at an average weight of 10g each, to a maximum of 11kg $m^{-3}$ when the fish reach 500g (Fujiya, 1979), although, under ideal conditions, stocking densities of up to 20kg $m^{-3}$ can be achieved (Kafuku and Ikenoue, 1983). However, today, the tendency is to keep stocking levels as low as possible (Nose, 1985). The effects of temperature and oxygen on feeding and survival are shown in *Table 8.2*.

Most yellowtail farmers use a wide range of cheap trash fish such as

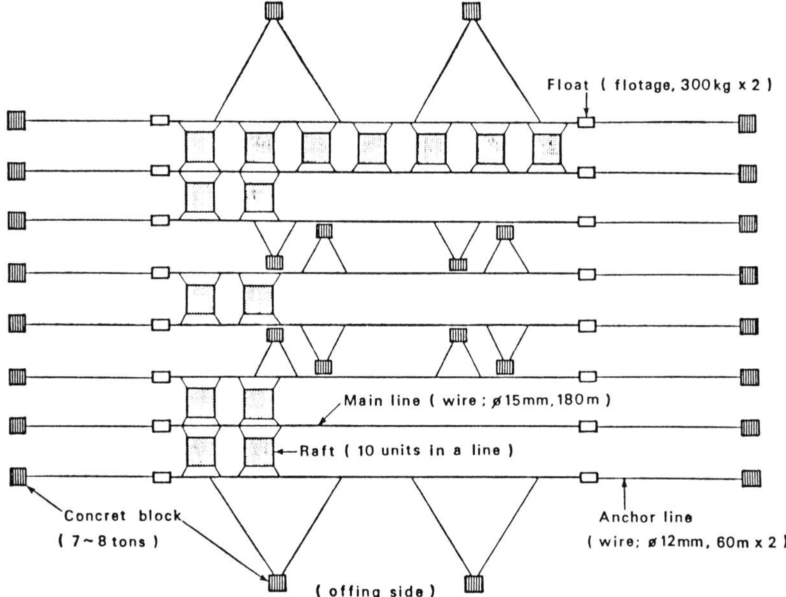


*Fig 8.3* Arrangement of floating net cages for yellowtail culture (from Kafuku and Ikenoue, 1983).

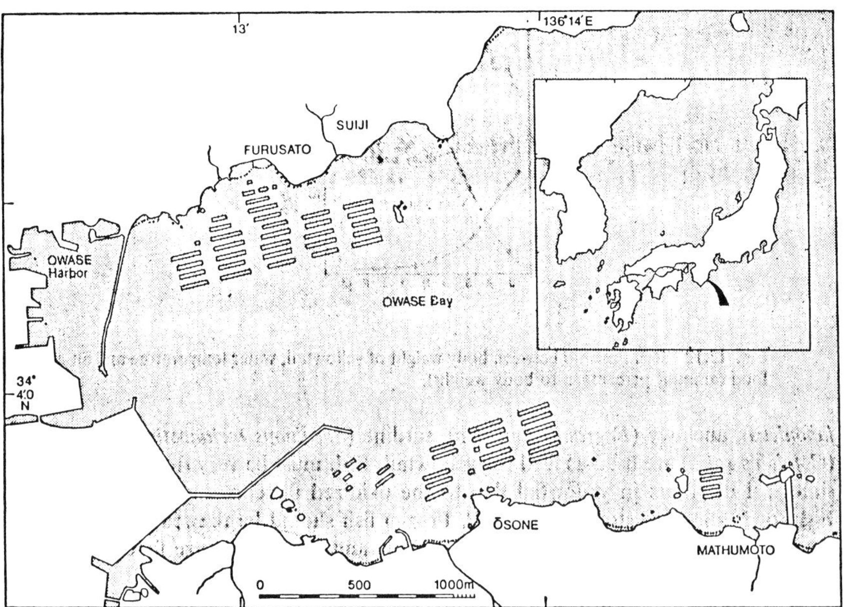


*Fig 8.4* Distribution of net cages in Owase Bay, Mie Prefecture (from Kafuku and Ikenoue, 1983).

mackerel (*Scomber japonicus*), anchovy (*Engraulis japonica*), sardine (*Sardinops melanosticta*), sand eel (*Ammodytes personatus*) and saury (*Cololabis saira*) to feed their fish (Kafuku and Ikenoue, *ibid*). The composition varies with location and season, and is primarily governed by cost and availability. The FCR is, on average, 7-8:1 (Brown 1983). However, nutritional diseases in farmed yellowtails are widespread, and are usually due to fish being fed diets high in unsaturated fats, or low in vitamins (Furukawa, 1976; Park, 1978; Ishihara *et al*, 1978a,b). Work carried out in Korea by Min *et al* (1981) has shown that when yellowtail were fed a variety of trash fish diets, growth and survival were best with high protein/low lipid (*ie* largely white-fleshed fish) formulations. It is imperative that trash fish diets are used when fresh or stored at low temperatures ($< -20$°C) prior to use, otherwise fats may oxidise leading to a range of physiological disorders (Furukawa, 1976; Kafuku and Ikenoue, 1983). Moreover, recent research work has implicated trash fish quality in streptococcal disease outbreaks, and some workers recommend

*Table 8.1* Recommended stocking densities for yellowtail culture.

| *Size (g)* | *No m$^{-3}$* | *Stocking density (kg m$^{-3}$)* |
|---|---|---|
| 10 | 90 | 0.9 |
| 50 | 60 | 3 |
| 100 | 50 | 5 |
| 200 | 35 | 7 |
| 400 | 25 | 10 |
| 500 | 22 | 11 |
| 700 | 14 | 10 |
| 900 | 11 | 10 |
| 1,000 | 10 | 10 |

Data from Fujiya (1979).

*Table 8.2* Temperature and oxygen requirements for culture of yellowtail.

| | |
|---|---|
| (a) *Temperature* (°C) | |
| < 9 | death |
| 9 – 11 | no feeding, no growth |
| 11 – 13 | feeding poorly, no growth |
| 14 – 17 | feeds, but poor growth |
| 18 – 27 | feeds and grows well. OPTIMUM CONDITIONS |
| 28 – 31 | feeding poorly, poor growth |
| > 31 | death |
| (b) *Oxygen* (ppt) | |
| < 2 | death |
| 2 – 3 | troubled respiration, abnormal swimming, poor appetite |
| 3 – 4 | appetite normal |
| 4 + | swimming normal, appetite normal. OPTIMUM CONDITIONS |

From Fujiya (1979) and Kafuku and Ikenoue (1983).

that it may be safer to feed yellowtail frozen blocks of trash fish than to use defrosted fish (Taniguchi, 1983). Vitamin deficiencies, such as those described in Furukawa (1976) and Ishihara *et al* (1978a,b) are best avoided by feeding a variety of trash fishes at any one time. Moist, pelletised feeds, consisting of 70% white fish meal, 5-10% gluten binder, plus minerals, vitamins and trace metals are gaining in popularity, since although more expensive, they are believed to be less polluting (Nose, 1985).

Feeding is carried out once or twice per day, although care must be taken not to overfeed. Fish are fed to 80% satiation level (Fujiya, 1979) and the feeding rate during the first year of growth decreases from 40% body weight per day when stocked in July, to 4-5% in January (*Table 8.3*). During this period the water temperature falls thus reducing food intake (Sakamoto *et al*, 1979). As temperature increases during July and August of the second year of growth (*Fig 8.5*), appetite improves, growth accelerates and the feeding rate increases to ~8% body weight per day.

At some sites fouling is a problem and synthetic net bags may have to be changed as frequently as every 8 days in summer and 20 days in winter. Synthetic net bags must also be regularly changed as the fish grow, the mesh sizes used during a two-year growth cycle ranging from 1cm to 6cm (Kafuku and Ikenoue, 1983). Growth is rapid, the fish attaining anything up to 2kg after 12 months and 4-6kg after 18 months (*Fig 8.5*). On average, 80% of the fish survive one year of culture, although this figure falls to 50-60% of the initial numbers stocked by the end of the second year and may be much lower if affected by a disease outbreak (Kafuku and Ikenoue, 1983). Yellowtail are susceptible to a number of parasitic and fungal diseases and parasitic infections. Until 15 years ago large losses were often associated with infestations of ectoparasites such as *Benedenia seriolae*, *Axine heterocerca* and *Caligus*. However, these parasites can be readily controlled through bathing fish in chemical solutions (Fujiya, 1979) and are no longer major causes of mortalities (Kufuku and Ikenoue, 1983). Of much greater significance today are bacterial diseases such as vibriosis, pseudotuberculosis, *Nocardia* and streptococcal disease, most of which first appeared in the late 1960s and early 1970s (*Table 8.4*). In the four years between 1973 and 1977, losses through these diseases increased 8-fold, accounting in 1977 for 13% of total fish

*Table 8.3* Feeding rates for yellowtail (after Fujiya 1979).

| *Body weight* (*g*) | *Water temperature* (*°C*) | *Daily feeding rate* (*% body wt per day*) |
|---|---|---|
| 10–50 | 15–22 | 40–60 |
| 50–100 | 18–24 | 30–50 |
| 100–200 | 20–25 | 20–30 |
| 200–400 | 23–26 | 15–20 |
| 400–600 | 25–28 | 11–15 |
| 600–800 | 24–28 | 9–11 |
| 800–1,000 | 18–25 | 8–10 |
| 1,000–1,200 | 13–20 | 6–8 |

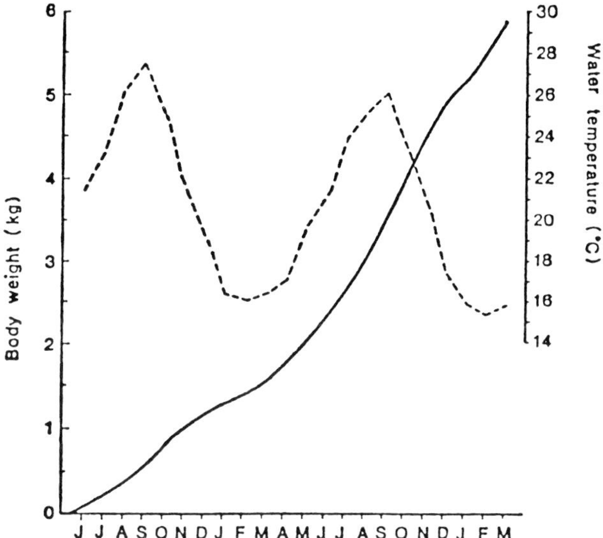


*Fig 8.5* Growth of cultured yellowtail in Owase Bay, Mie Prefecture (from Kafuku and Ikenoue, 1983).

stocked (Honma, 1980). Currently, streptococcosis accounts for most of the disease losses at yellowtail farms. The pathology of vibriosis and pseudotuberculosis is well known (Sugimoto *et al*, 1976), and both diseases respond to treatment with nitrofuran drugs. However, despite intensive research efforts the strain of *Streptococcus* responsible for mortalities in yellowtail has not yet been identified and vaccines to control the infection are still under development. Preventative measures include maintenance of good water quality and low stocking densities and ensuring that fresh trash fish is always used (Taniguchi, 1983).

Other major factors responsible for mortalities of fish are industrial pollution and dinoflagellate blooms (*Fig 8.6*). On average 50-100 oil-spill incidents occur in Japanese inshore waters each year, causing many millions of pounds of damage. The principal dinoflagellate species involved in blooms are *Gymnodinium* sp. (Kyushu, Shikoku and other areas), *Chatonella antiqua* (Naikai inland sea) and *C. marina* (Kyushu area) (T Sano, pers. comm). According to Shimada *et al* (1983) the presence of large numbers of *Chatonella* causes damage to the gill epithelia which impairs gaseous exchange. Doi *et al* (1981) report that during a bloom of *C. antiqua*, the yellowtail swam around their cages in a frenzied state, thus increasing oxygen demand and leading to death by asphyxiation. Losses incurred during a toxic dinoflagellate bloom can be catastrophic. For example, in 1984 a single bloom episode of *Gymnodinium* sp. in Kumano-nada in Wokayama Prefecture resulted in $4 \times 10^9$ Yens damage to aquaculture interests (T Sano, pers. comm). Research into the cause and amelioration of blooms is currently very active (see Section 4.2.2).

*Table 8.4* Bacterial diseases in cultured yellowtails, their importance and treatment.

| *Disease* | *Symptoms* | *Importance* | *Prevention and treatment* | *Reference* |
|---|---|---|---|---|
| Bacterial ulcer disease | Caused by *Cytophaga* sp. | + | | Miyazaki *et al*, 1975 |
| Vibriosis | Infect body openings (eyes, mouth, anus) or wounds, causing suppuration. Muscle and internal organs develop. *Vibrio* sp. | ++ | Oxytetracycline, sulfa and furan drugs administered in feed. Bathing in furan solution | Kafuku and Ikenoue, 1983 |
| Pseudotuberculosis (Pasteurellosis) | Large number of pseudo-tubercules in liver, spleen and kidney. Highly infectious, high mortality rate. Caused by *Pasteurella piscida* | +++ | Remove infected fish, treat with nitrofuran compounds (*eg* sodium nifurstyrenate) | Matsuzato, 1975<br>Sugimoto *et al*, 1976 |
| Streptococcal disease | Septicaemia of inner surface of gill opercula, liver and intestine. Causes persistent mortalities. *Streptococcus* sp. still unidentified | +++ | Remove infected fish. Antibiotics of limited use. Use fresh feed | Kusuda *et al*, 1982<br>Taniguchi, 1983 |
| *Nocardia* disease | First symptoms anorexia and skin discoloration. Tubercules appear on skin and in younger fish on gills. In older fish lesions also form on internal organs (spleen, kidneys). Caused by *Nocardia kampachi* | + | Streptomycin and sulphasoxizole appear to be active, but no effective treatment. Remove infected fish | Kariya *et al*, 1968<br>Egusa, 1980 |

### 8.2.4 HARVESTING AND MARKETING

When harvested, fish are removed from the cages by dip net, killed by knife, transferred to ice-filled polystyrene boxes (*Figs 8.7* and *8.8*) or iced water and transported as swiftly as possible by boat or truck to the large city markets in eastern Honshu Prefecture and Tokyo, and sold fresh. Some are also sold live. Most fish (90%) are harvested during October-December in the first year of growth when they are between 600g and 800-1200g in weight are known as 'hamachi'. Most of the remaining fish are sold during the same period of the second year of growth, when they are known as 'warasa' (5-7kg). However, only farms in the southernmost parts of Japan which have suitable over-wintering temperatures practice this (Fujiya, 1979). There is no farmed equivalent of the large 'buri' — the wild, 8kg+ fish which fetch highest market prices.

During the winter months, the numbers of cultured and wild yellowtail arriving on the market increases dramatically (*Fig 8.9*). There are several reasons for this. Yellowtail is regarded as a festive fish, which attains its best flavour during mid-winter and it is caught in largest numbers at this time of year. Much of the fish is sold directly to restaurants and the remainder retailed fresh, frozen or dried, through supermarkets. There are a few differences between the flesh of wild and cultured fish. Cultured fish have significantly higher lipid content, higher calorific value, and lower ash, protein and moisture levels than wild fish (Saeki and Kumugai, 1979) and usually fetch higher prices.

Recent figures for the economics of yellowtail culture are difficult to find. Production costs in the 1970s were dominated by the price of fry (15%) and feed (60%) (Honma, 1980; Brown, 1983) and considering the rise in demand for fry and the increase in prices of trash and pelleted feeds since then, the situation today must be similar. During the mid-1970s, it was the most profitable type of fish farming in Japan, enabling a farmer owning two 500m$^3$ cages to produce 10-20,000 fish at 1-1.5kg each per annum, and to earn a typical middle-class income (Fujiya 1979; Brown, 1983). Since then, the

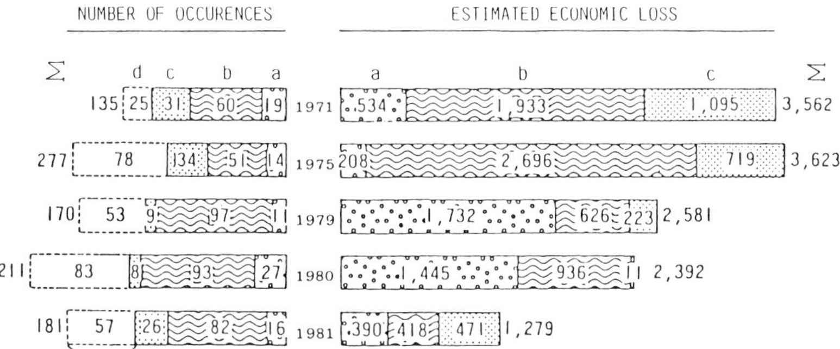


*Fig 8.6* Reported damages to Japanese aquaculture and fishery caused by marine pollution in coastal waters during the decade 1971-1981. (a) = red tides; (b) = oil pollution; (c) = others; (d) unidentified causes; Σ = total (from Nose, 1985).

*Fig 8.7* Harvesting yellowtail, Kagashima Prefecture (courtesy Professor T Sano).

*Fig 8.8* Yellowtail packed in insulated shipping boxes (courtesy Professor T Sano).

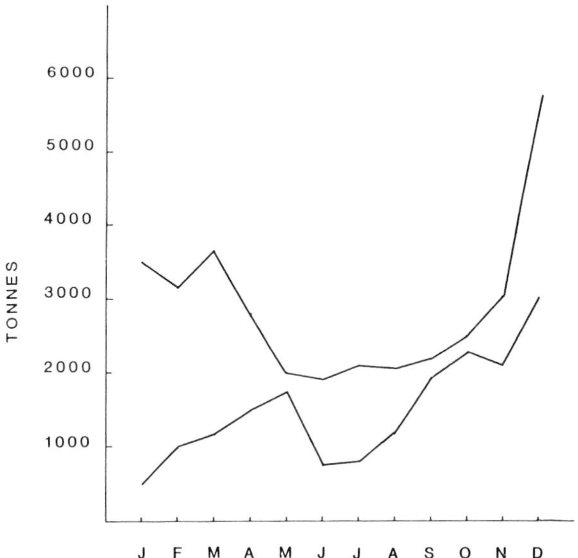


*Fig 8.9* Quantities of farmed (top line) and wild (bottom line) yellowtail arriving at central wholesale markets in six large cities. (redrawn from Yamaha Fishing Journal, 1980).

average number of cages per farm has increased by 50% and the average size of the cages has almost doubled, although there has been a reduction in production per unit cage area of around 35% (Nose, 1985). Today, a typical yellowtail farm produces 40 tonnes of fish per annum, a considerable increase over average production per farm 10 years ago. Wholesale prices have increased by 35% over the same period and the current price to farmers is £3.00 - £3.20 per kilo. Sashimi quality yellowtail retails at not less than £1.25 per 100g (N Endo, pers. comm).

Disease, pollution and, most significantly, self-pollution of sites, many of which were established 15 years or more ago, are the main factors restricting growth of the industry and forcing growers to move to new and increasingly more exposed offshore sites (see Kafuku and Ikenoue, 1983; Nose, 1985; Rosenthal, 1985, for reviews).

## 8.3 Atlantic salmon farming in Scotland

### 8.3.1 Introduction

Populations of the Atlantic salmon (*Salmo salar*) occur in almost all of the countries which border the northern Atlantic Ocean, including the north eastern states of the USA, eastern Canada, Greenland, Iceland, Scandanavia, UK, France and Spain, although in many of these countries, stocks have been sadly depleted (Went, 1980). It is an anadromous species which spends the juvenile stages of its life in freshwater and then migrates to the sea where most of its growth occurs and where it becomes sexually mature.

Sexually mature adults tend to return to the rivers of their origin to breed. Spawning typically occurs during the period October — January in the head waters of streams where the water is shallow, cold and clean and where there is a suitable substrate of gravel. The female or hen fish selects a site and excavates a shallow depression, or redd, in the gravel with her tail into which she deposits her eggs — some 1,500-20,000, depending on the size of the fish (1,200-1,600 eggs $kg^{-1}$ body weight). The male, or cock fish, then discharges his milt (sperm) over the eggs, which are subsequently covered over with several centimetres of gravel. Many of the adults — particularly the males — die after spawning, although some, known as kelts, survive and return to the sea and may even spawn again following a period of two or more years at sea (see Netboy, 1973; Sedgwick, 1982 for reviews).

Egg development is dependent upon temperature (see later) and in Scotland the young do not generally hatch until March/April. The newly hatched fry or alevins remain in the redds for a period of three or four weeks until most of the yolk reserves have been utilised. They then migrate to the surface of the redd, emerging in April/May.

The juveniles remain in freshwater for two, three or even four years and continue to feed and grow, and develop the typical vertical bar markings which distinguish salmon parr from the juveniles of other salmonids. The young fish adopt and defend territories against conspecifics and prefer boulder-strewn, fast-flowing reaches where they orient themselves into the current, holding station with the aid of their pectoral fins (Keenleyside and Yamamoto, 1962; Symons, 1976). They feed on a wide range of invertebrate prey items, whose size is a function of fish body length, and which are carried passively or semi-passively in the drift. (Thorpe and Wankowski, 1979; Wankowski and Thorpe, 1979).

Prior to commencing the marine phase of their life cycle, the parr undergo a process of transformation known as smoltification which involves a wide number of changes in appearance and behaviour and in the rates of growth, metabolism and osmoregulation which prepares the fish for life in saltwater. Not surprisingly, such a complex process is not yet fully understood, although it is known that environmental cues such as light and temperature play a major role, influencing rhythms via various endocrine pathways (see Langdon, 1985 and Thorpe *et al*, 1985, for reviews). Smoltification can occur as early as the spring of the first year after hatching in rivers at the southern-most end of their range, when the fish are just over one year old (Sl). However, in Canada and arctic Norway, smoltification may not take place until the fish are five or six years old. In Scotland, most salmon smolt at 2+ or 3+ years of age (Thorpe *et al*, *ibid*).

When the smolts enter seawater, they spend a period of time in home coastal waters before migrating to the feeding grounds off the south and south west coasts of Greenland. Initially they feed on small crustaceans, such as amphipods, euphausids (krill), and pteropods, although at a later stage they adopt a largely piscivorous diet, feeding on small pelagic species such as capelin (*Mallotus villosus*), sandeels (*Ammodytes* spp.) and herring (*Clupea harengus*). They will also feed on prawns (*Pandalus borealis*) and squid.

The growth of Atlantic salmon in seawater is rapid and after one sea winter they can attain a weight of 1.5 to 3.5kg (50-65cm). A proportion of these fish will become sexually mature and return to fresh water as grilse, whilst the remainder stay for a second or even third year at sea before returning to spawn. The two and three sea winter fish are known as salmon and attain 4-6kg (70-90cm) and 8-14kg (90-105cm) or more respectively. Males are generally considerably larger than females.

### 8.3.2 Fishing and farming industries

There have been salmon fisheries in Scotland for many hundreds of years, and today ownership of fishing rights in much of the country's rivers is in private hands, although the right to fish for salmon along the greater part of the Scottish coast still resides with the Crown (Association of Scottish District Salmon Fishery Boards, 1977). Permissible methods of catching salmon under the Salmon Fishery Acts are net fishing, rod fishing and cruive (trap) fishing. Detailed catch statistics have only been compiled since 1952 (*Fig 8.10*). Whilst much of the variation apparent in the record appears to be

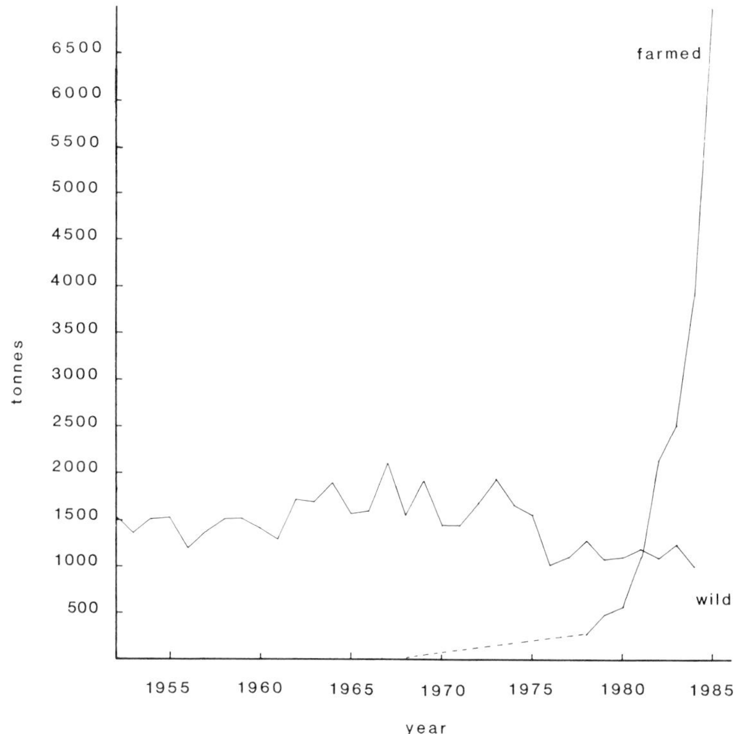


*Fig 8.10* Farmed and wild Atlantic salmon production in Scotland (data from DAFS).

due to changes in grilse numbers, nevertheless, during the period 1976-present, total salmon and grilse catches by all methods have been on average much lower than those recorded for 1952-1975, although the reasons for this are unclear. Debate rages around whether the decrease in Scottish catches is due to the dramatic increase in intervention fisheries pressure (*ie* exploitation of the stocks around Greenland and Faroes) or merely reflects part of a long term environmentally-induced cyclic variation in numbers (Went, 1980; Shearer, 1984). Today catches of wild salmon are around 1,100 tonnes per annum.

The concept of farming Atlantic salmon in Scotland dates back to the 1870s, when Sir James Maitland of Howietoun, Stirling, demonstrated that it was possible to capture fish from the rivers at spawning time, strip the eggs and sperm, successfully fertilise and hatch the eggs, and rear the young until they smoltified to smolt (Maitland, 1887). By the turn of the century it had been shown that captive smolts could be reared in saltwater ponds, although it was not until the late 1960s that the possibility of farming an anadromous fish, such as the salmon, began to emerge in Scotland (Glen, 1974). Marine Harvest Ltd, a subsidiary of Unilever, built the first marine fish farm in Scotland at Lochailort, Invernesshire, in 1966. Encouraged by Norwegian efforts and their own early attempts to culture rainbow trout in seawater, the company began to farm salmon in 1968. A number of other companies also became involved around this time, including Highland Trout Company (Scotland) Ltd, and Golden Sea.

Despite the industry in Scotland having begun only a few years after that in Norway, growth has for the most part been comparatively slow. Unlike Norway, where salmon farming was envisaged as a means of increasing the incomes of local fishermen/farmers, in Scotland the industry has, until recently, been largely developed by national or multinational companies, albeit employing local labour. However, in the 1980s the rates of growth in the two countries have been similar and today Scotland is the second largest producer of farmed Atlantic salmon after Norway. According to the Department of Agriculture and Fisheries for Scotland (DAFS) in 1985 the industry produced 6,900 tonnes (*Fig 8.10*), employed 750 people and had a turnover of £27 million per annum. Some 65 companies are involved, operating 90 seawater and 46 freshwater sites (*Fig 8.11*). Only half a dozen or so of the seawater sites are land-based, the rest all employed cages. Areas such as Lewis, Harris, Orkney and Shetland have only recently begun to be exploited. Most of the production, however, is accounted for by five companies (*Table 8.5*), and one of these — Marine Harvest Ltd — produces in excess of 1,000 tonnes per annum (Anon, 1985c). Norwegian interests in the Scottish industry have also increased dramatically during the 1980s contributing towards the rapid growth in production. Conservative estimates suggest that by the end of the decade the industry will be producing 10,000 tonnes per annum, although optimists predict that the figures could be as high as 20,000 tonnes.

Much of the credit for the current healthy state of the industry belongs to the Highlands and Islands Development Board (HIDB) who by 1983 had

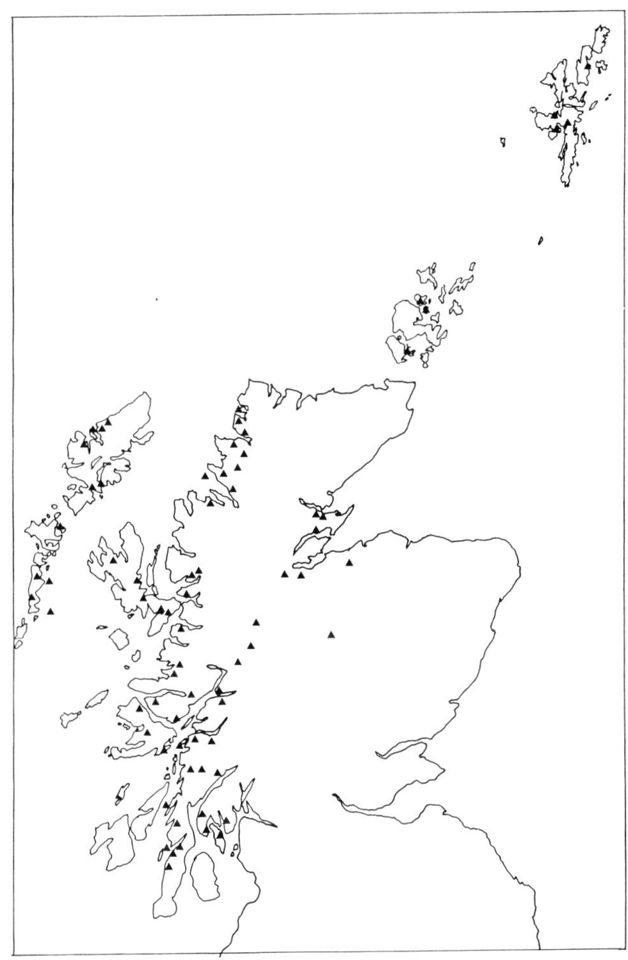

*Fig 8.11* Distribution of salmon farms in Scotland, 1984 (Data from the HIDB).

*Table 8.5* The distribution of production among salmon farming companies, Scotland, 1984.

| *Production (tonnes)* | *0* | *10* | *10–25* | *26–50* | *51–100* | *101–250* | *250* | *Total* |
|---|---|---|---|---|---|---|---|---|
| No. of companies | 18 | 11 | 9 | 4 | 7 | 3 | 5 | 57 |
| Cumulative tonnage | 0 | 48.8 | 146.8 | 165 | 493.5 | 486.7 | 2,570.8 | 3,911.6 |
| % total production | 0 | 1.3 | 3.8 | 4.2 | 12.6 | 12.4 | 65.7 | |

Source: DAFS.

invested £11.7 million, nearly 30% of which had been spent on research and development. In recent years the EEC's Integrated Development Programme (IDP) has helped greatly in the funding of fish farming projects.

### 8.3.3 Smolt rearing

The rearing of Atlantic salmon begins with the stripping of eggs and milt from sexually mature adults in October and November (see Edwards, 1978, for details). Today most of the larger hatcheries rear their own broodstock and operate a broodstock improvement programme, although some smaller units still rely on buying in eggs. Fertilised eggs are usually hatched in plastic trays held in shallow fibreglass troughs through which water flows by gravity. The rate at which the eggs develop depends upon temperature and is usually calculated on the basis of 'degree-days', the product of the number of days after fertilisation times the temperature. Eggs typically become eyed at 245 degree-days and hatch after 510 degree-days at temperatures around 8°C. At hatching the water supply is increased to match increased oxygen consumption, and the newly- hatched fry (alevins) are left in the egg trays until the yolk sac is at least two-thirds absorbed. First feeding usually begins at 290 degree-days after hatching.

The first-feeding stage may occur at any time from early February until late March, depending on the time of stripping and the temperature of the water supply to the hatchery, and around this time the fry are moved into the nursery. Traditionally, an indoor facility equipped with 1-2m diameter glass fibre tanks, holding 25cm of water, is favoured. The initial stocking density is high — around 10,000 fish $m^{-2}$ — and the flow adjusted so that the fish can comfortably maintain station. Salmon starter feeds are 0.3-0.6mm in size and have a high protein content (54%), and the fry are fed at a rate of 3-6% body weight $day^{-1}$. Mortalities at this stage can be high and most farmers expect to lose 10%. After a few weeks the fry are thinned out or moved to larger tanks outdoors where they remain until the end of the summer. In late August or September the parr are graded. At this stage it is possible to differentiate between those fish which will smolt the following April/May ($S_1$) and those which will not be ready to transfer to seawater until they have spent a second winter in fresh water ($S_2$), on the basis of size differences (*Fig 8.12*). The proportion of the stock that is potentially $S_1$ is highly variable and depends upon stock origins and prevailing environmental conditions, as well as management (Langdon, *ibid*). There are thus enormous variations between hatcheries and between years (see later). In a typical hatchery in an average year in Scotland, around 65-75% of the stock could be expected to smolt at the end of the first year.

Potential $S_1$ fish will be stocked at densities of around 1-2,000 $m^{-2}$ in 4-7m diameter tanks held outside, whilst at some farms, as many as possible of the designated $S_2$ fish will be sold off for stocking purposes. Small $S_2$ salmon parr currently have a market value of 10-40p each. The $S_1$ fish are fed according to manufacturers recommendations with the appropriate feed (2-3mm, 54% protein) at a rate of 0.6-1.4% body weight $day^{-1}$, depending on size and temperature. Unless there is a disaster (*eg* pipes freezing, a sudden fall in pH, *etc*), mortalities during the winter are usually extremely small, and by April-May the smoltifying salmon will be 35-45g each.

Although the majority of smolts are still produced in land-based systems, smolt rearing in cages is becoming increasingly popular ( *Fig 8.13*) partly because of the rapid expansion in salmon farming in areas with poor flowing water resources, such as Orkney and Shetland. The HIDB, which tries to maintain a regional balance between smolt production and ongrowing operations, has assisted in the development of cage smolt production in these areas. Typically, Kames or other wooden-collar cages, 20-36m$^2$, are used and net characteristics are determined by fish size. Usually 12-15g potential $S_1$ parr would be stocked, following autumn grading, in cages fitted with 4m deep bags fabricated from 9mm bar length knotless nylon square mesh netting. Around 10,000 fish per cage would be stocked initially. In the

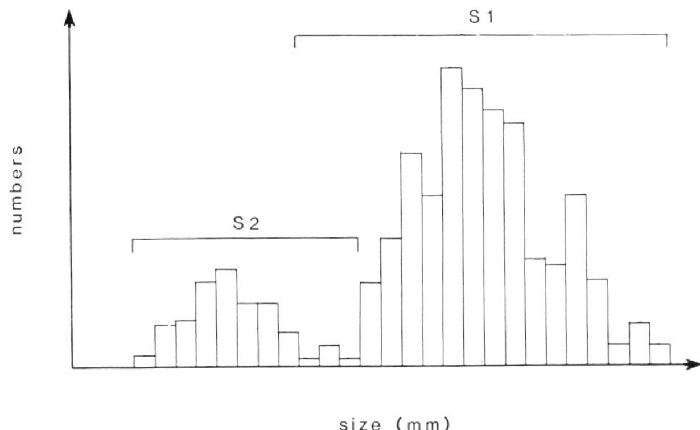


*Fig 8.12* Example of bimodal distribution of salmon parr sizes apparent during grading in September or October.

*Fig 8.13* Salmon parr cages, Lewis, Scotland (courtesy J Smith).

spring, when the parr are around 20g, the bags are changed for 12mm stretch mesh diamond netting bags in which the fish remain until transfer to sea cages. Potential $S_2$ fish are managed in a similar manner. However, some farmers have begun to transfer alevins when only 0.2-0.5g in size, directly from the hatchery to cages, with considerable success. The cages are equipped with bags fabricated from 4mm bar length square mesh netting, and are only 1m deep. Between 40 and 50,000 fish would be stocked in each cage. Fouling of such small mesh net bags is, however, a problem.

Statistics published by DAFS indicate that at present $15 \times 10^6$ ova are laid down each year, around 60% of which come from farms' own broodstock, 20% from other Scottish farms' broodstock, 6% from wild fish and 15% imported from Norway. Smolt production increased steadily during the 1980s to over $6 \times 10^6$ per annum in 1985 and, throughout the same period, the proportion of $S_1$ smolts has also steadily risen thanks largely to better management. In 1982 50% of smolts stocked by ongrowers were $S_1$ fish whilst in 1985 the proportion was 85%. (*NB* Figures for 1986 are expected to be poor due to the unusually cold, wet summer of 1985). However, although smolt sales now account for more than 40% of total salmon sales, there are still insufficient smolts available to meet the demand from Scottish salmon farmers, with the result that the price has risen dramatically to around £1.60 per smolt, or £40 per kilo (1986 prices)! The demand for smolts in the home market is further exacerbated by increasing pressure from Norwegian salmon producers, who are prepared to pay even higher prices. The current large profit margins enjoyed by smolt producers and the comparatively low capital costs required to establish a business is attracting growing numbers of people into the industry and preliminary assessments suggest a three-fold increase in smolt production over the next 3-4 years.

### 8.3.4 Ongrowing

Most cage salmon farms are sited in sheltered sea lochs (*Fig 8.14*) along the west coast of mainland Scotland, the Hebrides, Orkney and Shetland. Smolts are usually delivered to sea sites in early May-June by road and/or well boat (see *Fig 6.3*). The timing of transfer depends as much on management considerations as on environmental factors. Fish are normally stocked directly into sea cages without acclimation and initial stocking densities of 15-20 smolts $m^{-3}$ are common. Fish are normally left for a day or so before commencing feeding.

Cages currently used in Scotland are much bigger than those favoured at the beginning of the 1970s. Today most cages are at least $250m^3$ in volume and whilst some of the larger, more established companies still design and build their own cages, most farmers buy either Scottish-built designs largely fabricated from pressure-treated larch or imported steel collar cages from Norway. Almost all nets, however, are fabricated in Scotland where there is a flourishing, specialist cage net industry that has grown and evolved with the fish farming industry.

The juvenile salmon are fed largely on pelleted, commercial feeds with a

*Fig 8.14* Typical sea loch cage salmon farm site (courtesy M J Phillips).

protein content of 46-50%. Fish are fed several times per day and most farms rely on automatic feeders augmented by some hand feeding. Pigmented feeds are usually administered for several months prior to harvesting. Sample weighings of fish are taken every month or so to monitor growth and nets are checked regularly for damage and to assess the degree of fouling. A careful check is also made of fish health.

As the fish grow they may be moved from 12mm to 18mm and ultimately to 25mm stretch mesh diamond net bags although many farms do not use all three net sizes. It is claimed by some, for example, that the 25mm mesh nets can cause skin lesions on the fish. A few farms also prefer to use more expensive hexagonal mesh nets as they are easier to handle. In late spring/early summer after their first winter at sea the fish are graded and stocks divided on the basis of size differences into potential grilse and potential salmon. (On Orkney and Shetland, this operation is carried out in late summer). Grilsing is partly determined by genetics and partly by environment. However, both scientists and fish farmers have demonstrated that it is possible to select for late maturing stock with a low grilsing rate (Alderson, 1984; Gjedre, 1984). In mainland Scotland and the Hebrides, approximately 40-50% of the stocks are harvested as grilse during the period April-July, at an average weight of around 1.7kg. In Orkney and Shetland the grilsing pattern is similar to that in Norway and most of the grilse are harvested in September when fish weigh over 3kg on average.

Salmon are harvested from late October through to January or February although there is a growing trend to sell most of the fish prior to Christmas.

According to official statistics released by DAFS, the average weight of salmon produced by Scottish farms is 2.6kg. However, fish harvested in December can attain 4kg or even 5kg in weight. Some of the larger, faster growing salmon may be kept for broodstock.

The major problems faced by cage farmers at sea sites are a result of predators, disease and bad weather conditions. Seals, diving birds (cormorants, guillemots, *etc*) and dogfish are the principal predatory species involved and can cause not only significant mortalities amongst stock, but can also damage nets, spread disease and adversely affect growth and disease resistance (see Section 7.4). The usual preventative methods employed include anti-predator nets (3-4cm square mesh) and top nets. Bacterial diseases (fin rot, vibriosis, furunculosis) and parasites (*eg* sea lice) are responsible for most of the economically serious losses in the industry. Salmon diseases, their prevention and treatment have recently been reviewed by Roberts and Shepherd (1986). Many diseases can be treated effectively if diagnosed at an early stage and both DAFS, Aberdeen, and the Institute of Aquaculture, Stirling, offer comprehensive diagnostic services. Some of the larger farms employ trained graduate biologists to help in disease monitoring. All drugs, however, must be obtained from veterinarians. Vaccines against vibriosis and furunculosis are currently being developed in Scotland and elsewhere.

### 8.3.5 Harvesting and marketing

The Scottish Salmon Growers Association (SSGA) lay down specifications covering the harvesting procedure and product quality for all Scottish farmed salmon, whether marketed fresh or frozen, whole, gutted, or headed and gutted (Neusinger, 1983). At most salmon farms in Scotland fish are starved for 24 hours until the gut is empty, and killed in the traditional way by knocking the fish on the back of the head with a wooden club, although there is some pressure from Norwegian interests in the industry to change to killing by bleeding (see Section 6.4). The majority of fish are sold whole and fresh, and are transported to market or to fish smokers on ice in polystyrene boxes (*Fig 8.15*).

Farmed Atlantic salmon is sold directly to fish smokers, sent to the wholesale markets or wholesalers in Aberdeen, Glasgow, Manchester or Billingsgate (London), exported, sold to other processsors, or sold directly to retail and catering outlets where it must compete against wild Atlantic and Pacific salmon, and imported farmed fish (Shaw and Rana, 1985a, 1985b).

The wholesale price for fresh Scottish farmed salmon at Billingsgate in 1984 varied from £3.87 to £4.91 per kg, depending on size, quality and time of year (from Fish Farmer). Real prices for fresh salmon in the UK have in fact declined since the late 1970s due to a marked increase in the availability of farmed fish, and wild salmon still fetches up to twice the price of farmed fish, despite favourable comparisons having been widely made in the press.

Smoked salmon is currently enjoying a boom in demand. Approximately 40% of the UK domestic salmon consumption is smoked and more than half of this is sold to the catering industry via wholesalers. The remaining 1,900 tonnes is sold via non-multiple stores, such as fishmongers (700 tonnes), multiple stores (900 tonnes), mail order (160 tonnes), contract smokers (60 tonnes) and direct sales from stalls, *etc* (80 tonnes) (Shaw and Rana, 1985a). Retail prices vary enormously depending on whether a whole side or a pack of sliced salmon is purchased.

The future of salmon farming in Scotland looks promising, although a number of factors are likely to have a moderating effect on the growth of the industry:-

(i) Smolt supplies. The comparatively slow growth of this sector of the industry, coupled with the poor summer of 1985 and increased pressure on smolt producers to sell their products abroad are likely to have an adverse effect on the growth of farmed salmon production in the mid-late 1980s.

(ii) Ongrowing sites. Most of the best sites for cage rearing of salmon have already been taken and in future farmers will be forced to utilise more exposed sites and use expensive, sophisticated cage designs, or opt for land-based ongrowing systems. Moreover, there may be increased competition for existing sites from other users. Currently the site rental is a nominal £20 per annum, and permission is granted for three years at a time. A more realistic site fee is probably desirable in the long term.

Much of the growth of the industry in recent years has been at existing sites. However, continued expansion should be done with care. Although it is at present impossible to give hard and fast rules

*Fig 8.15* A good quality, farmed salmon, ready for sending to market (courtesy D A Robertson).

concerning carrying capacity of seawater sites (see Chapter 5), undoubtedly some of the older and more heavily exploited areas are troubled with poor water quality, chronic disease problems and disappointing growth. The provision of development guidelines is therefore of major importance.

(iii) Markets. Consumer demand for fresh and smoked salmon products must continue to be stimulated if the industry is to maintain its growth and remain profitable, and it is believed that price-led growth through a significant reduction in price is probably necessary (Shaw and Rana, 1985b).

## 8.4 Tilapia farming in the Philippines

### 8.4.1 Introduction

Much has been written about the tilapias and there are a number of excellent bibliographies, text books and reviews available (Bardach *et al*, 1972; Balarin and Hatton, 1979; Hepher and Pruginin, 1981; Pullin and Lowe-McConnell, 1982; Balarin and Haller, 1982; Jauncey and Ross, 1982; Macintosh and Sampson, 1986). For this reason, little will be said here regarding their ecology or biology and many of the broader aspects of their farming will also be omitted.

### 8.4.2 Taxonomy, biology and culture

The tilapias are members of the large tropical Cichlidae family of inland water fishes and all seventy-five or so species are believed to have come originally from continental Africa and Palestine (Philippart and Ruwet, 1982). Although there is still much debate on the issue, the current concensus of opinion suggests that the group may be divided into three genera, *Tilapia, Sarotherodon* and *Oreochromis*, on the basis of differences in feeding and reproductive habits and anatomical and electrophoretic characteristics (Trewavas, 1982, 1983; McAndrew and Majumdar, 1984). Members of the genus *Tilapia* are substrate spawners whilst those belonging to the genera *Sarotherodon* and *Oreochromis* are biparental and maternal mouthbrooders respectively.

Interest in the tilapias has increased tremendously in recent years as they have a number of important attributes which predispose them to farming. All species feed at the base of the aquatic food web on plankton, macrophytes or detritus (Chimitz, 1955; Moriarty and Moriarty, 1973a, 1973b; Bowen, 1982), they breed readily in captivity (see later), and are resistant to handling, disease and poor water quality (Roberts and Sommerville, 1982). The fish have a fine-textured, white flesh and have an excellent taste which has proved universally popular.

Around half-a-dozen species and various hybrids are currently important in commercial culture (*Table 8.6*), although there are a number of other species whose potential has not yet been fully evaluated. *O. mossambicus* was

*Table 8.6* Tilapia species currently important in aquaculture.

| |
|---|
| *Oreochromis niloticus* |
| *O. mossambicus* |
| *O. aureus* |
| *Tilapia rendalli* |
| *T. zillii* |

formerly the most popular species but it has a number of traits which today are recognised as undesirable for most farming situations. It usually attains sexual maturity in 2-3 months when only 8-9cm in size and thereafter continues to breed every 4-6 weeks throughout the year in tropical conditions, females producing 100-500 eggs per spawn (Hepher and Pruginin, 1981; Macintosh and Sampson, 1986). Subsistence farming of this species in ponds frequently results in overcrowding, insufficient food and stunting. Although most other species of tilapias share similar reproductive traits, they tend to mature a little later and at a larger size.

Problems of early maturation, stunting and disruption of growth in females through partitioning of energy into reproduction and cessation of feeding during fry incubation can be tackled by:-

(i) Manually sexing fry at an early stage in their growth and subsequent ongrowing of males.
(ii) Use of predators to control over-population.
(iii) Sex-reversal of fry through the use of hormones to produce all males
(iv) Hybridisation of selected species to produce a high proportion of males.

However, not all of these methods are practical for large-scale production of fish and most require well-organised hatcheries operated by trained personnel (see Guerrero, 1982; Lovshin, 1982; Macintosh and Sampson, 1986, for reviews).

Tilapia culture is not just restricted to the tropics but is practised on a commercial scale in many subtropical parts of the world such as Israel (Hepher and Pruginin, 1981) and the southern states of the USA (Suffern, 1980) and is being experimented with wherever suitable supplies of warm water are available, such as in thermal power station effluents (Melard and Philippart, 1980) and in recycle systems (Meske and Manthe, 1983). Cage farming of tilapias began at Auburn University, Alabama, USA in the late 1960s with the experimental culture of *O. aureus* in small ($0.7m^3$) cages in ponds (Pagan, 1969; Suwanasart, 1972). Interest quickly spread, and to date some three dozen countries have experimented with rearing different species in cages in a range of environments. Cage culture of tilapias has proved attractive since apart from the usual advantages cited for cages (see Chapter 2), it has been discovered that using cages can overcome the major disadvantage of tilapias — the overcrowding and stunting that can occur in land based systems. Pagan-Font (1975), working with *O. aureus*, and Rifai (1980), using *O. niloticus*, demonstrated that tilapias fail to rear offspring successfully in cages with mesh sizes greater than 0.6cm, for although some eggs were produced, these fell through the net floor of the cages.

Reviews of cage culture of tilapias may be found in Coche (1977, 1979, 1982) Balarin and Haller (1982) and Campbell (1985).

### 8.4.3 CAGE FARMING IN THE PHILIPPINES

#### *8.4.3.1 Introduction*

*O. mossambicus* was introduced by the Bureau of Fisheries (BFAR) from Thailand in 1950, but pond culture of the species failed through inappropriate management (Guerrero, 1985). The species spread, however, to many parts of the country and although it became an important component of inland fisheries production, it was widely regarded as a pest by fish farmers as it invaded milkfish ponds, proliferated and competed for food. *O. niloticus* was first introduced from Israel in 1972 and subsequent introductions over the next 12 years included *O. aureus* from Singapore, *T. zillii* and a hybrid 'red' strain from Taiwan (Guerrero, *ibid*).

With the introduction of these new species and better management methods, through research and dissemination of results by scientists and not least through the effects and ingenuity of local fish farmers, the 1970s witnessed renewed interest in tilapia farming in the Philippines. Although per capita fish consumption was falling, both the sales and the price of tilapia increased steadily. Ponds and pens were given over to tilapia culture and in 1975 the first commercial cage culture began in Lake Bunot in the San Pablo lakes area of Laguna Province, Luzon. Over the next few years it was eagerly introduced and adopted at other neighbouring lakes, including Laguna de Bay. In Mindanao in the south of the country it began in lakes Lanao and Sebu in 1977 and at Lake Mainit in 1981 (Oliva, 1985). There are currently several dozen lakes and reservoirs and a number of rivers being used for commercial culture of tilapias and these vary enormously in size and productivity and in the extent of development (*Table 8.7*). By early 1984, some 10,000 tonnes of tilapia per annum were being produced from cages, which corresponded to ~20% of the total fisheries and aquaculture production of tilapias in the country (Guerrero, 1985).

#### *8.4.3.2 Hatchery industry*

In all but a few areas, such as Mindanao, where the darker *O. mossambicusis* is still popular, *O. niloticus* is the species currently preferred by both commercial growers and buyers. 'Natural' (*ie* wild *O. niloticus* x *O. mossambicus*) tilapia may be used by some small-scale operators or in areas where *O. niloticus* is unavailable (Escover and Claveria, 1985). Imported and locally developed 'red' strains of tilapia are also gaining in popularity, although their share of the market remains small. In the early 1970s BFAR supplied most of the fingerlings required by the small grow-out industry and today they operate over thirty hatcheries in all regions of the country.

However, in the past five years, hundreds of small, back-yard operations and dozens of larger commercial hatcheries have also become established. Moreover, many ongrowing units have discovered that it is more profitable to operate their own hatcheries. Current (1984) fingerling requirements are probably in the region of 600 million per year, assuming 25% mortality and 10 pieces per kilo at harvest, and over 75% of these are produced from small ($<10^6$ fingerlings $annum^{-1}$), privately-owned units.

Unfortunately, the quality of *O. niloticus* fingerlings produced by the private sector is highly variable (Smith and Pullin, 1984). Most stocks are contaminated with *O. mossambicus* and fingerlings often have poor grow-out performance. The role of government hatcheries is crucial in promoting the production of high quality fingerlings through the culture and dispersal of good broodstock to hatcheries.

Hatcheries may be either land-based or water-based, the land-based operations using either open pond or hapa-in-pond production methods (Beveridge, 1984a). Details of open pond hatchery systems and management can be found in Smith *et al* (1985). Hapas give greater flexibility in the use of pond space but add to production, maintenance and labour costs. A commercial hapa-in-pond hatchery is illustrated in *Fig 8.16*. The hapas are small (3 × 3 × 1m) fixed cages fabricated from fine mesh (1-3mm) nylon mosquito netting attached to bamboo or 'ipil-ipil' (*Leucaena leucocephala*) poles driven into the bottom of ponds so that the hapa remains 10-20cm clear of the pond bottom. They are stocked with broodstock fish at densities of 4-7$m^{-2}$ at ratios of 3, 5 or 7 females to each male. Fry are collected once per fortnight or so by concentrating them in a corner of the hapa and scooping them out with a small net (*Fig 8.17*). They may then be transferred to nursery hapas or ponds where they are kept until sold. Using a good quality locally-compounded feed (*eg* 25% fish meal, 75% rice bran), production rates of 240 fry $m^{-2}$ $month^{-1}$ can be achieved in land-based systems, although this is at least an order of magnitude greater than production typically achieved in small back-yard operations where less feeding is likely to occur and where broodstock and management techniques are often poor (Beveridge, 1984a; Yater and Smith, 1985). Under experimental conditions, Hughes and Behrendts (1984) have achieved in excess of 2,000 fry $m^{-1}$ $month^{-1}$ with the hapa-in-pond method in the USA (Stocking density = 5$m^{-2}$; ratio of 2 females to each male).

During the late 1970s and early 1980s, researchers from the Southeast Asian Fisheries Development Center (SEAFDEC), Laguna Lake Development Authority (LLDA), and various universities experimented with lake-based hatcheries and nurseries with the result that today there are several commercial lake-based fry and fingerling units in operation, particularly in the San Pablo Lakes and in Laguna de Bay in southern Luzon. Most operators use large (typically 5 × 5 × 2m) hapa net bags attached to bamboo or 'anahaw' posts driven into the lake substrate (*Fig 8.18*). These are stocked with broodstock and managed in a similar manner to hapa-in-pond systems (*Fig 8.19*). Trials conducted by SEAFDEC staff have determined that optimum stocking densities are around 2.5-5.0 broodstock $m^{-2}$ that a

*Table 8.7* Utilisation of Philippine inland waters for cage culture of tilapia. Data on surface areas, mean depths and primary production from PCARRD (1981).

| *Region and province* | *Lake/reservoir/ river* | *Area (ha)* | *Mean depth (m)* | *Gross primary production* ($g\ Cm^{-2}\ day^{-1}$) | *Year cages introduced* | *Present*[1] *area (ha)* | *% lake area* | *No. operators* | *Av. area per operator* ($m^2$) | *References* |
|---|---|---|---|---|---|---|---|---|---|---|
| I<br>Ilocos Norte | Paoay | 402 | — | 3.4 | | | | | | Coche, 1982 |
| III<br>Neuva Ecija | Pantabangan | 8,420 | 23.6 | — | 1981 | E | | 1 | | Guerrero *et al*, 1985 |
| IV<br>Laguna | Bunot | 29.6 | 16.1 | 1.9 | 1975 | 8.75 | 29.56 | | | Garcia, 1979; Guerrero, 1982 |
| | Calibato | 43.0 | — | 7.8 | 1976 | | | | | Aragon *et al*, 1985 |
| | Laguna de Bay | 89,076.3 | 3.8 | 0.5–9.0 | 1976 | 100 | 0.11 | 1,500–2,000 | 530–870 | Santiago, 1982; Lazaga and Roa, 1985; Escover and Claveria, 1985 |
| | Mohicap | 14.0 | 28.8 | 1.1 | 1977 | | | | | Aragaon *et al*, 1985 |
| | Palacpaquen | 43.5 | 6.9 | 1.8 | 1977 | | | | | Aragon *et al*, 1985 |
| | Sampaloc | 105.4 | 17.6 | 1.8 | 1977 | 2.5 | 2.5 | | | Aquino, 1982 |
| Quezon | Gunao | | | | | | | | | Guerrero, 1982 |
| Batangas | Taal | 24,356.4 | 45.3 | 0.7 | | | | | | Guerrero, 1985 |
| Mindoro | Naujan River | — | — | — | 1982 | | | 9 | | |

*Table 8.7 (continued)*

| Region and province | Lake/reservoir/ river | Area (ha) | Mean depth (m) | Gross primary production (g Cm$^{-2}$ day$^{-1}$) | Year cages introduced | Present[1] area (ha) | % lake area | No. operators | Av. area per operator (m$^2$) | References |
|---|---|---|---|---|---|---|---|---|---|---|
| V | | | | | | | | | | |
| Camarines Sur | Bato | 3,792.5 | 3.6 | 15.2 | 1981 | 7.1 | <0.01 | | | J. Bisuña, pers comm. |
| | Buhi | 1,015.8 | 7.3 | 2.3 | 1979 | 7.9 | <0.01 | | | Gindelberger, 1982 Guerrero, 1985 |
| X | | | | | | | | | | |
| Surigao-del Norte | Mainit | 17,430.2 | | 13.4 | | 4.1 | <0.01 | | | Guerrero, 1985 |
| XI & XII | | | | | | | | | | |
| Lanao | Lanao | 33,999.7 | 60.3 | 0.5–0.8 | 1977 | | | | 500–1,000 | Oliva, 1985 |
| Cotabato | Buluan | 6,134.2 | | 5.0 | 1981 | 5.5 | <0.01 | 1* | 5.5ha | Oliva, 1985 |
| | Lahit | 248.8 | | | 1979 | | | | | PCARRD, 1981 |
| | Sebu | 964.1 | | | 1977 | 45–50 | 0.05 | 400 | | Oliva, 1985 |
| | Sultan | 197.3 | | | | | | | | PCARRD, 1981 |

1 = 1984
E = Experimental
* = Southern Philippines Development Authority

*Fig 8.16* A commercial hapa-in-pond hatchery, Laguna de Bay, Philippines.

*Fig 8.17* Harvesting tilapia fry from a pond hapa, Laguna de Bay, Philippines.

*Fig 8.18* Lake-based hatchery, Laguna de Bay, Philippines. Fixed cages constructed from anahaw palms and fitted with top nets for storm protection (see text).

*Fig 8.19* Harvesting tilapia fry from a lake-based hatchery hapa, Laguna de Bay, Philippines.

ratio of 5-7 females per male results in highest fry production per unit cage area (SEAFDEC, 1981, 1983) although more recent work suggests 2 or 3 females per male should be used.

A number of modified designs have also been developed to improve fry production by minimising cannibalism of young fry by older fry and broodstock (*Fig 8.20*) (Guerrero and Garcia, 1984; Beveridge, 1984a). In *Fig 8.20b*, the broodstock are confined to a large mesh (30mm) net enclosure within the hapa net bag, so that the movement of fry into the outer area of the unit is facilitated. However, the flow of water through such a cage may eventually become restricted as the broodstock do not have access to the walls of the outer cage and so cannot graze epiphytic algae and detritus. An alternative design has been proposed by Dr A Bautista of SEAFDEC (*Fig 8.20c*) in which good flow conditions to the broodstock are maintained.

Trials conducted by Guerrero and Garcia (1984) comparing the designs shown in *Figs 8.20a* and *b* concluded that although there were no significant differences in fry production per unit area, production per breeder was greatly increased, although the cost of the modified design offset this advantage.

Lake-based hatcheries have a number of advantages over land-based operations; they are cheaper to construct, since they do not involve land purchase or excavation, and they allow non-land owning sectors of the community access to the tilapia culture industry. Moreover, the fish can derive a large proportion of their food requirements from plankton (see Guerrero and Garcia, 1984). However, lake-based hatcheries also suffer from a number of disadvantages. They are vulnerable to pollution, over-exploitation of the site, theft, vandalism and storms and typhoons (Beveridge, 1984a). Although typhoon damage can be reduced by fitting top nets to the cages so that they can be submerged if necessary (see Chapter 7), some hatchery operators in Laguna de Bay have decided that lake-based systems involve too great a risk and so have relocated their businesses to the comparative safety of land.

#### *8.4.3.3 Ongrowing operations*

The fingerlings required by cage farmers are between 0.5 and 20g in size. The smaller and hence cheaper fish are preferred by most farmers, although larger, commercial farms operating their own hatcheries and nurseries often use larger fish. Fingerlings are usually transported in sealed, 8 litre plastic bags, partially filled with oxygenated water (3,000-5,000 fish per bag). Fish are usually starved for 12-24h prior to transport (*Fig 8.21*). Stocking of cages is carried out in the early morning or late afternoon when temperatures are slightly lower.

Two types of cages are in common use in the Philippines — fixed and floating. Coloured knotless nylon or PE nets of variable mesh size are employed (*Table 8.8*). In some lakes such as Laguna de Bay, cage and pen culture is restricted to designated areas in an attempt to maintain navigation routes and to protect the right of access to the lake by other users such as

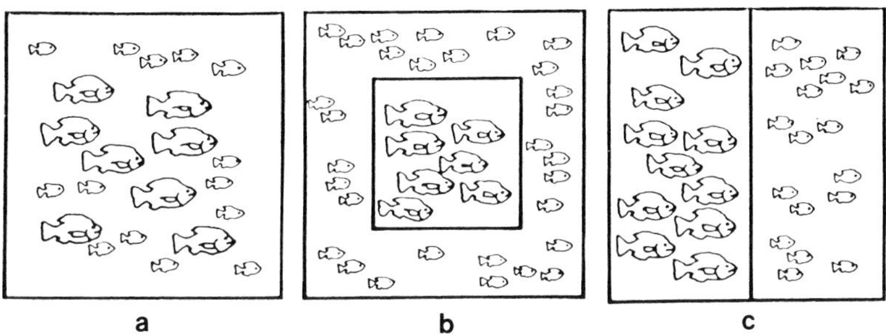


*Fig 8.20* Designs for lake-based tilapia hatcheries (a) simple hapa; (b) broodstock restrained in central area of hapa; (c) broodstock restrained in one half of the hapa (see text) (from Beveridge, 1984a).

*Fig 8.21* Sealed plastic bags of oxygenated water plus tilapia fry ready for distribution. Note outer protective bag made of palm leaves.

*Table 8.8* Recommended mesh sizes for cage culture of tilapia.

| *Size of fish* | *Purpose* | *Mesh size (mm)* |
|---|---|---|
| Fry (<12g) | Nursery | 1 – 3 |
| Fingerling (12–30g) | Grow-out | 4 – 8 |
| 30–200g | Grow-out | 10 – 20 |
| 200g + | Grow-out | 20 – 25 |
| Breeding (150g +) | Breeding | 1 – 3 |

*Table 8.9* A summary of cage tilapia production in Philippine lakes.

| *Lake* | *Date* | *Cage size (m)* | *Species* | *Stocking density* $(-m^3)$ | *Size at stocking (g)* | *Culture period (month)* | *Size at harvest (g)* | *Feed* | *Production* $(kg\ m^{-3}$ $(month^{-1}))$ | *Reference* |
|---|---|---|---|---|---|---|---|---|---|---|
| Bunot | 1975 | 50×25×5 | *O. mossambicus* | 16 | 5–10 | 6 | 100–240 | rice bran | 0.26–0.64 | Guerrero, 1981 |
| | 1979 | 50×25×5 | *O. mossambicus* | 16 | 5–10 | 6 | 100–150 | rice bran | 0.30–0.40 | Guerrero, 1981 |
| | 1980 | 20×25×5 | *O. niloticus* | 4 | — | 4 | 250 | no feed | 0.24 | Alvarez, 1981 |
| Laguna de Bay | 1978 | 5×10×3–10×20×5 | *O. niloticus* | 4–8 | 21 | 4–5 | 100 | no feed | 0.07–0.18 | Mane, 1979 |
| | 1978 | 5×10×3–10×20×5 | *O. niloticus* | 10–16 | 21 | 4–5 | 100 | various feeds | 0.18–0.36 | Mane, 1979 |
| Pillila, LDB | 1981 | 3×3×1.5–5×20×1.5 | *O. niloticus* *O. niloticus* | 20 | 1 | 3 | 100–125 | 1:1 wheatmeal broiler feed | 0.73 | Lampa, 1981 |
| Los Banos, LDB | 1982 | 266×3 | *O. niloticus* | 4.3 | 20.2 | 10 | 2,112 | rice bran and other materials | 0.05 | Aragon *et al*, 1985 |
| Binangonan, LDB | 1982 | 248×3–1,440×5 | *O. niloticus* | 6.8 | — | 6.6 | 118 | limited feeding | 0.12 | Lazaga and Roa, 1985 |
| Cardona, LDB | 1982 | 138×3–2,900×5 | *O. niloticus* | 7.4 | — | 6.3 | 119 | limited feeding | 0.14 | Lazaga and Roa, 1985 |
| Sampaloc | 1983 | 10×10×9–25×20×9 | *O. niloticus* | 1.6–2.0 | 12.5–16.0 | 6–9 | 225–300 | no feed | 0.05–0.08 | Guerrero, 1985 |
| Taal | 1983 | 10×5×3 | — | 50 | — | 4 | 100 | no feed | 1.25 | Guerrero, 1985 |

*Table 8.9 (continued).*

| *Lake* | *Date* | *Cage size (m)* | *Species* | *Stocking density* ($m^{-3}$) | *Size at stocking (g)* | *Culture period (month)* | *Size at harvest (g)* | *Feed* | *Production (kg* $m^{-3}$ *(month*$^{-1}$*))* | *Reference* |
|---|---|---|---|---|---|---|---|---|---|---|
| Bato | 1982 | 6×3×3–10×8×3 | *O. mossambicus* × *O. niloticus* | 50 | — | 4 | 70–200 | no feed or rice bran and shrimp | 0.18–0.50 | Escover and Claveria, 1985 |
| Buhi | 1982 | 2×3×2.5–22×3×2.5 | *O. mossambicus* × *O. niloticus* | 10 | — | 4 | 70–200 | rice bran and fish meal/ dried shrimp | 0.18–0.50 | Escover and Claveria, 1985 |
| Lanao | 1982–83 | 5×10×5 or smaller | *O. niloticus* × *O. mossambicus* | 1–5×$10^3$ cage$^{-1}$ | 1 | 7 | 125 | rice bran and fish meal and ipil-ipil | 0.18 | Oliva, 1985 |
| Buluan | 1982–83 | 5×10×5 | *O. niloticus* | 10 | 21 | 5 | 200 | no feed | 0.40 | Oliva, 1985 |
| Sebu | 1982–83 | 5×10×5 (most) | *O. niloticus* and *O. mossambicus* | 10 | 1.5 | 5.8 | 160 | tiki-tiki, ipil ipil, copra | 20.28 | Oliva, 1985 |

fishermen and snail gatherers (*Fig 4.21*). Lake-side municipalities may charge a small annual rental fee based on cage area. Regulations apart, cages are usually sited in sheltered areas or as near as possible to the operators home for security reasons. Some operators also construct their cages inside or on the leeward side of fishpens which offer a measure of protection during stormy weather.

Cages vary enormously in size (*Table 8.9*) and in general, the larger the farm, the larger the cages that are used (Escover and Claveria, 1985). Stocking densities range from 5 to 50 fingerlings per $m^3$ depending on site, cage size, feeding practices, and desired size at harvest. Stocking densities may also change with season (Bautista, 1984) and from year to year. Some supplementary feeding is carried out by the majority of farmers at most sites, although when first exploited, several lakes such as Laguna de Bay and the San Pablo lakes, were sufficiently productive to support high densities of fast-growing fish without recourse to the use of feeds (PCARRD, 1981; Coche, 1982). However these lakes were rapidly over-exploited (Beveridge, 1984b). At several of the more recently developed sites such as Lake Buluan in Mindanao and Lake Taal in Batangas, extensive culture is still widely practised (Oliva, 1985; Guerrero, 1985), whilst at other sites, feeding is only carried out during the unproductive winter months (*Table 8.9*).

Rice bran ('cono' or fine grade) is probably the most widely used feed in the country. Other materials include composted water lilies, chicken and pig manure, dried fish and shrimp meal, copra meal, ipil-ipil leaf meal, kitchen scraps and wastes from various food processing industries (Smith *et al*, 1985). Several cheap but good compounded feeds have been developed, the most widely used consisting of 77% rice bran and 23% fish meal (Pantastico and Baldia, 1979, Guerrero, 1980). Recommended feeding rates with this feed are 3-5% body weight per day, depending on fish size, the feed being either broadcast on the water surface or moistened into balls and fed to the fish several times per day. However, recent surveys instigated by PCARRD and ICLARM have shown that most farmers, especially the poorer ones, feed their fish as and when they can afford (Smith *et al*, 1985).

Details of cage tilapia production from a number of lakes are summarised in *Table 8.9*. Not only are there tremendous differences in production between sites, but there are also year-to-year variations and differences between sites within a lake. In lakes Bunot and Laguna, for example, there has been a decline in production over the years and a corresponding increase in the use of supplementary feeds. In Laguna de Bay, sites in Rizal Province (*eg* Pillila and Binangonan) have higher production figures than those in Laguna Province (*eg* Los Baños) due largely to differences in productivity within the lake. Highest production is at Lake Taal, where 1.25kg $m^{-3}$ $month^{-1}$ has been achieved without recourse to supplementary feeding (Guerrero, 1985). In addition to pollution, typhoons and poaching, many farmers cite problems of overcrowding at existing sites (*Fig 8.22*) and lack of available capital for development (Smith *et al*, 1985).

Quality of fry and fingerlings is of concern not only to cage growers but also to all tilapia farmers (Smith and Pullin, 1984). The economics of cage

*Fig 8.22* Overcrowding of cages at the outflow of Lake Buhi, Philippines, November 1983.

farming in different regions of the country have been evaluated by Gonzales (1984), Oliva (1985), Aragon *et al* (1985), Lazaga and Roa (1985) and Escover and Claveria (1985). Not surprisingly, profitability was found to vary enormously, some farms even incurring losses on paper due to high, non-cash labour costs. However, Gonzales (1984) demonstrated that even small, family-run, 6 unit cage operations could realise sufficient profit to increase typical rural incomes more than six-fold.

## 8.5 Whitefish farming in northern Europe

### 8.5.1 Introduction

The Coregonidae, or whitefishes as they are more commonly known, are a cold water-loving family of fishes which are widely distributed throughout freshwater lakes of Northern Europe, Asia and North America (Wheeler, 1978). It is now widely accepted that there are three genera within the family — *Stenodus*, *Prosopium* and *Coregonus* — although the number of species within each genus is still a matter of considerable debate, since many of the characteristics used by taxonomists to diagnose species (*eg* shape, size, number of gill rakers, scales, *etc*) are highly variable and are governed by environmental conditions. There are probably between 20 and 30 species belonging to the genus *Coregonus*, six of which occur in Europe: *C. albula* (cisco or vendace), *C. peled* (peled), *C. pidschian* (arctic whitefish), *C. nasus* broad whitefish), *C. lavaretus* (common whitefish or powan) and *C. oxyrhynchus* (houting) (Wheeler, 1969).

Most species of whitefish are pelagic, at least for part of their life, and feed largely on zooplankton, although some, like the common whitefish, may as adults adopt a benthophagic lifestyle during the colder winter months. Many species possess long, fine gillrakers which are believed to be specialist adaptations for filter feeding (Nilssen, 1978). If surface water temperatures become too high, they may remain below the thermocline in stratified lakes. Spawning occurs in the colder months of the year and the fish often congregate in huge numbers in the spawning areas. Eggs are laid in open water, usually over sand or gravel substrates, and hatch after 100-150 days or so. The fry remain pelagic for the first year at least. Details of breeding in the commercially more important species are summarised in *Table 8.10*.

Commercial coregonid fisheries have been in existence in many parts of northern Europe, Asia and North America for centuries and, according to FAO, catches of European species currently total around 45,000 tonnes per annum, with USSR and Finland accounting for 85% of production. The fishes are eaten fresh — they are the preferred ingredient for gefilte fish (Bardach *et al*, 1972) — dried or smoked, but although popular, have a comparatively low market value. Today, however, many of the major fisheries are in decline through a combination of over-fishing, pollution and damming of migratory routes.

Culture of coregonids dates back to the late 19th Century in North America and Europe (Bardach *et al*, *ibid*), although today the principal exponents are the Russians and Finns who raise large numbers of vendace, arctic and common whitefish for replenishing depleted fisheries and for stocking reservoirs. Common whitefish and peled are also apparently used as predators in carp ponds (Bardach *et al*, ibid).

### 8.5.2 Cage culture

Cage rearing of coregonid species — common whitefish, peled, vendace — has been carried out experimentally since the early 1970s in Poland, the Federal Republic of Germany, the German Democratic Republic, the USSR,

*Table 8.10* Details of reproduction in the three commercially important cultured species of European whitefish (from various sources).

| *Species* | *Max. size (cm)* | *Age at maturity (y)* | *Spawning time* | *No. eggs* | *Hatching time (days)* |
|---|---|---|---|---|---|
| *C. lavaretus* | 20 (usual)<br>70 (Baltic pops.) | 3–4 | Oct.–Jan. | 1,000–<br>28,000 | 100 |
| *C. albula* | 20–26 | 2–3 | Oct.–Dec. | 1,700–<br>4,800 | 100–200 |
| *C. peled* | 27 (lakes) | — | Sept.–Nov. | 29,000–<br>105,000 | — |

Finland and Switzerland (Bronisz, 1979; Marciak, 1979; Uryn, 1979; Jäger and Nellen, 1981; Mamcarz, 1984; Mamcarz and Szczerbowski, 1984; Szczerbowski and Mamcarz, 1984; Zaugg and Pedroli, 1984; Mamcarz and Worniallo, 1985; Worniallo and Mamcarz, 1985). However, despite considerable success, no commercial operations have yet been established other than the small-scale rearing of juveniles for restocking purposes in the Federal Republic of Germany (T. Jäger, pers. comm.) and Poland (Mamcarz and Worniallo, 1985).

Fry are produced by capturing wild fish, stripping and fertilising the eggs, and incubating them in Zougg or MacDonald hatching jars (see Huet, 1971 for details). Hatching occurs in late March/early April. Usually the larvae must be fed for at least the first few weeks on specially-reared live foods (*eg Artemia*) before they will take dry diets (Gunkel, 1979). However, larvae have been successfully transferred from the hatchery to cages, immediately after the yolk has been absorbed, when only 2-5 days old, thus avoiding the difficulties associated with live food rearing and feeding (Mamcarz and Szczerbowski, 1984).

Early attempts to rear coregonid larvae and fry in cages simply relied on the passive transport of planktonic food through the cages by external currents (Dembinski and Zuromska, 1973 and Marciak *et al* 1976, in Marciak, 1979). The feeding behaviour of larval coregonids has been studied and described in detail by Braum and others (Braum, 1978). Reactions to prey are apparently triggered by prey items up to $1mm^2$, and in aquaria it has been shown that the larvae can scan volumes of 11-18*l* $h^{-1}$. Using small, floating cages, stocked with 2,000-10,000 fish $m^{-3}$, differences in food preferences were found between species and between size groups (Marciak, 1979). When first introduced into cages in April there were striking differences between larval peled and common whitefish in diet composition. The former's diet consisted largely of *Cyclops* sp. (75%) with relatively small numbers of *Daphnia* sp. (23%) and *Leptadera kindti* (1%) whilst the diet of the latter was largely composed of *Daphnia* sp. (80%), the balance being made up of *Cyclops* sp. (16%) and *Bosmina* sp. (4%).

Five weeks after stocking, a comparison was made between the food preferences of the smallest and largest fishes within a cage. Large fish were found to congregate around the windward cage wall and preyed on relatively large plankters such as copepodite stages and *Cyclops* sp. (58%), whilst the smallest fish tended to occupy the deep water in the centre of the cage and fed largely on rotifers (54%).

However, selectivity and differences between species in food preferences diminished with time. Moreover, Marciak (*ibid*) observed that the availability of planktonic food to the caged fish decreased markedly during the summer months. Initially, plankton density inside the cages was around 40% of that recorded outside. However, by late summer, densities inside the cages were only 2% of those outside. The food requirements of coregonid fry are discussed by Braum (1978) and Gunkel (1979) who have shown that they increase many fold during the first few months. The shortage of available food in Marciak's experiments was reflected in the presence of periphytic algae and sedentary stages of animals from the cage walls in the diet of older fish.

The first experiments with illuminated cages began in Poland in 1974 and because growth and survivorship at all stages were found to considerably improve (Uryn, 1979), all subsequent work on cage rearing of coregonids has involved the use of lights. Small (2 × 2 × 2m), submersible cages are typically used and are suspended from a variety of types of floating raft (*Fig 8.23*). Mains electricity supply (220v) is generally transformed to 24v and is used to operate 60 or 100w bulbs which are sealed in waterproof PVC housings (Jäger and Nellen, 1981; Mamcarz and Szczerbowski, 1984).

Details of stocking densities and cage mesh sizes for common whitefish are summarised in *Table 8.11*. Values for peled and vendace are similar. The cages are submerged to a depth of 3-5m partly to avoid high summer temperatures and partly to minimise the rate of algal fouling. The cages are only lifted in order to clean the netting or check on fish growth (*Fig 8.24*) although during the winter the cages may remain suspended under ice for

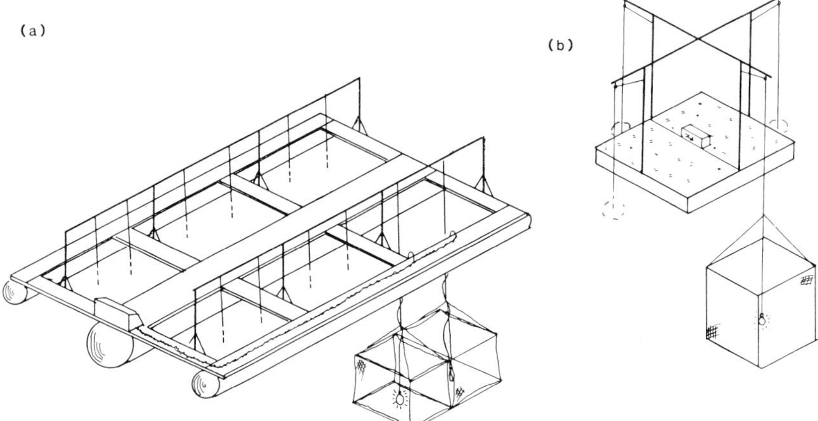


*Fig 8.23* Two types of floating raft/platform used to support submersible illuminated cages. (a) Poland; (b) Federal Republic of Germany. (Redrawn from Jäger and Kiwus, 1980; Mamcarz and Szczebowski, 1984.)

*Table 8.11* Stocking densities and mesh sizes (bar length, square mesh) used in cage common whitefish culture, Poland.

| *Age* | *Stocking density* | *Cage mesh size* |
|---|---|---|
| 0 – 6–9 weeks | 400–6,000 fish $m^{-3}$ | 0.8–1.0mm |
| 9 – 26 weeks | 300–4,000 fish $m^{-3}$ | 1.8mm |
| 26 – 52 weeks | 120–1,000 fish $m^{-3}$ | 5.5mm |
| 1+ years | 95– 430 fish $m^{-3}$ | 5.5mm |
| 2+ years | 80 fish $m^{-3}$ | 5.5mm |

*Fig 8.24* Inspecting cage for fouling and examining fish stocks. Note heavy duty zips on walls to facilitate observation of fry.

several months. The effect of the light bulbs is to induce a positive phototactic response in the zooplankton, thus attracting them into the cages (*Fig 8.25*), although it has been found that the larger individuals from amongst the biggest taxons (*eg Diaptomidae, Daphnidae, Leptadera kindti*) are effectively selected (Bronisz, 1979; Szcerbowski and Mamcarz, 1984). Unfortunately, there are no published accounts on the effects of different light levels or wavelengths on the attraction of zooplankton. In some studies lights have only been switched on at night whilst in others the lights remain on continuously, the proponents of the latter strategy claiming that not only is fry growth improved but also that the life of the lights is considerably extended. Thus by using lights, both the duration and availaility of planktonic food to the caged coregonids is increased.

Bronisz (1979) estimated that there was an approximate five-fold increase in the consumption of prey in illuminated cages, although the intake of prey at night is only about 80% of the daytime rate, and that the caged fish tended to predate on the larger zooplankters. However detailed studies of food preferences in three species of coregonid by Marciak (1979) have shown that when first stocked, vendace tend to feed on a diverse range of prey items, whereas the common whitefish feed on large mobile plankters (*eg Daphnia* sp., *Cyclops* sp.) and peled choose small sluggish items such as rotifer eggs and juvenile *Daphnia*. Feeding differences between species, however, were much less marked in the illuminated than in the unilluminated cages, and differences were also much less pronounced as the fish grow older (Marciak, *ibid*).

Growth and survival of caged coregonids is highly variable, depending on species, age and site characteristics. Mortalities are lowest in all species during the early stages. In one set of experiments, survival and growth of newly stocked larvae were excellent until June, when high temperatures and poor oxygen conditions depressed growth significantly (Mamcarz and Szczerbowski, 1984). Mortalities over the first winter were high (32-77%), but again growth was excellent during the period March-June of the second year. Culture of 1+ and 2+ fishes resulted in no further increase in stock weight, as survivership and growth were poor. At the end of the third year mean individual weights were 52-144g.

Mortalities and growth were also correlated with stocking density and trophic state. Mortalities increased with stock density and growth was highest in the mostly productive lakes (Mamcarz and Szczerbowski, *ibid*). In the above experiments no single specific cause was given for the mortalities. However, in trials conducted in the Federal Republic of Germany 61% mortalities of fry recorded over a six month growing period were largely attributed to bacterial gill disease (Jäger and Nellen, 1981).

In conclusion, it seems that the extensive cage culture of coregonid fry for restocking purposes is a viable proposition, and trials with rearing of planktivorous juveniles of other species such as vimba (*Vimba vimba*), pike (*Esox lucius*), pike perch ( *Lucioperca lucioperca*) tench (*Tinca tinca*) and even salmonids (*Salmo salar*, *S. trutta*) have been successful (Jäger and Kiwuș, 1980; Jäger and Nellen, 1981; Holm and Moller, 1984; Mamcarz and Szczerbowski, 1984; Zaugg and Pedroli, 1984). However, the extensive rearing of coregonids for the table does not look promising as survival and growth beyond the first few months are poor. Apart from the causes given

*Fig 8.25* Swarms of feeding coregonid fry gather round the light.

above the use of lights may also adversely affect the fish. Whilst the juvenile stages are phototactic, the adults are not, and the continuous illumination may contribute towards stress and disease susceptibility. Moreover, disruption of gonads was observed in both caged peled and common whitefish (Mamcarz and Worniallo, 1985; Worniallo and Mamcarz, 1985) and both species failed to attain sexual maturity despite being held for four years. A combination of overcrowding and poor nutrition was held to be largely responsible.

The environmental impact of extensive culture of zooplanktivorous fishes may at first appear minimal and indeed Szczerbowski and Mamcarz (1984) found no significant difference in physical or chemical parameters between cage and control sites in their study. However, in several studies a significant decline in zooplankton biomass and production, with a resultant fall in caged fish production, was observed (Bronisz, 1979; Szczerbowski and Mamcarz, 1984). Illuminated cages have also been found to disrupt the feeding behaviour and distribution of the wild fish community by attracting dense shoals of planktivorous species.

It seems, therefore, that although extensive culture of planktivores in illuminated cages has some potential, it should be carried out with caution. The economics of the system require to be properly evaluated and research into environmental effects must be a priority.

## 8.6 Tiger prawn farming in the Indo-west Pacific region

### 8.6.1 Introduction

There are several dozen species of prawn belonging to the family Penaeidae and around 15-20 of these occur in the marine and coastal brackish waters around the Indo-west Pacific region. However, only a few of these species are of commercial importance and even fewer have been farmed. Species which have been cultured to date include *Penaeus indicus*, *P. japonicus*, *P. merguiensis* and *P. monodon* and it is the latter species, commonly known as the giant tiger prawn or sugpo, which is most widely cultured.

*P. monodon* is distributed throughout the greater part of the Indo-west Pacific region, from 30°E to 155°E and from 35°N to 35°S. It occurs in southern Japan, Taiwan, Korea, Hong Kong, Philippines, Malaysia, Singapore, Thailand, Sri Lanka, India, Pakistan, Tanzania, South Africa and Australia (Motoh, 1981). Juveniles and adolescents tend to inhabit surface estuarine or inshore waters whilst adults occur in deeper water in bays or other coastal areas. The life cycle of all penaeid species is similar. *P. monodon* males become sexually mature when they attain a carapace length of 37cm+, females when they are somewhat larger (around 47cm+)(Motoh, *ibid*). During mating the male transfers and deposits a spermatozoa-containing spermatophore into the seminal receptacle of a female, which she stores until ready to spawn. Spawning takes place in deep water at night at temperatures of 25-30°C and in high salinities (Ling, 1977; Kafuku and Ikenoue, 1983). During spawning the female sheds both eggs and sperm simultaneously into the water whilst swimming. A gravid female will produce between 200,000 and 800,000 eggs, depending on body size (Motoh, 1981).

The eggs hatch at around 12h after fertilisation. The newly hatched nauplii subsist on the food reserves stored in the yolk sac and undergo a series of six moults within 36h or so. At this stage they are termed zoea, and begin to feed on phytoplankton. The zoea then moult three times within four days and become myses, approximately 2.5mm long. The myses begin to feed on small zooplankton and continue to grow rapidly and moult. After three moultings in five days, they metamorphose into post-larvae, 5.0mm long.

The various post-larval stages are indicated by the letters PL, followed by a subscript indicating their age in days from metamorphosis. At around $PL_{20}$, the prawns, now termed juveniles, leave the deeper, high salinity waters for shallower, inshore areas, where they settle on the sea-bed and adopt a sedentary life, feeding on small benthic invertebrates, detritus and algae. Gradually, they become nocturnal, burrowing in the sand during the day and emerging to feed after sunset. They also migrate to deeper water as they grow. By the time they are one year old, the males are around 12cm and the females 15cm in length. The males live for two years, the females three years (see *Table 8.12* for summary).

### 8.6.2 FISHING AND FARMING INDUSTRIES

Penaeid prawn fisheries exist in many countries in the region (Gulland and Rothschild, 1984), and according to recently published FAO statistics catches are around 900,000 tonnes per annum, most of which are landed in India, China and Indonesia. However, over-exploitation and pollution have led to a serious depletion of stocks in many areas, and the increasing pressure caused by capture of gravid females for aquaculture purposes (see later) has contributed towards the worsening situation.

In many countries in the region the farming of prawns is well established, having been incidental to the extensive pond culture of brackishwater species, such as milkfish (Bardach *et al*, 1982). Modern prawn farming has its origins in Japan where in 1934 Fujinaga successfully managed to spawn and partially rear *P. japonicus* (Bardach *et al*, *ibid*), although it was not until the 1960s that the commercial prawn farming industry began to develop. Today culture of various penaeid prawn species is practised in most countries in the region and methods range from traditional, extensive farming to highly intensive. In extensive culture the prawns are usually reared with fish — either concurrently or sequentially — in tidally flushed coastal ponds with little or no feed inputs. Low stocking densities and production figures of a few hundred kg per hectare per year are typical of such methods. However, in Taiwan and the Philippines, *P. monodon* is grown in monoculture in intensively managed ponds which utilise pumped seawater, aeration and specially formulated high protein feeds and yields of 5-10 tonnes per hectare per annum have been claimed.

World farmed prawn production is currently estimated at around 130,000 tonnes per annum, 70% of which is produced in the Indo-west Pacific region

*Table 8.12* The life history of the giant tiger prawn (modified from Motoh, 1981).

| *Stage* | *Commences at* | *Duration* | *Carapace length (mm)* *Male* | *Female* | *Model of life* | *Habitat* |
|---|---|---|---|---|---|---|
| Embryo | Fertilisation | 12 h | 0.29[1] | | Planktonic | Outer littoral area |
| Larvae | Hatching | 20 days | 0.5–2.2 | | Planktonic | Outer/inner littoral area |
| Juvenile | Completion of gill system | 15 days | 2.2–11.0 | | Benthic | Estuarine area |
| Adolescent | Stability of body proportion, development of outer genitalia | 4 months | 11–30 | 11–37 | Benthic | Estuarine area |
| Subadult | Commencement of sexual maturity, first copulation | 4 months | 30–37 | 37–47 | Benthic | Inner/outer littoral area |
| Adult | Completion of sexual maturity | 10 months | 37–71 | 47–81 | Benthic | Outer littoral area |

1 = egg diameter.

(from FAO, 1984) and approximately 25-30% of this will be accounted for by *P. monodon* culture in Taiwan, Indonesia, India, Philippines, Thailand and Malaysia.

### 8.6.3 Cage culture

Research into the feasibility of culturing prawns in cages seems to have originated in Japan in the late 1960s or early 1970s, although there are few published accounts available (Bardach *et al*, 1972). According to Tomiyama (1973), 3 × 3m floating bamboo collar cages were successfully used to culture *P. japonicus* at marine sites. A few years later Sampath and Menon (1975) succeeded in rearing several species, including *P. monodon*, in small (4.5 × 2 × 1m) fixed cages in estuarine areas near Cochin, India and since then penaeid species have been successfully cultured in cages in other parts of India (Krishnan *et al*, 1984; Maheswari, 1984; Venkatasamy, 1984), the Philippines (Anon, 1979; Peña and Prospero, 1984), Singapore, and even Italy (Lumare and Palmegiano, 1980), although few commercial cage-based operations have been established.

As in other methods of rearing *P. monodon*, cage culture involves stocking with wild or hatchery-reared postlarvae. In India and elsewhere, most trials have involved using $PL_{20}$ — $PL_{35}$ stages. However, in the Philippines it has recently been reported that commercial operators stock postlarvae at $PL_5$ stage in floating or fixed nursery cages with considerable success (Peña and Prospero, 1984). The fixed cage nurseries measure 3.7 × 2.7m and are 1.3m deep. They are fabricated from fine mesh (0.5mm) nylon hapa netting supported by bamboo posts which are driven into the bottom of a brackish water pond. Alternatively, floating bamboo slatted cages, 4 × 3 × 1.5m, supported by bamboo poles and ferrocement floats are used. These are anchored 4m apart in sheltered coastal areas with a 15-20kg block anchor at each corner.

The optimum stocking density is 30,000 $PL_5$ prawns per cage. A feeding frame constructed from 4 bamboo poles and galvanised iron wire fitted with net panel sides and with monofilament nylon (N° 60) lines and hooks (N° 27) strung at 10cm intervals across the base is necessary, as are net feeding plates (*Fig 8.26*). Feeding is carried out each morning. Initially the postlarvae are fed on a paste of finely ground trash fish spread on the smaller of the feeding plates. After two or three days, the larger plate can be used. From stage $PL_{11}$ onwards they are fed on fresh brown mussel (*Modiolus metcalfei*) meat hung from the hooks on the feeding frame and by the time they are ready to be transferred to ongrowing systems, 5*l* of mussel meat will have been used. Excess feed must be removed daily to prevent fouling of the water.

The postlarvae remain in the nursery cages until they become juveniles ($PL_{20}$) or until $PL_{30}$. During this period the nets will have to be changed once or twice. The postlarvae are carefully transferred from the old to the new net by using a fine mesh scoop. When being transferred to the ongrowing system, a small floating cage (1.3 × 0.6 × 0.6m) is tied alongside the nursery cage, and the postlarvae transferred by scoop net (mesh size = 2mm). The small

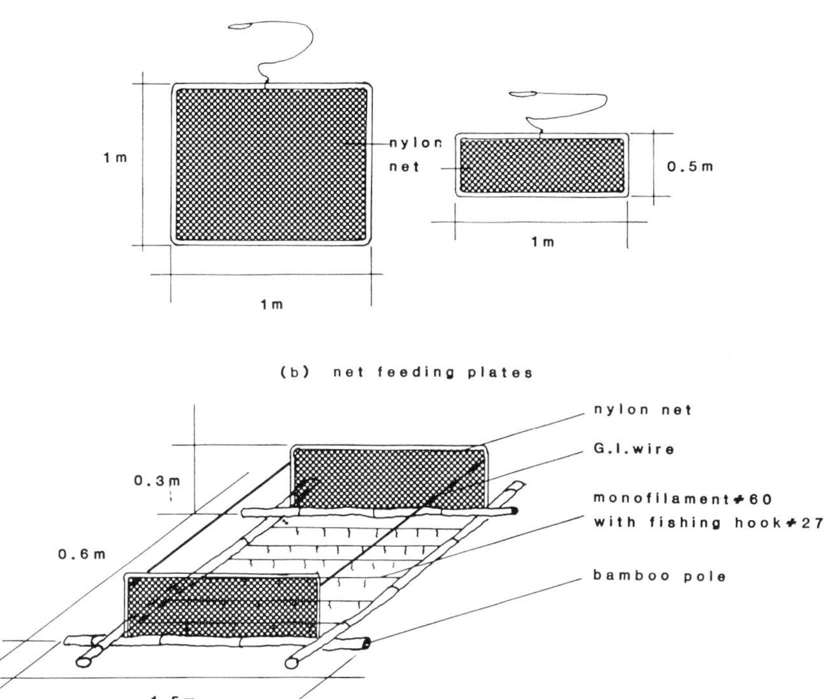


*Fig 8.26* Feeding frame and plates for nursery rearing of *P. monodon* (redrawn from Peña and Prospero, 1984).

harvesting cage can then be towed to the new site.

According to Peña and Prospero (1984) survival rates in the nursery cages are at least 80%.

A number of trials involving the cage rearing of *P. monodon* juveniles were carried out in Tamil Nadu, India, during the early 1980s and these set out not only to examine the production capabilities of cage systems for prawn culture, but also attempted to determine the influences of stocking density, cage size, cage type and mesh size on production (Krishnan *et al*, 1984; Maneeswaran, 1984). The results are summarised in *Table 8.13*.

There appears to be no marked difference between fixed and floating cages in terms of survival, growth or production, but because of the way in which the data is presented in the papers, it is impossible to determine the influence of cage size. Detailed information on size of postlarvae at stocking, feeding rate and feed quality are also absent. However, survival is shown to be surprisingly high and in most cases was greater than 80%. Production was shown to increase with stocking density to a maximum of 0.09kg $m^{-2}$ $month^{-1}$ at stocking densities of around 10-12 individuals $m^{-2}$, although there

*Table 8.13* Summary of experimental cage *P. monodon* production, Tamil Nadu, India.

| Cage type | Cage size (m) | Initial size | Stocking density ($m^{-2}$) | Growth period | Feed | Feed rate (% body wt $day^{-1}$) | Survival (%) | Size at harvest | Production (kg $m^{-2}$ month) | Reference |
|---|---|---|---|---|---|---|---|---|---|---|
| Floating | 3×2×1 | 63mm (1g) | 15 | 4 months | 40.5% protein 20.4% ash 13.4% fibre 11.6% carbohydrate 8.5% moisture | 5–50 | 100 | 111mm (9.6g) | 0.03 | Maneeswari, 1984 |
| | | 59mm (0.3g) | 10 | 5 months | | 5–50 | 38 | 144mm (28.9g) | 0.02 | |
| | | 45mm (0.6g) | 1 | 3 months | | 5–25 | 100 | 170mm (41g) | 0.01 | |
| Floating | 2×5×1–12×5×1 | 0.7–2.5g | 4.5 | 90–120 days | Clam meat, rice bran, oil cake, frog fleshwaste, dried filamentous algae. Also, dry pellets | 5–15 | 84.4 | 33.7g | 0.03–0.04 | Krishnan *et al*, 1984 |
| | | | 5.0 | | | | 98.5 | 30.7g | 0.04–0.05 | |
| | | | 10.0 | | | | 96.8 | 24.0g | 0.06–0.08 | |
| | | | 11.3 | | | | 100.0 | 22.0g | 0.07–0.09 | |
| | | | 15.0 | | | | 69.3 | 20.0g | 0.05–0.07 | |
| Fixed | 2×5×1–12×5×1 | 0.7–2.5g | 3 | 73–134 days | Clam meat, rice bran, oil cake, frog fleshwaste, dried filamentous algae. Also, dry pellets | 5–15 | 86.5 | no data | 0.01–0.02 | Krishnan, *et al*, 1984 |
| | | | 4 | | | | 88.0 | | 0.01–0.02 | |
| | | | 5 | | | | 61.5 | | 0.02–0.03 | |
| | | | 6 | | | | 76.0 | | 0.02–0.04 | |
| | | | 7 | | | | 100.0 | | 0.02–0.03 | |
| | | | 8 | | | | 56.0 | | 0.01–0.02 | |
| | | | 10 | | | | 100.0 | | 0.05–0.09 | |

was a corresponding decrease in size at harvest. Krishnan *et al* (1984) demonstrated that growth, survival and production were higher when nets of mesh size 10 meshes $cm^{-2}$ were used compared with 40 meshes $cm^{-2}$.

In most cases where cage culture of penaeids has been attempted, the cages have been sited in seawater or highly saline brackish water conditions. However, in the Philippines freshwater cage culture of *P. monodon* has been successfully developed by research scientists at SEAFDEC, Binangonan (Anon, 1979; Dalagan, 1980; Pantastico and Oliveros, 1980). Highest survival (98-99%) was achieved using $PL_{35}$ stages and an acclimation period of one to three days. During acclimation the juveniles are placed in tanks containing 35-50% seawater (12-18‰) at densities of $20m^{-2}$, and after 24h the salinity reduced at a rate of 2‰ every nine hours. The tanks are kept aerated and the prawns fed on algal cake (see later). When the salinity is nearly at zero the prawns are transferred to cages in sheltered, pollution free areas of Laguna de Bay.

Both fixed and floating cages of variable size (3 × 1 × 1m −10 × 10 × 4.5m) have been employed, although smaller designs are usually used during the initial stages (*eg Fig 8.27*). In most cases sites with clay substrates are chosen and the cage bottoms placed in contact with the mud, partly in order to increase food availability to the prawns, and partly to reduce the risks of cannibalism. Elsewhere artificial seaweed, threaded through the cage meshes

*Fig 8.27* Small (4 × 1 × 1m) fixed cages stocked with juvenile *P. monodon*, Laguna de Bay, Philippines.

*Fig 8.28* Harvesting cage-grown *P. monodon*.

have been used for this purpose (Tomiyama, 1973).

The prawns are usually stocked at $40m^{-2}$ and during the first month no supplemental feed is given. From $PL_{65}$ — $PL_{125}$, the prawns are fed on a 1:1 ratio of starter algae cake (40% protein) and trash shrimp at a rate of 5% body wt. $day^{-1}$, one third of which is given in the morning and the balance fed in the afternoon. At $PL_{125}$, the prawns are transferred to large (5 × 10 × 3m) cages fitted with 5.5mm stretch mesh nets at stocking densities of 15 - $30m^{-2}$. From $PL_{125}$ — $PL_{155}$ the juveniles are fed on 1:1 grower algal cake (30% protein) and trash shrimps at 5% body wt. $day^{-1}$. During the fifth month the feeding rate remains the same, although an algal cake finisher feed (20% protein) is substituted for the grower feed. The prawns are harvested at the end of the fifth month, when they weigh around 40g each, by simply lifting the cage net and picking the prawns out (*Fig 8.28*).

The SEAFDEC pilot studies were carried out at a 0.5ha site adjacent to

Tapao Point, Talim Island, Laguna de Bay, during July-September 1979 (Anon, 1979). Nineteen hapa nets and 35 larger mesh cages were used to produce 1.7 tonnes of marketable prawns at a rate of return of 117% on the initial capital investment of $5,000.

The rearing of prawns in cages looks promising, with rates of production comparable to those reported from intensive pond culture. The advantages of such a system for producing *P. monodon* in freshwater lakes have been summarised by Dalagan (1980). There are far fewer predators of prawns in freshwater and the use of cages results in the virtual exclusion of any that may be there. Water quality is excellent due to the high exchange rate and there is a good — though unquantified — supply of natural feed which the prawns can take advantage of. Even the algae fouling the cages are believed to be beneficial, contributing towards growth (Rensel and Prentice, 1979). Moreover, harvesting of 100% of the stock is facilitated.

Despite the above advantages, the cage culture of prawns has hardly progressed beyond the experimental stage. Limited cage production from Singapore (R Chou, pers. comm), India and the Philippines are reported. However, cage prawn farms are, like other water-based culture operations, vulnerable to adverse weather conditions and pollution. On several occasions SEAFDEC's pilot operations in Laguna de Bay were destroyed by typhoons (Dalagan, 1980). More importantly, the economics of cage prawn farming have yet to be fully evaluated.

# References

ADCP, 1983. Fish feeds and feeding in developing countries. United Nations Development Programme and Food and Agriculture Organization of the United Nations. ADCP/REP/83/18. FAO, Rome. 395 pp.

ADRON, J W, GRANT, P T and COWEY, C B, 1973. A system for the quantitative study of the learning capacity of rainbow trout and its application to the study of preferences and behaviour. J. Fish Biol., *5*, 625-636.

ALABASTER, J S (ed) 1982. Report of the EIFAC workshop on fish farm effluents. Silkeborg, Denmark, 26-28 May, 1981. EIFAC Tech. Pap., *41*, 166 pp.

ALABASTER, J S and LLOYD, R, 1980. Water quality criteria for freshwater fish. Butterworths, London, 1980. 361 pp. 2nd Edition.

ALDERSON, R, 1984. Broodstock selection. Selection for late age at maturity. *In*: Proc. Int. Symp. Salmonid reproduction, Bellevue, Washington. 31. Oct. — 2 Nov., 1983. University of Washington Press, Seattle 8.

ALDRIN, J F, MESSAGER, J L, MEVEL, M, 1979. Essai sur le stress de transport chez le saumon coho juvenile (*Oncorhynchus kisutch*). Aquaculture, *17*, 279-289.

ALFEREZ, V N, 1982. Fishpen design and construction. *In*: Report of the training course on small-scale pen and cage culture for finfish. South China Sea Fisheries Development and Co-ordinating Programme SCS/GEN/82/34, 23-51.

ALVAREZ, R C, 1981. Growing tilapia in floating cages. Greenfields, *11*, 8-12.

AMEND, D F, CROY, T R, GOVEN, B A, JOHNSON, K A, McCARTHY, D H, 1982. Transportation of fish in closed systems: Methods to control ammonia, carbon dioxide, pH and bacterial growth. Trans. Am. Fish. Soc., *111*, 603-611.

ANDREWS, J W, MURAI, T and CAMPBELL, C, 1973. Effects of dietary calcium and phosphorous on growth, food conversion, bone ash and haematocrit levels in catfish. J. Nutr., *103*, 766-771.

ANDREWS, J Y and MATSUDA, Y, 1975. Influence of various culture conditions on the oxygen consumption of channel catfish. Trans. Am. Fish. Soc., *104*, 322-327.

ANON, 1979. Cage farming of *Penaeus monodon* in freshwater. Asian Aquacult., *2* (7), 4-5.

ANON, 1980. Pink trout unit is on target for 40 tonnes a year. Fish Farmer *3* (5), 30-31.

ANON, 1981. Beautiful valley trout farm developing cage culture projects in Northern Maine. Aquacult. Mag., *7* (4).

ANON, 1982a. Big freeze strikes chill in salmon producers. Fish. Farming Int., *9* (1), 15.

ANON, 1982b. Antifouling suspect in Japanese yellowtail deformities. Fish Farming Int., *9* (6), 15.

ANON, 1983a. Salmon feed promise in fish silage. Fish Farming Int., *10* (4), 16.

ANON, 1983b. Fish farming in Japan. Fish Farming Int., *10* (7), 5.

ANON, 1984a. Alarm uses mining safety cable. Fish Farming Int., *11* (1), 14.

ANON, 1984b. Breaking the ice with bubbles. Fish Farmer 7 (6), 24.

ANON, 1985a. Dutch company develops high energy pellets. Fish Farming Int., *12* (3), 21.

ANON, 1985b. Seafish for business and pleasure. Fish Farmer, *8* (4), 19-20.

ANON, 1985c. An industry in the making. Fish Farmer, *8* (4), 8-9.

ANSUINI, F J and HUGENIN, J E, 1978. The design and development of a fouling-resistant marine cage system. Proc. World Maricult. Soc., *9*, 737-745.

APHA, 1980. Standard methods for examination of water and wastewater. American Public Health Association, Washington, DC 1134 pp. 15th Edition.

AQUINO, L V, 1982. Some ecological considerations relative to cage culture in Sampaloc lake. Unpublished MSc thesis, College of Fisheries, University of Philippines. 77 pp.

ARAGON, C T, COSICO, J and SALAYO, N, 1985. Tilapia marketing in Laguna Province. *In*: Proc. PCARRD-ICLARM Tilapia Economics Workshop, U.P. Los Baños, Laguna, Philippines. August 10-13, 1983. I R Smith, E B Torres and E O Tan (eds). ICLARM, Philippines (in press).

ARAGON, C T, TORRES, E B, LIM, M M de, and TIOSECO, G L, 1985. Economics of cage culture in Laguna Province. *In*: Proc. PCARRD/ICLARM Tilapia Economics Workshop, U.P. Los Baños, Laguna, Philippines. August 10-13, 1983. I R Smith, E B Torres and E O Tan (eds). ICLARM, Philippines (in press).

ARAI, S, NOSE, T and KAWATSU, H, 1975. Effects of minerals supplemented to the fishmeal diet on growth of eel, *Anguilla japonica*. Bull. Freshwater Fish. Res. Lab., Tokyo, *24*, 95-99.

ARIZONO, M and SUIZU, H, 1977. Environmental deterioration in yellowtail farms. Bull. Yamaguchi Pref. Gaikai Fish. Exp. Sta., *15*, 43-56.

ASSOCIATION OF SCOTTISH DISTRICT SALMON FISHERY BOARDS, 1977. Salmon fisheries of Scotland. Fishing News Books Ltd., Farnham, Surrey, 79 pp.

AVAULT, J Jr, 1981. Feeding methods affect production and profit. Aquacult. Mag., 7 (4), 38-59.

AWANG KECHIK, I, 1985. Preliminary observations on the cage culture of bighead carp, *Aristichthys nobilis*, at the Durian Tunggal Reservoir, Melaka. *In*: Proc. Development and Management of tropical living aquatic resources, U P M, Selangor, Malaysia. August 1-5, 1983. Universiti Pertanian Malaysia Press, Kuala Lumpur (in press).

BACKIEL, T, KOKUREWICZ, B and OGORZALEK, A, 1984. High incidence of skeletal anomalies in carp, *Cyprinus carpio*, reared in cages in flowing water. Aquaculture, *43*, 369-380.

BAKER, J P, 1984. Introduction to the proceedings of a symposium on mitigation techniques for acidified surface waters. Fisheries, *9* (1), 2-3.

BALARIN, J D and HALLER, R D, 1982. The intensive culture of tilapia in tanks, raceways and cages. *In*: Recent Advances in Aquaculture, Vol:1. J F Muir and R J Roberts (eds). Croom Helm, London.

BALARIN, J D and HATTON, J D, 1979. Tilapia: A guide to their biology and culture in Africa. Institute of Aquaculture, University of Stirling. 174 pp.

BARDACH, J E, RYTHER, J H and McLARNEY, W O, 1972. Aquaculture. The farming and husbandry of freshwater and marine organisms. John Wiley and Sons, New York, 868 pp.

BARICA, - J, 1976. Nutrient dynamics in eutrophic inland waters used for aquaculture. South China Sea Development Programme, Manila. SCS/76/WP/24. 29 pp.

BARRETT, E C, 1974. Climatology from satellites. Methuen, London. 418 pp.
BASCOM, W, 1964. Waves and beaches. Anchor Books, Doubleday and Company, Inc., New York. 267 pp.
BAUTISTA, A, 1984. Management of tilapia cage farms. Asian Aquacult., *6* (6), 3-4.
BEAMISH, F W H, 1964. Respiration of fishes with special emphasis on standard oxygen consumption II. Influence of weight and temperature on respiration of several species. Can. J. Zool., *42*, 177-194.
BEHRENDT, A, 1984. Eastern Europe solves some problems. Fish Farmer, 7 (3), 28-29.
BEN YAMI, M, 1974. Gnawing at fish netting — a problem in cage raising of herbivorous fish. Aquaculture, *3*, 199-202.
BEN YAMI, M, 1978. Tuna fishing with pole and line. Fishing News Books, Ltd., Farnham, Surrey, 168 pp.
BERRY, T, 1981. Mariculture equipment data file, with special reference to raft and cage design. National Board for Science and Technology, Dublin. 73 pp.
BERTEAUX, H O, 1976. Buoy engineering. Wiley Interscience, New York. 314 pp.
BEVERIDGE, M C M, 1984a. Lake-based or land-based tilapia hatcheries? ICLARM Newsl., 7 (1), 10-11.
BEVERIDGE, M C M, 1984b. Cage and pen fish farming. Carrying capacity models and environmental impact. FAO Fish. Tech. Pap., *255*, 131 pp.
BEVERIDGE, M C M and MUIR, J F, 1982. An evaluation of proposed cage culture in Loch Lomond, an important reservoir in central Scotland. Can. Wat. Res. Journ., 7, 107-113.
BEVERIDGE, M C M, BEVERIDGE, M, and MUIR, J F, 1982. Cage culture and Loch Lomond. Report commissioned by Central Scotland Water Development Board. Institute of Aquaculture, University of Stirling. 68 pp.
BEVERTON, R J H and HOLT, S J, 1957. On the dynamics of exploited fish populations. Fish. Invest., London, Ser. II, *19*, 1-533.
BISHOP, C T, 1980. Design and construction manual for floating tyre breakwaters. Hydraulic Division, National Wat. Res. Inst., Burlington, Ontario. 95 pp.
BLAIR, A, CAMPBELL, R and GRANT, P T, 1982. A submersible fish cage that can be rotated on the surface to remove biofouling and for other purposes. Aquaculture, *29*, 177-184.
BOEUF, G and HARACHE, Y, 1980. Present status of salmonid cage rearing in western France and prospects for development. *In*: Proc. IFM Cage Rearing Symposium, Reading University. 26-27 March, 1980. Janssen Services, London, 50-57.
BOVEN, C J P VAN, 1968. Design considerations for permanent-type offshore structures. *In*: Offshore Europe, 1968. Scientific Surveys (Offshore), Ltd., London, 51-55.
BOWEN, S H, 1982. Feeding, digestion and growth-qualitative considerations. *In*: The Biology and Culture of the Tilapias. R S V Pullin and R H Lowe-McConnell (eds.). ICLARM, Manila, 141-156.
BOYD, C E, 1979. Water quality in warm water fishponds. Auburn University Agricultural Experimental Station, Alabama. 359 pp.
BOYD, C E, 1980. Reliability of water analysis kits. Trans. Am. Fish. Soc., *109*, 239-243.
BOYDSTUN, L B and HOPELAIN, J S, 1977. Cage rearing of steelhead rainbow trout in a freshwater impoundment. Prog. Fish. Cult., *39*, 70-75.
BOYDSTUN, H and PATTERSON, T, 1982. The pros and cons on the use of self-feeders. Salmonid., Jan-Feb, 14-16.
BRAATEN, B R and SAETRE, R, 1973. Rearing of salmonids in Norwegian coastal

waters — environment and types of installations. Transl. Ser. Fish. Res. Board Can., *2759*, 100 pp.
BRAATEN, B, ERVIK, A and BOJE, E, 1983. Pollution problems on Norwegian fish farms. Aquacult. Ireland, *14*, 6-10.
BRANDAL, P O and EGIDIUS, E, 1979. Treatment of salmon lice (*Lepeophtherius salmonis* Kroyer) with Nevugon — description of method and equipment. Aquaculture, *18*, 183-188.
BRAUM, E, 1978. Ecological aspects of the survival of fish eggs, embroyos and larvae. *In*: Ecology of freshwater fish production. S D Gerking(ed). Blackwell Scientific Publications Oxford, 102-136.
BRETT, J R and ZALA, C A, 1975. Daily pattern of nitrogen excretion and oxygen consumption of sockeye salmon (*Oncorhynchus nerka*) under controlled conditions. J. Fish. Res. Bd. Can., *32*, 2479- 2486.
BRONISZ, D, 1979. Selective exploitation of lake zooplankton by coregonid fry in cage culture. Spec. Publ. Eur. Maricult. Soc., *4*, 301-307.
BRONZI, P and GHITTINO, P, 1981. Floating cages and/or raceways: two different systems of utilising warm waters discharged by thermal power stations in inland fish culture. Schr. Bundesforschungsanst. Fisch. Hamb., (16/17) vol 1, 289-300.
BROWN, E E, 1983. World fish farming: Cultivation and economics. AVI Publishing Co., Inc., Westport, Connecticut. 397 pp. 2nd Edition.
BRYLINSKY, M, 1980. Estimating the productivity of lakes and reservoirs. *In*: The Functioning of Freshwater Ecosystems. E D Le Cren and R H Lowe-McConnell (eds). Cambridge University Press, England. 411-454.
BULLOCK, A M, 1982. The pathological effects of ultraviolet radiation on the epidermis of teleost fish with reference to the solar radiation effects in higher animals. Proc. Roy. Soc. Edinb., *81B*, 199-210.
BULLOCK, A M, 1984. The skin response of fish to ultraviolet radiation: A histological study. Unpubl. PhD thesis, University of Stirling, 206 pp.
BULLOCK, A M and COUTTS, R R. 1985. The impact of solar ultraviolet radiation upon the skin of rainbow trout, *Salmo gairdneri* Richardson, farmed at high altitude in Bolivia. J. Fish Dis., *8*, 263-272.
BURNETT, R, 1984. Island smolt shuttle. Fish Farming Int., *11* (8), 4-5.
BUSCH, C D and GOLDMAN, R K, 1981. Water circulation — an alternative to emergency aeration. Proc. World Maricult. Soc., *12* (1), 13-19.
BUTLER, M J A, 1982. Plight of the blue fin tuna. Nat. Geogr. Mag., *162*, 220-239.
BUTZ, I and VENS CAPPELL, B, 1982. Organic load from the metabolic products of rainbow trout fed with dry food. *In*: Report on the EIFAC Workshop on fish farm effluents. J S Alabaster (ed). Silkeborg, Denmark, 26-28 May 1981. EIFAC Tech. Pap., *41*, 73-82.
CALKINS, J (ed), 1982. The role of solar ultraviolet radiation in marine ecosystems. Plenum Press, New York. 724 pp.
CAMPBELL, D, 1985. Large scale cage farming of *Sarotherodon niloticus*. Aquaculture, *48*, 57-70.
CANFIELD, D E Jr and BACHMANN, R W, 1981. Prediction of total phosphorus concentrations, chlorophyll a and Secchi depths in natural and artifical lakes. Can. J. Fish. Aquat. Sci., *38*, 414-423.
CARMICHAEL, W W (ed), 1982. The water environment: algal toxins and health. Plenum Press, New York. 320 pp.
CHACON-TORRES, A, 1984. Water movement and stirring effects in fish cages. Unpubl. M.Sc. thesis, University of Stirling, Stirling, Scotland. 74 pp.
CHAMBERLAIN, G, 1978. The use of caged fish for mariculture and environmental monitoring in a power plant cooling water system. Unpubl. M.Sc. thesis, Texas A.

and M. University College Station. 135 pp.
CHAMBERLAIN, G and STRAWN, K, 1977. Submerged cage culture of fish in supersaturated thermal effluent. Proc. Ann. World Maricult. Soc., *8*, 625-645.
CHAPRA, S C, 1975. Comment on 'An emperical method of estimating the retention of phosphorus in lakes' by W B Kirchner and P J Dillon. Water Resource Res., *11*, 1033-1034.
CHEAH, S H and CHUA, T E, 1979. A preliminary study of the tropical marine fouling organisms on floating net cages. Malay. Nat. J., *33*, 39-48.
CHEN, F-Y, 1979. Progress and problems of netcage culture of grouper (*Epinephelus tauvina*) in Singapore. Proc. World Maricult. Soc., *10*, 260-271.
CHIMITS, P, 1955. Tilapia and its culture: A preliminary bibliography. FAO Fish. Bull., *8*, 1-33.
CHONG, C V Y, 1977. Properties of materials. McDonald and Evans, Ltd., Plymouth. 318 pp.
CHOW, K W (ed), 1978. Fish feed technology. Report of the FAO/UNDP Training Course in Fish Feed Technology, University of Washington, Seattle. ADCP/REP/79/8. FAO, Rome.
CHOW, V T (ed), 1959. Handbook of applied hydrology. McGraw Hill, New York. 700 pp.
CHUA, T E, 1979. Site selection, structural design, construction, management and production of floating cage culture systems in Malaysia. *In*: Proc. IDRC/SEAFDEC Int. Workshop on Pen and Cage Culture of Fish. Tigbanan, Iloilo, Philippines, 11-22 Feb. 1979. SEAFDEC, Iloilo, Philippines. 65-80.
CHUA, T-E and TENG, S K, 1977. Floating fish pens for rearing fishes in coastal waters, reservoirs and mining pools in Malaysia. Fish. Bull. Minist. Agric. Malays., *20*, 36 pp.
CHUA, T-E and TENG, S K, 1978. Effects of feeding frequency on the growth of young estuary grouper *Epinephelus salmoides* Maxwell, cultured in floating net cages. Aquaculture, *14*, 31-47.
CHUA, T-E and TENG, S K, 1980. Economic production of estuary grouper *Epinephelus salmoides* Maxwell, reared in floating cages. Aquaculture, *20*, 187-228.
CIESLUK, A F, 1974. Dissolved gas supersaturation associated with the thermal effluent of an electric generating station and some effects on fish. Unpubl. M.Sc. thesis, Texas A & M University, College Station, Texas. 142 pp.
COCHE, A G, 1977. Premiers résultats de l'élevage en cages de *Tilapia nilotica* (L) dans le Lac Kossou, Côtes d'Ivoire. Aquaculture, *10*, 109-140.
COCHE, A G, 1979. A review of cage culture and its application in Africa. *In*: Advances in Aquaculture. Papers presented at FAO Tech. Conf. on Aquaculture, Kyoto, Japan, 26 May — 2 June, 1976. T V R Pillay and W A Dill (eds). Fishing News Books Ltd., Farnham, Surrey. 428-441.
COCHE, A G, 1982. Cage culture of tilapias. *In*: Biology and culture of tilapias. R S V Pullin and R H Lowe-McConnell (eds). ICLARM, Philippines. 205-246.
COCHE, A G, 1983. The cultivation of fish in cages. A general bibliography. FAO Fish Circ. 714 (Rev. 1), 61 pp.
COLLINS, I, 1983. A study of the environmental impact of particulate matter, derived from a salmonid cage culture system on L. Fad, Isle of Bute. Unpublished B.Sc. thesis, University of Stirling, Stirling, Scotland. 49 pp.
COLLINS, R A, 1975. The Hungarian People's Republic. Implementation of cage culture research at the Fish Culture Research Institute, Szarvas. FI: P.P. Hun/71/512/1. FAO, Rome. 11 pp.
COLT, J E and TCHOBANOGLOUS, G, 1981. Design of aeration systems for aquaculture. *In*: Proc. Bioeng. Symp. Fish Cult., Traverse City, Michigan, 16-18

October, 1979. L J Allen and E C Kinney (eds). Fish Culture Section of the American Fisheries Society, Bethesda, MA. 138-148.

CONNELL, J J, 1980. Control of fish quality. Fishing News Books, Ltd., Farnham, Surrey. 240 pp. 2nd Edition.

COOK, J-M, CRISTIAL, G and LAMARE, J-P, 1984. Technologie des structures d'élevage en mer (cages flottantes et immergées). *In*: L'aquaculture du Bar et des Sparides. G Barnabe and R Billard (eds). INRA, Paris. 395-402.

COUPER, A D, 1983. The Times Atlas of the Oceans. Times Books, Ltd., London. 272 pp.

CSANADY, G T, 1975. Hydrodynamics of large lakes. Ann. Rev. Fluid Mech., 7, 357-385.

DALAGAN, C B, 1980. Prawns can also be grown in freshwater. Greenfields, *10* (9), 29-33.

DARBYSHIRE, M and DRAPER, L, 1963. Forecasting wind-generated sea waves. Engineering, 482-484.

DELA CRUZ, C R, 1980. Capture and culture fisheries in Chinese lakes. ICLARM Newsl., *3* (4), 8-9.

DE YOUNG, B, 1978. Enhancing wave protection with floating tyre breakwaters. New York College of Agriculture and Life Sciences, Cornell University, Ithaca, New York. 30 pp.

DILLON, P J and RIGLER, F H, 1974. A test of a simple nutrient budget model predicting the phosphorus concentration in lake water. J. Fish. Res. Bd. Can., *31*, 1771-1778.

DILLON, P J and RIGLER, F H, 1975. A simple method for predicting the carrying capacity of a lake for development, based on lake trophic status. J. Fish. Res. Bd. Can., *32*, 1519-1531.

DISNEY, J G, HOFFMAN, A, OLLEY, J, CLUCAS, I J, BARRANCO, A, FRANCIS, B J, 1978. Development of a fish silage/carbohydrate animal feed for use in the tropics. Trop. Sci. *20*, 129-146.

DOI, A, HATASE, O, SHIMADA, M, MURUKAMI, T H and OKAICHI, T, 1981. Ultrastructural changes in gill epithelia of a yellowtail, *Seriola quinqueradiata*, exposed to sea bloom.

DUGDALE, R C, 1967. Nutrient limitation in the sea: dynamics, identification and significance. Limnol. Oceanogr., *12*, 685- 695.

DYER, K R, 1973. Estuaries — a physical introduction. Wiley Interscience, New York. 140 pp.

EDWARDS, A and EDELSTEN, D J, 1976. Marine fish cages — the physical environment. Proc. R. Soc. Edinb. (B), *75*, 207-211.

EDWARDS, D J, 1978. Salmon and trout farming in Norway. Fishing News Books, Ltd., Farnham, Surrey. 195 pp.

EDWARDS, P *et al*, 1984. Re-use of cesspool slurry and cellulose agricultural residues for fish culture. ODA project R3677. Final Report. Asian Institute of Technology, Bangkok, Thailand. 338 pp.

EFFIRD, K D, 1975. Interrelation of corrosion and fouling for metals in seawater. *In*: Proc. Corrosion 75. Natl. Assoc. Corrosion Engineers, Houston, Texas. 14 pp.

EGUSA, S, 1980. Disease problems in Japanese yellowtail, *Seriola quinqueradiata*, culture: a review. *In*: Specl. Meet. Diseases of commercially important marine fish and shellfish, 9, 1-20.

EIRAS, J C, 1982. A case of sunburn in farmed brown trout *Salmo trutta* L. Publ. Inst. Dr Augusto Nobre, *162*, 1-12.

ELEY, R L, CARROLL, J H and DE WOODY, D DE, 1972. Effect of cage catfish culture on water quality and community metabolism of a lake. Proc. Okla. Acad. Sci.

*52*, 10-15.

ELLER, L L, 1975. Gill lesions in freshwater teleosts. *In*: The pathology of fishes. W E Ribelin and G Migaki (eds). University of Wisconsin Press, USA 305-330.

ELLIS, M M, 1944. Water purity standards for freshwater fishes. Spec. Sci. Rep. US Fish. Wildl. Serv., *2*, 68 pp.

ENELL, M, 1982. Changes in sediment dynamics caused by cage culture activities. *In*: 10th Nordic Symp. Sediment, Otaniemi, Finland, 1982. I Bergstrom, J Kettunen and M Stenmark (eds). Div. Water Engineering, Helsinki University of Technology, 72-88.

ENELL, M and LOF, J, 1983. Environmental impact of aquaculture — sediment and nutrient loadings from fish cage culture farming. Vatten, *39*, 364-375.

ENELL, M, LOF, J and BJORKLUND, T-L, 1984. Fiskkasseodling med rening teknisk beskrivning och reningseffekt. Institute of Limnology, University of Lund, 34 pp.

ENGLE, C R, 1982. Growth of fed and unfed bighead carp in cages at two stocking densities. Prog. Fish Cult., *44*, 216-217.

ESCOVER, E M and CLAVERIA, R L, 1985. Economics of cage culture in Bicol freshwater lakes. *In*: Proc. PCARRD/ICLARM Tilapia Economics Workshop, U.P. Los Banos, Laguna, Philippines. August 10-13, 1983. I R Smith, E B Torres, E O Tan (eds). ICLARM, Philippines.

FANGE, R and GROVE, D J, 1978. Fish digestion. *In*: Fish Physiology Vol. VIII. D J Randall, W Hoar and J Brett (eds). Academic Press, London, 162-260.

FAO, 1977. Joint SCSP/SEAFDEC Workshop on Aquaculture Engineering. South China Sea Fisheries Development and Coordinating Programme, Manila, Philippines. 452 pp.

FAO, 1983. Freshwater aquaculture development in China. FAO Fish. Tech. Pap., *215*, 125 pp.

FEDORUK, A, SRISUWANTACH, V, 1984. Sand goby culture in Thailand. Dept. Fisheries, Bangkok. FAP/WP-17. 59 pp.

FRERICHS, G N, 1984. Isolation and identification of fish bacterial pathogens. Institute of Aquaculture, University of Stirling, Scotland. 48 pp.

FRIEDMAN, L and SHIBKO, S I, 1972. Non-nutrient components of the diet. *In*: Fish Nutrition. J E Halver (ed). Academic Press, London. 182-255.

FUJIYA, M, 1979. Coastal culture of yellowtail (*Seriola quinqueradiata*) and red sea bream (*Pagrus major*) in Japan. *In*: Advances in Aquaculture. T V R Pillay and W A Dill (eds). Fishing News Books, Ltd., Farnham, Surrey. 453-458.

FURUKAWA, A, 1976. Diets of yellowtails. *In*: Proc. 1st Int. Conf. on Aquaculture Nutrition. Delaware University, Newark, DE. 85-104.

GAIGHER, I G and KRAUSE, J B, 1983. Growth rate of mozambique tilapia (*Oreochromis mossambicus*) and silver carp (*Hypophthalmichthys molitrix*) without artificial feeding in floating cages in plankton-rich waste water. Aquaculture, *31*, 361-367.

GARNER, J, 1962. How to make and set nets. Fishing News Books Ltd., Farnham, Surrey. 95 pp.

GEUSKENS, G (ed), 1975. Degradation and stabilisation of polymers. Applied Science Publishers, 203 pp.

GINDELBERGER, B, 1982. After sinarapan: The planned small-scale fish cage industry of Lake Buhi. ICLARM Newsl., *5*, 6-7.

GJEDRE, B, 1984. Response to individual selection for age at sexual maturation in Atlantic salmon. Aquaculture, *38*, 229-240.

GLEN, D, 1974. Unilever salmon farm in a Scottish sea loch. Fish Farming Int., *1* (3), 12-23.

GODDARD, S and SCOTT, P, 1980. Understanding appetite. Fish Farmer, *3* (6), 40-41.

GONZALES, E, 1984. Small-scale tilapia cage technology adopted in fishing villages in Laguna Lake, Philippines. Aquaculture, *41*, 161-169.

GOOD, M L, KULKARNI, V H, MONAGHAN, C P and HOFFMAN, J F, 1979. Antifouling marine coatings and their long term environmental impact. *In*: Proc. 3rd coastal marsh and estuary management symposium. J W Day Jr, D D Culley Jr, R E Turner and and A J Mumphrey Jr (eds). LSU, Baton Rouge. 19-35.

GOWEN, R J, 1984. The ecology of phytoplankton in Scottish coastal waters with particular reference to toxic species and their importance to mariculture. Scottish Marine Biological Association, Dunstaffnage, Oban, Scotland. 92 pp.

GOWEN, R J, BRADBURY, N B and BROWN, J R, 1985. The ecological impact of salmon farming in Scottish coastal waters: a preliminary appraisal. Proc. ICES C.M. 1985/F.35/Ref. E+C/ Sess. W. 18 pp.

GRAVE, H, 1975. A new type of net cage for fish culture used in Kiel Fjord, Meeresforsch., *24*, 209-211.

GRAVE, H, 1981. Net cage culture under extreme hydrological and meteorological conditions. Spec. Publ. Eur. Maricult. Soc., *6*, 25-36.

GROVE, D J, LOIZIDES, L G and NOTT, J, 1978. Satiation amount, frequency of feeding and gastric emptying rate in *Salmo gairdneri*. J. Fish. Biol., *12*, 507-516.

GUERRERO, R D III, 1979. Cage culture of tilapia in the Philippines. Asian Aquacult., *2* (11), 6.

GUERRERO, R D III, 1980. Studies on the feeding of *Tilapia nilotica* in floating cages. Aquaculture, *20*, 169-175.

GUERRERO, R D III, 1981. An introduction to fish culture in the Philippines. Philippine Education Co., Inc., Manila, Philippines. ˇ 70 pp.

GUERRERO, R D III, 1982. Development, prospects and problems of the tilapia cage culture industry in the Philippines. Aquaculture, *27*, 313-315.

GUERRERO, R D III, 1985. Tilapia farming in the Philippines: Practices, problems and prospects. *In*: Proc. PCARRD-ICLARM Tilapia Economics Workshop, U.P. Los Banos, Laguna. August 10-13, 1983. I R Smith, E B Torres and E O Tan (eds). ICLARM, Philippines (in press).

GUERRERO, R D III and GARCIA, A M 1984. Studies on the fry production of *Oreochromis niloticus* in a lake-based hatchery. *In*: Proc. Int. Symp. on Tilapia in Aquaculture, Nazareth, Israel. May 8- 13, 1983. Tel Aviv University, Israel. 388-393.

GUERRERO, R D III, SEVILLEJA, R C and TOLENTINO, A M, 1985. Studies on the cage culture of *Oreochromis niloticus* in Aya Reservoir and Laguna de Bay, Philippines. *In*: Proc. Regional Workshop on Limnology and Water Resources Management in Developing Countries of Asia and the Pacific. University of Malaysia, Kuala Lumpur. December 1982. (In press).

GULARTE, R C and HUGENIN, J E, 1984. Mesh hydrodynamic data. *In*: Design guide for the use of copper alloy expanded metal mesh in marine aquaculture. INCRA Project 268B. Woods Hole Engineering Associates, Inc., Woods Hole, MA. 54-58.

GULLAND, J A, 1970. Food chain studies and some problems in world fisheries. *In*: Marine food chains. J H Steele (ed). University of California Press, Berkeley. 296-315.

GULLAND, J A and ROTHSCHILD, B J (eds), 1984. Penaeid shrimps. Their biology and management. Fishing News Books Ltd., Farnham Surrey. 312pp.

GUNKEL, G, 1979. Laboratory experiments on the cultivation of young Whitefish *Coregonus fera J*: Food intake, gross growth efficiency and growth of the fry. Spec. Publ. Eur. Maricult. Soc. *4*, 211-242.

HALVER, J E and TIEWS, K (eds), 1979. Proc. World Symp. Finfish Nutrition and

Fishfeed Technology. Hamburg 20-23 June, 1978. Vol I and II. Schr. Bundesforschungsanst, *16*, 593 pp.

HASSE, J, 1974. Utilisation of rabbit fishes (Siganidae) in tropical oyster culture. Prog. Fish. Cult., *36*, 160-162.

HASSELROT, B and HULTBERG, H, 1984. Liming of acidified Swedish lakes and streams and its consequences for aquatic ecosystems. Fisheries, *9* (1), 4-9.

HASTEIN, T, SALTE, R and KITTELSEN, A, 1980. Et tilfelle av solfarbrenning hos laksyngel *Salmo salar* L. Nordsk. Vet., *92*, 721-724.

HATTINGH, J, FOURIE, F Le R and VAN VUREN, J H J, 1975. The transport of freshwater fish. J. Fish. Biol., 7, 447-450.

HEAL, S C, 1983. Marine farm prospects in British Columbia. Fish Farming Int., *10* (5), 10-11.

HENDERSON, E B, 1980. Salmonid cage rearing trials in the Bay of Fundy. *In*: Proc. Inst. Fisheries Management Cage Fish Rearing Symposium, University of Reading. 26-27 March, 1980. Janssen Services, London. 76-97.

HENRICKSEN, A, 1979. A simple approach for identifying and measuring acidification of freshwater. Nature, *278*, 542-545.

HEPHER, B and PRUGININ, Y, 1981. Commercial fish farming with special reference to fish culture in Israel. John Wiley and Sons, New York. 261 pp.

HERBERT, D W M and MERKENS, J C, 1961. The effect of suspended mineral solids on the survival of trout. Int. J. Air Wat. Poll., *5*, 46- 55.

HERBERT, D W M and RICHARDS, J M, 1963. The growth and survival of fish in some suspensions of solids of industrial origin. Int. J. Air Wat. Poll., 7, 297-302.

HEY, R J and PAGAN, A V, 1983. Metals. *In*: Civil engineering materials. N. Jackson (ed). McMillan Publishers, Ltd., London. (3rd edition). 1-65.

HICKLING, C F, 1962. Fish Culture. Faber and Faber, London. 295 pp.

HILL, G and RAI, H, 1982. A preliminary characterisation of the tropical lakes of the Central Amazon by comparison with polar and temperate systems. Arch. Hydrobiol., *96*, 97-111.

HISOAKA, M, NOGAMI, K, TAKEUCHI, O, SUZUKI, M and SUGIMOTO, H, 1966. Studies on seawater exchange in fish farm. II. Exchange of sea water in floating net. Bull. Naikai Reg. Fish. Res. Lab., *115*, 21-43.

HOELZL, A and VENS CAPPELL, B, 1980. Profitability of food-fish production in net cages. Fisch. Teichwirt., *32*, 2-5.

HOFFMAN, G L and MEYER, F P, 1974. Parasites of freshwater fishes: A review of their control and treatment. Neptune, New Jersey. T F H Publishers. 327 pp.

HOGENDOORN, I H, 1984. Dutch multiple fish culture. Fish Farmer, 7 (1), 18-20.

HOLDGATE, M W, 1979. A perspective of environmental pollution. Cambridge University Press, Cambridge. 278 pp.

HOLM, J C and MOLLER, D, 1984. Growth and prey selection by Atlantic salmon yearlings reared on live freshwater zooplankton. Aquaculture, *43*, 401-412.

HOLT, R S, 1977. Culture of fish in cages and their use as biological monitors of water quality within heated discharge waters of a power station. Unpubl. Ph.D. Thesis, Texas A & M University, College Station. 405 pp.

HOLT, R S, COBB, B F III and STRAWN, K, 1978. Organoleptic and biochemical comparisons of cage-raised and wild striped mullet (*Mugil cephalus*). *In*: Proc. First Annual Tropical and Sub-tropical Fisheries Technical Conference. B F Cobb and A B Stockton (eds). Texas A & M University. 473-495.

HONMA, A, 1980. Aquaculture in Japan. Japan FAO Association, Tokyo. 81 pp.

HOPWOOD, A, 1983. How to choose an aeration system. Fish Farmer, 5, (7), 20-23.

HORBUND, H M and FREIBERGER, A, 1970. Slime films and their role in marine fouling. A review. Ocean Eng., *1*, 631-634.

HORVATH, L, TAMAS, G and TOLG, I, 1984. Special methods in pond fish husbandry. Halver Corporation, Seattle, USA. 148 pp.

HOWARD, K T and KINGWELL, S J, 1975. Marine fish farming: development of holding facilities, with particular reference to work by the WFA on the Scottish west coast. *In*: Proc. Oceanology International 1975, 16-21 March, 1975, Brighton, UK BPS Exhibitions Ltd., London. 183-190.

HOYER, M V and JONES, J R, 1983. Factors affecting the relation between phosphorus and chlorophyll a in mid-western reservoirs. Can. J. Fish. Aquat. Sci., *40*, 192-199.

HUET, M, 1956. Apercu de la pisiculture en Indonesie. Bull. Agric. Congo Belge, *47*, 901-957.

HUET, M, 1986. Textbook of fish culture; breeding and cultivation of fish. Fishing News Books Ltd., 436 pp.

HUGENIN, J E and ANSUINI, F J, 1978. A review of the technology and economics of marine fish cage systems. Aquaculture, *15*, 151-170.

HUGENIN, J E and ROTHWELL, G N, 1979. The problems, economic potentials and system design of large future tropical marine cage systems. Proc. Annu. Meet. World Maricult. Soc., *10*, 162-181.

HUGENIN, J E, FULLER, S C, ANSUINI, F J and DODGE, W T, 1981. Experience with a fouling-resistant modular marine fish cage system. *In*: Proc. Bioeng. Symp. for Fishculture, Traverse City, Mich. 16-18 October 1979. L J Allen and E C Kinney (eds). Fish Culture Section of the American Fisheries Society, Bethesda, MA. 201-209.

HUGHES, D G and BEHRENDTS, L L, 1984. Mass production of *Oreochromis* (Tilapia) *niloticus* seed in suspended net enclosures. *In*: Proc. Int. Symp. Tilapia in Aquaculture, Nazareth, Israel. May 8-13, 1983. Tel Aviv University, Israel. 394- 440.

HULL, S T and EDWARDS, R D, 1979. Progress in farming turbot, *Scophthalmus maximus*, in floating sea cages. *In*: Advances in Aquaculture. T V R Pillay and W A Dill (eds). Fishing News Books, Ltd., Farnham, Surrey. 466-472.

HUTCHINSON, G E, 1967. A treatise on limnology. Vol. 2. Introduction to lake biology and the limnoplankton. Wiley, New York. 1115p.

IBRAHIM, K H, NOZAWA, T and LEMA, R, 1976. Preliminary observations on cage culture of *Tilapia esculenta* (Graham) and *Tilapia zillii* (Gervais) in Lake Victoria waters at the Freshwater Institute, Nyegezi, Nigeria. Afr. J. Trop. Hydrobiol. Fish., *4*, 121-127.

ICE, 1979. Corrosion in civil engineering. Institute of Civil Engineering, London. 197 pp.

IDRC/SEAFDEC, 1979. International workshop on pen and cage culture of fish. SEAFDEC Aquaculture Dept., Tigbauan, Philippines. 164 pp.

IDRC, 1983. Fish quarantine and fish diseases in Southeast Asia. Rep. Workshop held in Jakarta, Indonesia, 7-10 Dec., 1982. IDRC, Ottawa, Ont. 79 pp.

INOUE, H., 1972. On water exchange in a net cage stocked with the fish Hamachi. Bull. Jap. Soc. Sci. Fish., *38*, 167-176.

ISHIHARA, T, HARA, K, NAKAYAMA, H and YASUDA, M, 1978. Studies on thiaminase-1 in marine fish. 7. Thiamine requirements of yellowtail fed saury. Bull. Jap. Soc. Sci. Fish. *44*, 653-657.

ISHIHARA, T, HARA, K, YAGI, M and YASUDA, M, 1978. Studies on thiaminase-1 in marine fish. 8. Thiamine requirements of yellowtail fed anchovy. Bull. Jap. Soc. Sci. Fish, *44*, 659-664.

JAGER, T and KIWUS, A, 1980. Aufzucht von tiechtzechtlingen in erleucheten netzgehegen Fisch. und Teichwirt., *11*, 323-326.

JAGER, T and NELLEN, W, 1981. Illuminated net cages for rearing fish larvae.

Poster paper presented at the World Conference on Aquaculture, Venice. 21-25 September, 1981.

JANGKURU, A and DJAJADIREDJA, R, 1979. Common carp in floating net cage culture. *In*: Proc. Int. Workshop on pen and cage culture of fish, Tigbauan, Iloilo, Philippines. Feb. 11-22, 1979. SEAFDEC, Philippines. 55-60.

JANSSEN, J J A, 1981. Bamboo in building structures. Unpublished Ph.D. thesis, University of Eindhoven. 235 pp.

JARRAMS, P, STARKIE, A, EASTON, K W and TEMPLETON, R G, 1980. Salmonid rearing in floating net cages in freshwater reservoirs owned by the Severn-Trent Water Authorities. Fish. Managemt., *11*, 63-79.

JAUNCEY, K and ROSS, B, 1982. A guide to tilapia feeds and feeding. Institute of Aquaculture, University of Stirling, Scotland. 111 pp.

JEE, A K, 1978. Some problems in the cage culture of marble goby (*Oxyeleotris marmorata* Bleeker). Provis. Rep. Int. Found. Sci. Stokh., *2*, 193-196.

JONES, J R and BACHMANN, R W, 1976. Prediction of phosphorus and chlorophyll levels in lakes. J. Water Pollut. Control Fed., *48*, 2176-2182.

JONES, K J, AYRES, P, BULLOCK, A M, ROBERTS, R J and TETT, P, 1982. A red tide of *Gyrodinium aureolum* in sea lochs of the Firth of Clyde and associated mortality of pond-reared salmon. J. Mar. Biol. Ass. UK, *62*, 771-782.

JONES, K J, GOWEN, R J and TETT, P, 1984. Tidal mixing and summer phytoplankton distribution in the Sound of Jura. J. Exp. Mar. Biol. Ecol., *78*, 269-290.

KADOWAKI, S, KASEDO, T, NAKAZONO, T and HIRATA, H, 1978. Continuous records of DO contents by cruising in the coastal culture farms. — II. Diffusion of suspended particles by feeding. Mem. Fac. Fish. Kagoshima Univ., *27*, 281-288.

KADOWAKI, S, KASEDO, T, NAKAZONO, T, YAMASHITA, Y and HIRATA, H, 1980. The relationship between sediment flux and fish feeding in coastal culture farms. Mem. Fac. Fish. Kagoshima Univ., *29*, 217-224.

KAFUKU, T, and IKENOUE, H, 1983. Modern methods of aquaculture in Japan. Elsevier Publishing Co., Amsterdam. 216 pp.

KALLE, K, 1971. General aspects of salinity. *In*: Marine Ecology Vol. 1. I.O. Kinne (ed). Wiley Interscience, New York. 683-688.

KANIDYEV, A N and LYOUKSHINA, V D, 1979. Development and test of pelleted feed for young whitefish (*Coregonus lavaretus baeri* Kess.). Schr. Bundesforschungsanst. Fish. Hamb., *14/15*, Vol 1, 549-554.

KATO, J, NOMA, T and UEKITA, Y, 1979. Design of floating breakwaters. *In*: Advances in Aquaculture. T V R Pillay and W A Dill (eds). Fishing News Books, Ltd., Farnham, Surrey.

KARIYA, T, KUBOTHA, S, NAKAMURA, Y and KIRA, K, 1968. Nocardial infection in cultured yellowtails (*Seriola quinqueradiata* and *S. purpurescens*). 1. Bacteriological Study. Fish Path., *3*, 16-23.

KAWAKAMI, T, 1964. The theory of designing and testing fishing nets in model. *In*: Modern Fishing Gear of the World, Vol. 2. H. Kristjansson (ed). Fishing News Books, Ltd., Farnham, Surrey. 471- 482.

KEENLEYSIDE, M H A and YAMAMOTO, F T, 1962. Territorial behaviour of juvenile Atlantic salmon (*Salmo salar*). Behaviour, *19*, 139-169.

KENNEDY, W A, 1978. A handbook on rearing pan-size Pacific salmon using floating seapens. Ind. Rep. Fish. Mar. Serv. Can., *107*, 111 pp.

KERR, N M, 1981. Design of equipment and selection of materials — an engineer's assessment. Schr. Bundesforschungsanst. Fisch. Hamb., *16/17*, Vol. 1, 151-181.

KERR, N M, GILLESPIE, M J, HULL, S T and KINGWELL, S J, 1980. The design, construction and location of marine floating cages. *In*: Proc. The Institute of Fisheries Management Cage Fish Rearing Symposium, University of Reading. 26-27 March, 1980. Janssen Services, London. 23-49.

KETCHUM, B H (ed), 1983. Estuaries and enclosed seas. Ecosystems of the world Vol. 26. Elsevier Publishing Company, Amsterdam. 500 pp.

KETOLA, H G, 1975. Requirements of Atlantic salmon for dietary phosphorus. Trans. Am. Fish. Soc., *104*, 548-551.

KIEKHAFER, H, 1983. Fishzuchtin gehegen. Paul Parey, Hamburg. 75 pp.

KILAMBI, R V, HOFFMANN, E E, BROWN, A V, ADAMS, T C and WICKIZER, W A, 1976. Effects of cage culture fish production upon biotic and abiotic environment of Crystal Lake, Arkansas. Arkansas University, Fayetteville. 155 pp.

KILS, U, 1977. The salinity effect on aeration in mariculture. Meeresforsch., *25*, 201-206.

KILS, U, 1979. Oxygen regime and artificial aeration of net cages. Meeresforsch., *27*, 236-243.

KINNE, O, 1976. Marine ecology: a comprehensive, integral treatise on life in oceans and coastal waters. Vol. 3: Cultivation. Wiley Interscience, London. 577 pp.

KINNE, O, (ed) 1980. Diseases of marine animals. Vol.l. John Wiley and Son, New York. 466pp.

KLUST, G, 1982. Netting materials for fishing gear. Fishing News Books Ltd., Farnham, Surrey, 175 pp. (2nd Edition).

KLUST, G, 1983. Fibre ropes for fishing gear. Fishing News Books, Ltd., Farnham, Surrey. 216 pp.

KOGAN, R F and ROMANYCHEVA, O D, 1979. The use of floating breakwaters in fish culture. Proc. All-Union Res. Inst. Mar. Fish. Oceanogr., *137*, 19-30.

KOMPIANG, I P, ARIFUDIN, R and RAA, J, 1980. Nutritional value of ensiled by-catch fish from Indonesia shrimp trawlers. *In*: Advances in Fish Science and Technology. J J Connell (ed). Fishing News Books, Ltd., Farnham, Surrey. 349-352.

KONIKOFF, M and LEWIS, W M, 1974. Variations in weight of cage-reared channel catfish. Prog. Fish. Cult., *36*, 138-144.

KOOPS, H, 1972. Experiments on the cage farming of trouts in the western Baltic. ICES C M, 7, 1-16.

KOWALSKI, T (ed), 1974. Floating tyre breakwater conference papers. University of Rhode Island Mar. Tech. Rep., *24*, 375 pp. Narranganset, R I.

KOWALSKI, T, 1976. Scrap tyre floating breakwaters for North Sea operations. Naval Architect., *6*, 186-187.

KORZENEWSKI, K and KORZENEWSKA, J, 1982. Changes in the composition and physiological properties of the bacterial flora of water and bottom sediments in Lake Letowo, caused by intensive trout culture. Pol. Arch. Hydrobiol., *29*, 671-682.

KRESTER, W A and COLQUHOUN, J A, 1984. Treatment of New York's Adirondack lakes by liming. Fisheries, *9* (1), 36-41.

KRISHNAN, P, SHAIK JALALUDDIN, R and JAYASUNDARI, K, 1984. Studies on penaeid prawn growth in fixed and floating cages in the backwaters of Kovalam. *In*: Proc. Natl. Seminar on Cage and Pen Culture, Fisheries College, Tamil Nadu Agricultural University, Tuticorin. March 18-19, 1983. Tamil Nadu Agricultural University. 89-94.

KURONUMA, K and FUKUSHO, K, 1984. Rearing of marine fish larvae in Japan. IDRC, Ottawa, Ontario. 109pp.

KUSUDA, R, KAWAI, K and SHIRAKAWA, T, 1982. Serological study of *Streptococcus* sp. pathogenic to cultured yellowtail. Bull. Jap. Soc. Sci. Fish., *48*, 1731-1738.

KUWA, M, 1983. Corrosion and protection of fish culturing floating cage made of wire netting. Bull. Jap. Soc. Sci. Fish., *49*, 165-175.
KUWA, M, 1984. Fouling organisms on floating cage of wire netting and the removal by *Oplegnathus* sp. cultured with other marine fish. Bull. Jap. Soc. Sci. Fish., *50*, 1635-1640.
LAFONT, R and SAVOEUN, D, 1951. Notes sur la pisiculture au Cambodge. Cybium, 6, 54-61.
LALL, S P, 1979. Minerals in finfish nutrition. Schr. Bundesforschungsanst. Fisch. Hamb., *14/15*, Vol. 1, 85-98.
LAMPA, R R, 1981. Raising tilapia in a big way. Greenfields, *11* (10), 6-8, 10, 12, 14.
LANDLESS, P J, 1985. Aeration in floating cages. Fish Farmer, *8* (6), 12-14.
LANDLESS, P J and EDWARDS, A, 1976. Economical ways of assessing hydrography for fish farms. Aquaculture, *8*, 29-43.
LANDLESS, P J and EDWARDS, A, 1977. Some simple methods for surveying a marine farm site. Fish Farming Int., *4* (1), 32-34.
LANGDON, J S, 1985. Smoltification physiology in the behaviour of salmonids. *In*: Recent Advances in Aquaculture Vol. 2. J F Muir and R J Roberts (eds). Croom Helm, London. 79-118.
LANGFORD, T E, 1983. Electricity generation and the ecology of natural waters. Liverpool University Press. 342 pp.
LARSEN, D P and MERCIER, H T, 1976. Phosphorus retention capacity of lakes. J. Fish. Res. Bd. Can., *33*, 1742-1750.
LAWSON, R, 1984. Economics of fisheries development. Francis Pinter, London. 283 pp.
LE CREN, E D and LOWE-McCONNELL, R H (eds), 1980. The functioning of freshwater ecosystems. IBP Vol. 22. Cambridge University Press. 588pp.
LIAO, P B, 1971. Water requirements of salmonids. Prog. Fish. Cult., *33*, 210-215.
LIBERT, L and MAUCORPS, A, 1973. Mending of fishing nets. Fishing News Books, Ltd., Farnham, Surrey. 116 pp.
LIEFFRIG, D V M, 1985. The effect of hydrogen sulphide ($H_2S$) on aquaculture production. Unpublished M.Sc. thesis, Institute of Aquaculture, University of Stirling. 91 pp.
LIKENS, G E, WRIGHT, R F, GALLOWAY, J N and BUTLER, T R, 1979. Acid rain. Sci. Am., *241*, 43-51.
LINDBERGH, J M, 1979. The development of a commercial Pacific salmon culture business. *In*: Advances in aquaculture. T V R Pillay and W A Dill (eds). Fishing News Books, Ltd., Farnham, Surrey 441-447.
LING, S W, 1967. Feeds and feeding of warm-water fishes in ponds in Asia and the Far East. FAO Fish. Rep., *44*, 291-309.
LING, S W, 1977. Aquaculture in Southeast Asia: A historical review. University of Washington, Seattle. 108 pp.
LIPTON, A P, 1984. Studies on the culture of *Heteropneustes fossilis* in cages. *In*: Proc. Natl. Seminar on cage and pen culture, Fisheries College, Tamil Nadu Agricultural University, Tuticorin. 18-19 March, 1984. Tamil Nadu Agricultural University. 51-54.
LISTER, A R and BOON, A G, 1973. Aeration in deep tanks: An evaluation of fine-bubble diffused air system. J. Int. Wat. Pollut. Cont. 5, 3-18.
LITTLE, A H and PARSONS, H L, 1967. The weathering of cotton, nylon and terylene fabrics in the UK J. Text. Inst., *58*, 449-462.
LIVERSIDGE, J and MUNRO, A L S, 1978. The virology of teleosts. *In*: The pathology of fishes. R J Roberts (ed). Bailliere Tindall, London. 114-143.
LOVEGROVE, T, 1979. Control of fouling in farm cages. Fish Farming Int., *6* (1),

33, 35-37.

LOVELL, R T, 1973. Absorption by channel catfish of earthy-musty flavour compounds synthesised by cultures of blue-green algae. Trans. Am. Fish. Soc., *102*, 774-777.

LOVELL, R T, 1978. Dietary phosphorus requirements of channel catfish. Trans. Am. Fish. Soc., *107*, 617-621.

LOVSHIN, L L, 1982. Tilapia hybridisation. *In*: The biology and culture of tilapias. R S V Pullin and R H Lowe-McConnell (eds). ICLARM, Manila, Philippines. 279-308.

LOWELL, F, NEWTON, S H and TORRANS, E L, 1982. Caged catfish production in private farm ponds. Arkansas Farm. Res., *31* (2), 14-15.

LOYACANO, H A and SMITH, D C, 1976. Attraction of native fishes to catfish culture cages in reservoirs. Proc. 29th Ann. Conf. S.E. Assoc. Game Fish. Comm., 1976. Columbia, S.C. 63-73.

LUMARE, F and PALMEGIANO, G B, 1980. Acclimatazione di *Penaeus japonicus* Bate nella Laguna di Lesina (Italia Sud-orientale). Riv. Ital. Piscic. Ittiopatol., *15*, 53-58.

MACKERETH, F J H, HERON, J and TALLING, J F, 1978. Water analysis: some revised methods for limnologists. Freshwater Biological Association Scientific Publication No. 36. FBA, Windermere. 112 pp.

MACINTOSH, D J and SAMPSON, D R T, 1985. Tilapia culture: hatchery methods for *Oreochromis mossambicus* and *O. niloticus* with special reference to all-male fry production. Institute of Aquaculture, University of Stirling. 154 pp.

MAHESWARI, R UMA, 1984. Studies on the cage culture of prawns, *Penaeus indicus* and *P. monodon*. *In*: Proc. Nat. Seminar on Cage and Pen Culture, Fisheries College, Tamil Nadu Agricultural University, Tuticorin. 18-19 March, 1983. Tamil Nadu Agricultural University. 95-98.

MAHNKEN, C V W, 1975. Status of commercial net pen farming of Pacific salmon in Puget Sound. Proc. Annu. Meet. World Maricult. Soc., *6*, 285-298.

MAITLAND, J R G, 1887. The history of Howietoun. Edinburgh University Press, 278 pp

MAMCARZ, A, 1984. Rearing of coregonid fishes (Coregonidae) in illuminated lake cages III. Phenotypic variability in body size of *Coregonus peled* Gmel. during the first year. Aquaculture, *40*, 241-250.

MAMCARZ, A and SZCZERBOWSKI, J A, 1984. Rearing of coregonid fishes (Coregonidae) in illuminated lake cages. I. Growth and survival of *Coregonus lavaretus* L. and *Coregonus peled* Gmel. Aquaculture, *40*, 135-145.

MAMCARZ, A and WORNIALLO, E, 1985. Rearing of coregonid fishes (Coregonidae) in illuminated lake cages V. Gonad Development in peled (*Coregonus peled* Gmel.) under conditions of cage culture. Aquaculture, *49*, 41-54.

MANE, A M, 1979. Cage culture of tilapia in Laguna de Bay. Technical Consultation for Available Technologies in Aquaculture. SEAFDEC, Tigbauan, Iloilo. 5 pp.

MANNING, R, 1891. On the flow of water in open channels and pipes. Trans. Inst. Civil Engrs. Ireland, *20*, 161-207.

MARCELLO, R A Jr, and STRAWN, K, 1973. Cage culture of some marine fishes in the intake and discharge canals of a steam-electric generating station, Galveston Bay, Texas. Proc. Annu. World Maricult. Soc., *4*, 97-112.

MARCIAK, Z, 1979. Food preference of juveniles of three coregonid species reared in cages. Spec. Publ. Eur. Maricult. Soc., *4*, 127-137.

MARTIN, M, 1982. Impact of predators on fish farming. Aquacult. Mag.,*8* (2) and *8* (3), 36-37 and 44-45.

MARTYSHEV, F G, 1983. Pond Fisheries (English Trans.). AA Balkema,

Rotterdam and Amerind Publishers Co. Pvt. Ltd., New Delhi. 454 pp.

MARUYAMA, T and ISHIDA, R, 1976. Effect of water depth in net cages on the growth and body shape of *Tilapia mossambica*. Bull. Freshwat. Fish. Res. Lab. Tokyo, *26*, 11-19.

MARUYAMA, T and ISHIDA, R, 1977. Effects of water depth on the growth and body shape of common carp raised in net cages. Bull. Freshwat. Fish. Res. Lab., Tokyo, *27*, 95-102.

MASON, E E, 1981. Biology of freshwater pollution. Longman Group Ltd., Harlow, Essex. 250 pp.

MATHESON, A, 1979. An investigation of the occurrence and control of pleurocercoids of *Diphyllobothrium* spp. in farmed Atlantic salmon (Salmo salar). Unpublished MSc thesis, University of Stirling, Scotland. 65 pp.

MATSUZATO, T, 1975. Bacterial tuberculosis in yellowtail. Spec. Publ. Jap. Sea Fish. Res. Lab. 115-118.

MAVANIC, D S and BEWTRA, J K, 1974. Mass transfer of oxygen in diffused air systems. Can. J. Civil Eng., *1*, 71-84.

McANDREW, B J and MAJUMDAR, K C, 1984. Evolutionary relationships within three Tilapiine genera (Pisces: Cichlidae). Zool. J. Linn. Soc., *80*, 421-435.

McCARTNEY, T H, 1969. The effect of dietary inorganic phosphate and Vitamin D supplementation on liver glycogen of fingerling brown trout. Fish. Res. Bull. N.Y. State Conserv. Dep., *31*, 5-7.

McCAVE, I N, 1979. Suspended sediment. *In*: Estuarine Hydrography and Sedimentation. K R Dyer (ed). Cambridge University Press. 131- 185.

McDONALD, D G, 1983. The effect of $H^+$ upon the gills of freshwater fish. Can. J. Zool., *61*, 691-703.

McGREGOR, R C, 1978. Design of scrap tyre floating breakwaters with special reference to fish farms. Proc. Roy. Soc. Edinb., *76B*, 115-133.

McGREGOR, R C and GILBERT, C H G, 1983. Floating tyre breakwaters: A case history. *In*: Proc. 18th Int. Conf. Coastal Engineering. Capetown, South Africa, Nov. 1982. 17 pp.

McGUIGAN, J B and SOMMERVILLE, C, 1985. Studies on the effects of cage culture of fish on the parasite fauna in a lowland freshwater loch in the west of Scotland. Z. Parisitenkd., *71*, 673-682.

McLARNEY, W, 1984. The freshwater aquaculture book. Hartley and Marks, Washington. 583 pp.

McLEAN, J, 1984. Red tide — a growing problem in the Indo-Pacific region. ICLARM Newsl., 7 (4), 20.

MELACK, J M, 1976. Primary productivity and fish yields in tropical lakes. Trans. Am. Fish. Soc., *105*, 575-580.

MELARD, Ch and PHILIPPART, J C, 1980. Pisciculture intensive de *Sarotherodon niloticus* dans les effluents thermique d'une centrale de Belgique. Schr. der Bundesforschungstalt. Fisch., *16/17* , 637-685.

MERICAN, Z O and PHILLIPS, M J, 1985. Solid waste production from rainbow trout, *Salmo gairdneri* Richardson, cage culture. Aquacult. Fish. Mgmt., *16*, 55-70.

MESKE, C H and MANTHEY, M, 1983. *Sarotherodon niloticum* — tropical cichlids as food fish. Int. Fischwirt., *30*, 30-34.

MEYER, J, 1980. Herons at fish farms. *In*: Proc. Institute of Fisheries Management 11th Annual Study Course, University of Sussex. 23-25 September, 1980. Janssen Services, London. 58-63.

MEYER, J, 1982. Easy pickings. Birds, *8* (6), 51-54.

MILLS, D, 1965. The distribution and food of the cormorant in Scottish inland waters. Freshwater and Salmon Fisheries Research, *35*, 31- 37.

MILLS, D, 1979. Bird predation — current views. *In*: Proc. Institute of Fisheries Management. 10th Annual Study Course, Nottingham University. 18-20th September, 1979. Janssen Services, London. 264-271.
MILNE, P H, 1970. Fish farming: a guide to the design and construction of net enclosures. Mar. Res., *1*, 31 pp.
MILNE, P H, 1972. Fish and shellfish farming in coastal waters. Fishing News Books, Ltd., Farnham, Surrey. 208 pp.
MILNE, P H, 1974. A visit to Japan's fish farming industry. Fish Farming Int., *1* (2), 38-55.
MILNE, P H, 1976. Engineering and the economics of aquaculture. J. Fish. Res. Bd. Can., *33*, 288-298.
MILNE, P H, 1979. Selection of sites and design of cages, fish pens and enclosures for aquaculture. *In*: Advances in Aquaculture. T V R Pillay and W A Dill (eds). Fishing News Books, Ltd., Farnham, Surrey. 416-423.
MITANI, F, 1979. The present status and perspectives of the yellowtail fisheries in Japan. Actes Colloques CNEXO, *8*, 195-200.
MIYAZAKI, T, KUBATA, S S and EGUSA, S, 1975. A histopathological study on gliding bacteria disease in yellowtail (*Seriola quinqueradiata*). Fish Path., *10*, 69-74.
MOK, T K, 1981. Design and construction of floating cages and rafts. *In*: Report of the training course on small-scale pen and cage culture for finfish. 26-31 October, 1981, Laguna, Philippines. South China Seas Programme, Manila, Philippines. SCS/GEN/82/34. 103-108.
MØLLER, D, 1979. Recent developments in cage and enclosure aquaculture in Norway. *In*: Advances in Aquaculture. T V R Pillay and W A Dill (eds). Fishing News Books, Ltd., Farnham, Surrey. 447-453.
MOORE, R E, 1982. Toxins from marine blue-green algae. *In*: The water environment. W W Carmichael (ed). Plenum Press, New York. 15-24.
MORIARTY, C M and MORIARTY, D J W, 1973a. Quantitative estimation of the daily ingestion of phytoplankton by *Tilapia nilotica* and *Haplochromis nigripinnis* in Lake George, Uganda. J. Zool., *171*, 15-24.
MORIARTY, D J W, 1973. The physiology of digestion of blue-green algae in the cichlid fish *Tilapia nilotica*. J. Zool., *171*, 25- 40.
MORIARTY, D J W and MORIARTY, C M, 1973b. The assimilation of carbon from phytoplankton by the herbivorous fishes: *Tilapia nilotica* and *Haplochromis nigripinnis*. J. Zool., *171*, 41-56.
MORING, J R, 1982. Fin erosion and culture-related injuries of chinook salmon raised in floating net pens. Prog. Fish Cult., *44*, 189- 191.
MORING, J R and MORING, K A, 1975. Succession of net biofouling material and its role in the diet of pen-cultured chinook salmon. Prog. Fish. Cult., *37*, 27-30.
MORTENSEN, A, 1985. Massive fish mortalities in Tjaldavik cove on the Faroe Islands, on 10-20 July, 1984. *In*: Proc. 3d, Int. Conf. on toxic dinoflagellates, St. Andrews, New Brunswick. 8-12 June, 1985. (In press).
MOTOH, H, 1981. Studies on the fisheries biology of the giant tiger prawn, *Penaeus monodon* in the Philippines. SEAFDEC Tech. Rep. 7. SEAFDEC, Iloilo, Philippines. 128 pp.
MUIR WOOD, A M and FLEMING, C A, 1981. Coastal Hydraulics. McMillan Press, Ltd., London. 280 pp. Second Edition.
MULLER, F and VARADI, L, 1980. The results of cage fish culture in Hungary. Aquacult. Hung., *2*, 154-167.
MULLER-FUEGA, A, PETIT, J and SABAUT, T J, 1978. The influence of temperature and wet weight on the oxygen demand of rainbow trout (*Salmo gairdneri* R.) in freshwater. Aquaculture, *14*, 355- 363.

MURUGESAN, V K and PARAMESWARAN, S, 1984. Culture of air-breathing fishes in cages and pens. *In*: Proc. Natl. Seminar on Cage and Pen Culture, Fisheries College, Tamil Nadu Agricultural University, Tuticorin. March 18-19, 1983. Tamil Nadu Agricultural University. 59- 62.

NATARAJAN, P, RAMANATHAN, N and VENKATARAMANI, V K, 1984. Cage and pen culture of carps and prawns. *In*: Proc. Natl. Seminar on Cage and Pen Culture, Fisheries College, Tamil Nadu Agricultural University, Tuticorin. March 18-19, 1983. Tamil Nadu Agricultural University. 33-36.

NEEDHAM, E, 1980. Parasite and disease control in pen-reared fish. *In*: Proc. Institute of Fisheries Management Cage Rearing Symposium, Reading University. 26-27 March, 1980. Janssen Services, London. 61- 67.

NETBOY, A, 1973. The salmon: their fight for survival. Houghton Mifflin Co., Boston. 613 pp.

NEUSINGER, P R, 1983. A study of external characteristics of Scottish farmed salmon (*Salmo salar*) for use in quality assessment. Unpublished MSc. thesis, Institute of Aquaculture University of Stirling. 103 pp.

NEWTON, S H, 1980. Review of cage culture activity indicates continuing interest. Aquacult. Mag., 7 (1), 32-36.

NILSSEN, N A, 1978. The role of size-based predation in competition and interactive segregation in fish. *In*: Ecology of freshwater fish production. S D Gerking(ed). Blackwell Scientific Publications, Oxford. 303-325.

NISHIMURA, A, 1982. Effect of organic matter produced in fish farms on the growth of red tide algae *Gymnodinium* type-65 and *Chattonella antiqua*. Bull. Pl. Soc. Japan, *29*, 1-7.

NISHIMURA, A, 1983. The process of decomposition and sedimentation of deposits in the pearl oyster and yellowtail culture grounds. Bull. Jap. Soc. Oceanogr., *44*, 44-50.

NOMURA, M and YAMAZAKI, T, 1977. Fishing Techniques. Japanese International Cooperation Agency, Tokyo. 206 pp.

NORDA, 1984. Rainbow trout cage farming for Northern Lake Huron: A pilot project. Ministry of Natural Resources, Canada. 69 pp.

NOSE, T, 1979. Diet compositions and feeding techniques in fish culture. Schr. Bundesforschungsanst., *14/15*, Vol. 1, 283-296.

NOSE, T, 1985. Recent advances in aquaculture in Japan. Geojournal, *10*, 261-276.

NOSE, T and ARAI, S, 1979. Recent advances in studies on mineral nutrition of fish in Japan. *In*: Advances in Aquaculture. T V R Pillay and W A Dill (eds). Fishing News Books, Ltd., Farnham, Surrey. 584-589.

NRC, 1981. Nutrient requirements of coldwater fishes and shellfishes. National Academy Press, Washington DC. 63 pp.

NRC, 1983. Nutrient requirements of warmwater fishes and shellfishes. National Academy Press, Washington DC. 102 pp. Revised Edition.

O'DONNELL, A R, MANCE, G, and NORTON, R, 1984. A review of toxicity of aluminium in freshwater. W.R.C. Tech. Rep. TR 197. 27 pp.

OECD, 1982. Eutrophication of waters. Monitoring, assessment and control. OECD, Paris. 154 pp.

OGINO, C and TAKEDA, T, 1976. Mineral requirements in fish. 3. Calcium and phosphorus requirements in carp. Bull. Jap. Soc. Sci. Fish., *42*, 793-799.

OGINO, C and TAKEDA, T, 1978. Requirements of dietary rainbow trout for dietary calcium and phosphorus. Bull. Jap. Soc. Sci. Fish., *44*, 1019-1022.

OJEDA, G M and STRAWN, K, 1980. Comparison of wire cages and net cages for the culture of black drum (*Pogonius cromis*). Proc. Annu. Meet. World Maricult. Soc., *11*, 185-191.

OKOYE, R N, 1981. Techniques for transportation of juvenile tilapia. Unpubl. MSc thesis. Institute of Aquaculture, University of Stirling. 39 pp.
OLIVA, L P, 1985. Economics of cage culture in Mindanao. *In*: Proc. PCARRD ICLARM Tilapia Economics Workshop. U.P. Los Banos, Laguna, Philippines. August 10-13, 1983. I R Smith, E B Torres and E O Tan (eds). ICLARM, Philippines (in press).
OSTROFSKY, M L, 1978. Modification of phosphorus retention models for use with lakes with low areal loading. J. Fish. Res. Bd. Can., *35*, 1532-1536.
OTHMAN, M A, AWANG KECHIK, I, RASHID, Z A and KHALIL, R, 1985. A study on the physicochemical properties of Tengi River with respect to its suitability for cage fish culture. *In*: Proc. Int. Workshop on Development and Management of Tropical Living Aquatic Resources, Universiti Pertanian Malaysia, Selangor, Malaysia. August 1-5, 1983. Universiti Pertanian Malaysia (in press).
OVERREIN, L N, SEIP, H M and TOLLAN, A, 1981. Acid precipitation — effects on forest and fish. SNSF, Norway. 175 pp. (2nd Edition).
PAETSCH, B, 1977. Untersuchungen uber die fischhalterung von karpfen in netzkaefigen. Z. Binnenfisch., DDR, *24*, 364-365.
PAGAN, F A, 1970. Cage culture of tilapia. FAO Fish Cult. Bull., *3* (1), 6.
PAGAN-FONT, F A, 1975. Cage culture as a mechanical method for controlling reproduction of *Tilapia aurea* (Steindachner). Aquaculture, *6*, 243-247.
PALMEN, O, 1948. On the formation and structure of tropical hurricanes. Geophysica, *3*, 26-39.
PANKHURST, R C and HOLDER, D W, 1952. Wind-tunnel techniques. Sir Isaac Pitman and Sons, Ltd., London. 702 pp.
PANTASTICO, J B and BALDIA, J P, 1979. Supplemented feeding of *Tilapia mossambica*. Schr. Bundesforschungsanst., *14/15* (1), 587-593.
PANTASTICO, J B and BALDIA, J P, 1981. An assessment of algal growth in net cages in Laguna Lake. Fish. Res. J. Philipp., *6*, 19-25.
PANTASTICO, J B and OLIVEROS, E N, 1980. Acclimation of *Penaeus monodon* post larvae to freshwater. Fish. Res. J. Philipp., *5*, 33-38.
PANTULU, V R, 1979. Floating cage culture of fish in the lower Mekong River Basin. *In*: Advances in Aquaculture. T V R Pillay and W A Dill (eds). Fishing News Books Ltd., Farnham, Surrey. 423-427.
PARK, S-I, 1978. Nutritional liver diseases of cultured yellowtail *Seriola quinqueradiata*, caused by feed deficiency. Bull. Korean Fish. Soc., *11*, 1-4.
PARKER, M (ed), 1981. Red tides. Fisheries Seminar 1. Fisheries Research Centre, Dublin, Ireland. 44 pp.
PARRISH, J A, ROX ANDERSON, R, URBACH, F and PITTS, O, 1978. The spectrum of electromagnetic radiation: UV-A in perspective. *In*: UV-A-Biological effects of ultraviolet radiation with emphasis on human responses to longwave ultraviolet. Wiley, Plenum Press, New York. 37-66.
PARSONS, T R, MAITA, Y and LALLI, C M, 1984. A manual of chemical and biological methods for seawater analysis. Pergamon Press, Oxford. 173 pp.
PATHMASOTHY, S, 1985. A review of feeds, their processing and feeding techniques in Malaysian aquacultural systems. *In*: Proc. Int. Workshop on Development and Management of Tropical Living Aquatic Resources, Universiti Pertanian Malaysia, Selangor, Malaysia. August 1-5, 1983. Universiti Pertanian Malaysia (in press).
PCARRD, 1981. State of the art: Lakes and reservoirs research. Fisheries Series *1*, PCARRD, Philippines. 70 pp.
PENA, R and PROSPERO, O 1984. Floating nursery cage for higher sugpo surival. Asian Aquacult., *6* (3), 6-8.

PENCZAK, T, GALICKA, W, MOLINSKI, M, KUSTO, E and ZALEWSKI, M, 1982. The enrichment of a mesotrophic lake by carbon, phosphorus and nitrogen from the cage aquaculture of rainbow trout *Salmo gairdneri*. J. Appl. Ecol., *19*, 371-393.

PERSSON, P E, 1980. Muddy odour in fish from hypereutrophic waters. Dev. Hydrobiol., *2*, 203-208.

PHILIPPART, J Cl and RUWET, J Cl, 1982. Ecology and distribution of tilapias. *In*: R S V Pullin and R H Lowe-McConnell (eds). The Biology and Culture of Tilapias. ICLARM, Manila, Philippines. 15-60.

PHILLIPS, A M Jr, 1956. The nutrition of trout. I. General feeding methods. Prog. Fish. Cult., *18*, 113-118.

PHILLIPS, M J, 1985. Behaviour of rainbow trout, *Salmo gairdneri* Richardson, in marine cages. Aquacult. Fish. Mgmt., *16*, 223-232.

PHILLIPS, M J, MUIR, J F, BEVERIDGE, M and STEWART, J A, 1983. Cage farm management. Fish Farmer, *6* (4), 14-16.

PHILLIPS, M J, ROBERTS, R J, STEWART, A J and CODD, G A, 1985a. The toxicity of the cyanobacterium *Microcystis aeruginosa* to rainbow trout *Salmo gairdneri* Richardson. J. Fish Diseases, *8*, 339-344.

PHILLIPS, M J, BEVERIDGE, M C M and ROSS, L G, 1985b. The environmental impact of salmonid cage culture on inland fisheries: present status and future trends. J. Fish. Biol. (in press).

PHILLIPS, M J, BEVERIDGE, M C M and MUIR, J F, 1985c. Waste output and environmental effects of rainbow trout cage culture. Proc. ICES C.M. 1985/F: 21. 18 pp.

PICCINETTI, C, PICCINETTI, W and MANFRIN, G, 1980. The reproduction of blue fin tuna (*Thunnus thynnus* L.). Mem. Biol. Mar. Oceanogr., *10* (6), 255-258.

PICKERING, A D (ed), 1981. Stress and fish. Academic Press, London. 367 pp.

PINGREE, R D, HOLLIGAN, P M and MARDELL, G T, 1978. The effects of vertical stability on phytoplankton distribution in the summer on the northwest European shelf. Deep-Sea Res., *25*, 1011-1028.

PIPER, R G, McELWAIN, I B, ORME, L E, McCRACKEN, J P, FOWLER, L.G. and LEONARD, J R, 1982. Fish hatchery management. US Dept. of the Interior, Fish and Wildlife Service, Washington, D.C. 517 pp.

PITCHER, T J and HART, P J B, 1982. Fisheries ecology. Croom Helm, London. 414 pp.

PITT, R, TSUR, O and GORDIN, H, 1977. Cage culture of *Sparus aurata*. Aquaculture, *11*, 285-296.

POND, S and PICKARD, G L, 1978. Introductory dynamic oceanography. Pergamon Press, Oxford. 241 pp.

PORTER, C, 1981. Cage culture of gilthead bream (*Sparus aurata*) at an exposed site in the Red Sea. Spec. Publ. Eur. Maricult. Soc., *6*, 15-24.

POSTMA, H, 1967. Sediment transport and sedimentation in the estuarine environment. *In*: Estuaries. G H Lauff (ed). American Association for the Advancement of Science Publication No. 83. Washington, D.C. 158-179.

POUPARD, C J, 1978. Therapy of fish diseases. *In*: Fish Pathology. Bailliere Tindall, London. 268-275.

POWELL, M R, 1976. The resistance of copper-nickel expanded metal to fouling and corrosion in mariculture operations. Prog. Fish. Cult., *38*, 58-59.

POXTON, M G and ALLOUSE, S B, 1982. Water quality criteria for marine fisheries. Aquacult. Engineering, *1*, 153-192.

PPD, 1981. Manual on floating net-cage fish farming in Singapore's coastal waters. PPD, Ministry of National Development, Republic of Singapore. 23 pp.

PRESCOTT, G W, 1948. Objectionable algae with reference to the killing of fish and other communities. Hydrobiologia, *1*, 1-13.

PRIMAVERA, J H, 1983a. Prawn hatcheries in the Philippines. I. Asian Aquacult., 5 (2), 1-4.

PRIMAVERA, J H, 1983b. Prawn hatcheries in the Philippines. II. Asian Aquacult., 5 (3), 5-8.

PRITCHARD, D W, 1967. Observations of circulation in coastal plain estuaries. *In*: Estuaries. G H Lauff (ed). American Association for the Advancement of Science Publication No. 83. Washington, D.C. 37-44.

PRIVOL'NEV, T I, 1975. The prospects for raising rainbow trout in cages. Transl. Ser. Fish. Mar. Serv. Can., *3577*, 16 pp.

PULLIN, R S V and LOWE-McCONNELL, R H (eds), 1982. The biology and culture of tilapias. ICLARM, Manila, Philippines. 432 pp.

RAA, J and GILDBERG, A, 1982. Fish silage: A review. C R C Critical Review Series in Food Science and Nutrition, *16*, 383-420.

RAA, J, GILDBERG, A and STROM, T, 1983. Silage production — theory and practice. *In*: Upgrading waste for feeds and food. D A Ledward, A J Taylor and R A Lawrie (eds). Butterworths, London. 117-132.

RANSON, K, 1982. Bird predation at a cage fish farm. Unpublished B.Sc. thesis, University of Stirling. 44 pp.

RANSON, K and BEVERIDGE, M C M, 1983. Raiders from the skies. Fish Farmer, 6 (1), 22-23.

RAVAGHAVAN, S L, RAHMAN, M F and GOVIND, B V, 1979. 'Suspensoids', a factor for fish mortality. J. Inland Fish. Soc. India, *11*, 111- 112.

REAY, P, 1979. Aquaculture. Studies in Biology No. 106. Edward Arnold, Ltd., London. 60 pp.

RECKOW, K H, 1983. A method for the reduction of lake model prediction error. Water Res., *17*, 911-916.

REKSALEGORA, O, 1979. Fish cage culture in the town of Jambi, Indonesia. *In*: Proc. IDRC/SEAFDEC Int. Workshop on pen and cage culture of fish, Tigbauen, Iloilo, Philippines. 11-22 Feb., 1979. SEAFDEC, Iloilo. 51-53.

RENE, F, 1984. Essais d'—levage du loup (*Dicentrarchus labrax*) de la daurade (*Sparus aurata*) et du sar (*Diplodus sargus*) la Martinique. *In*: L'aquaculture du Bar et des Sparides. G. Barnabe and R. Billard (eds.). INRA, Paris. 403-418.

RENSEL, J E and PRENTICE, E F, 1979. Growth of juvenile spot prawn, *Pandalus platyceros*, in the laboratory and in net pens using different diets. Fish Bull. NOAA/NMFS, *78*, 886-890.

REYNOLDS, C S, 1984. The ecology of freshwater phytoplankton. Cambridge University Press. 384 pp.

REYNOLDS, C S and WALSBY, A E, 1975. Water blooms. Biol. Rev., *50*, 437-481.

REYNOLDS, F A and HAINES, T A, 1980. Effects of chronic exposure to hydrogen sulphide on newly hatched brown trout, *Salmo trutta* L. Environ. Pollut. (Ser. A), *22*, 11-17.

REYSSAC, J and ROUX, M, 1972. Communités phytoplanktoniques dans les eaux de Cote d'Ivoire. Groupes d'éspèces associées. Mar. Biol., *13*, 14-33.

RICHARDS, K, 1982. Rivers: form and process in alluvial channels. Methuen, London. 358 pp.

RICHARDS, R H, 1978. The mycology of teleosts. *In*: Fish pathology. R.J. Roberts (ed). Bailliere Tindall, London. 205-215.

RIFAI, S A, 1980. Control of reproduction of *Tilapia nilotica* using cage culture. Aquaculture, *20*, 177-185.

RIPPEY, S R and CABELLI, V J, 1980. Occurrence of *Aeromonas hydrophila* in the limnotic environment: relationship of organism to trophic state. Microb. Ecol., 6, 45-54.

ROBERTS, R J (ed), 1978. Fish pathology. Bailliere Tindall, London. 318 pp.

ROBERTS, R J and SHEPHERD, C J, 1986. Handbook of trout and salmon diseases. Fishing News Books Ltd., Farnham, Surrey. 1986. 2nd Edition.

ROBERTS, R J and SOMMERVILLE, C, 1982. Diseases of tilapias. *In*: Biology and culture of tilapias. R S V Pullin and R H Lowe-McConnell (eds). ICLARM, Philippines. 247-264.

ROBERTS, R J, BULLOCK, A M, TURNER, M, JONES, K and TETT, P, 1983. Mortalities of *Salmo gairdneri* exposed to cultures of *Gyrodinium aureolum*. J. Mar. Biol. Ass. U.K., *63*, 741-743.

ROBINSON, M A, 1982. Prospects for world fisheries to year 2,000. FAO Fish Circ., *722* (Rev.1), 16 pp.

ROSARIO, W R, 1985. Transfer of fish culture technology in Central Luzon, Philippines. *In*: Proc. PCARRD-ICLARM Workshop Philippine tilapia economics, Laguna, Philippines. August 10-13, 1983. I R Smith, E B Torres and E O Tan (eds). ICLARM, Philippines (in press).

ROSENTHAL, H, 1985. Constraints and perspectives in aquaculture development. Geo. Journal, *10*, 305-324.

ROSS, A J and JOHNSON, H E, 1962. Studies of the transmission of mycobacterial infections by chinook salmon. Prog. Fish. Cult., *24*, 147-149.

ROSS, B and JAUNCEY, K, 1981. A radiographic estimation of the effect of temperature on gastric emptying in *Sarotherodon niloticus* (L.) x *S. aureus* (Steindachner) hybrids. J. Fish Biol., *19* , 133-144.

ROSS, B and ROSS, L G, 1984. The oxygen requirements of *Oreochromis niloticus* under adverse conditions. *In*: Proc. Int. Symp. on Tilapia in Aquaculture, Nazareth, Israel. May 8-13, 1983. Tel Aviv University, Israel. 134-143.

ROSS, L G and ROSS, B, 1984. Anaesthetics and sedative techniques for fish. Institute of Aquaculture, University of Stirling, Scotland. 35 pp.

ROSSELAND, B O and SKOGHEIM, O K, 1984. Attempts to reduce effects of acidification on fishes in Norway by different mitigation techniques. Fisheries 9 (1), 10-16.

ROUND, F E, 1981. The ecology of algae. Cambridge University Press. 653 pp.

RUSSELL, E S, 1931. Some theoretical consideration on the overfishing problem. J. Cons. Int. Explor. Mer., *6*, 3-20.

SAEKI, K and KUMUGAI, H, 1979. Muscle components of wild and cultured yellowtail. J. Food Hyg. Soc. Jap., *20*, 101-105.

SAKAMOTO, S and YONE, Y, 1980. A principal source of deposited lipid in phosphorus deficient red sea bream. Bull. Jap. Soc. Sci. Fish., *46*, 1227-1230.

SAKAMOTO, S, ETO, A, FURUICHI, M and YONE, Y, 1979. Purified test for yellowtail — 1. Bull. Jap. Soc. Sci. Fish., *45*, 1179-1183.

SALMON, T P and CONTE, F S, 1982. Fighting the feathered foe. Fish Farming Int., *9* (3), 13.

SAMPATH, V and MENON, V R, 1975. Preliminary experiments in the cage culture of prawns at Kovelong in Tamil Nadu. Bull. Dep. Mar. Sci. Univ. Cochin, 7, 467-476.

SANTHANAM, R, NATARAJAN, P and KUTHALINGAM, M D K, 1984. Fouling problems in cages and pens. *In*: Proc. Natl. Seminar on cage and pen culture, Fisheries College, Tamil Nadu Agricultural University, Tuticorin. March 18-19, 1983. Tamil Nadu Agricultural University. 143- 147.

SANTIAGO, A E, 1982. Lake water quality and productivity assessment. *In*: Report of the SCS Training Course on Small-Scale Pen and Cage Culture for Finfish.

SCS/GEN/82/34. South China Sea Fisheries Development and Coordinating Programme, Manila, Philippines. 17-22.

SANTIAGO, C B, 1983. Artificial feeds and feeding of tilapias. *In*: Lecture notes. National Training Programme for Tilapia Hatchery and Cage Culture. SEAFDEC, Binangonan Research Station, Rizal. April-May 1983. SEAFDEC, Rizal, Philippines. 21 pp.

SARIG, S, 1971. Diseases of fishes III. The prevention and treatment of diseases of warm water fishes under subtropical conditions, with special emphasis on intensive fish farming. T F H Publications Inc. Ltd., Neptune City, New Jersey. 127 pp.

SCHANTZ, E J, 1982. Poisons produced by dinoflagellates: A review. *In*: The water environment. W W Carmichael (ed). Plenum Press, New York. 25-36.

SCHMITTOU, H R, 1969. Cage culture of channel catfish. Proc. Fish Farming Conf. Annu. Conv. Catfish Farmers, Tex., 72-75.

SCHRECK, C B, 1982. Stress and rearing of salmonids. Aquaculture, *28*, 241-249.

SEAFDEC, 1981. Tilapia fry rearing in cages. Asian Aquacult., *4* (3), 3, 6-7.

SEAFDEC, 1983. Tilapia cage farming: A new enterprise for small fishermen. Asian Aquacult., *5* (3), 1, 3.

SECRETAN, P A D, 1979. Too much stock escapes from nets and cages. Fish Farming Int., *6* (3), 23.

SECRETAN, P A D, 1980. Insurance and risk management as related to cage culture. *In*: Proc. I F M Cage Fish Rearing Symp., Reading University, 26-27 March, 1980. Janssen Services, London. 126-136.

SEDGWICK, S D, 1982. The salmon handbook. Andre Deutsch Ltd. London. 247 pp.

SERRUYA, C and POLLINGHER, U, 1983. Lakes of the warm belt. Cambridge University Press. 569 pp.

SHARMA, R N, 1979. Cage fish culture in Nepal. *In*: Proc. IDRC/SEAFDEC Int. Workshop on Cage and Pen Culture of Fish. Tigbauan, Iloilo, Philippines. 11-22 February, 1979. SEAFDEC, Philippines. 93- 96.

SHAW, S and MUIR, J F, 1986. Salmon economics and marketing. Croom Helm, London. (In press).

SHAW, S and RANA, J, 1985a. Salmon smoking in the UK Institute of Retail Studies, University of Stirling. Market Report No. 1.

SHAW, S and RANA, A J 1985b. The markets for Scottish-grown salmon. Institute of Retail Studies, University of Stirling Market Report No. 2.

SHEARER, W M, 1984. Scottish salmon catches 1952-1981. Scott. Fish. Bull., *48*, 49-54.

SHELKOV, Yu. A, SOLOMATOVA, V P and KUDENCOVA, P A; 1981. Some peculiarities of fish diseases in case of pen rearing. *In*: Fish pathogens and environment in European polyculture. S Olah, K Molnar and Z Jeney (eds). Muller, Szarvas, Hungary. 270-280.

SHELLARD, H C, 1965. Extreme wind speeds over the United Kingdom for periods ending 1963. Climatol. Memo Met. Off. Lond., *50*. 17 pp.

SHEPHERD, C J, 1978. Husbandry and management in relation to disease. *In*: Fish Pathology. R J Roberts (ed). Bailliere Tindall, London. 276-282.

SHILO, M, 1982. The toxic principles of *Prymnesium parvum*. *In*: The water environment. W W Carmichael (ed). Plenum Press, New York. 37-48.

SHIMADA, M, MURUKAMI, T H, IMAHAYASHI, T, OZAKI, H S, TOYOSHIMA, I, and OKAICHI, T, 1983. Effects of sea bloom, *Chattonella antiqua* on gill primary lamellae of the young yellowtail, *Seriola quinqueradiata*. Acta Histochem. Cytochem., *16*, 232-244.

SIGLER, J W, BJORNN, T C and EVEREST, F H, 1984. Effects of chronic

turbidity on density and growth of steelheads and coho salmon. Trans. Am. Fish. Soc., *113*, 142-150.
SIMPSON, J H, 1981. The shelf-sea fronts: implications of their existence and behaviour. Phil. Trans. R. Soc. London, Ser. A, *302*, 531-546.
SIRIKUL, B, 1982. Stocking and rearing of seabass in grow-out ponds and cages. *In*: Report of the Training Course on seabass spawning and larval rearing. SCS/GEN/82/39. South China Seas Fisheries Development and Coordinating Programme, Manila, Philippines. 42.
SKULBERG, O M, CODD, G A and CARMICHAEL, W W, 1984. Toxic blue-green algal blooms in Europe: A growing problem. Ambio, *13*, 244-247.
SLINGER, S J, RAZZAQUE, A and CHO, C Y, 1979. Effect of feed processing and leaching on the losses of certain vitamins in fish diets. Schr. Bundesforschungsanst., *14/15* (2), 425-434.
SMITH, I R, 1975. Turbulence in lakes and rivers. Freshwater Biological Association Publication No. 29. FBA, Windermere. 79 pp.
SMITH, I R and PULLIN, R S V, 1984. Tilapia production booms in the Philippines. ICLARM Newsl., 7 (1), 7-9.
SMITH, I R, TORRES, E B and TAN, E O, (eds) 1985. Proc. PCARRD-ICLARM Workshop on Philippine tilapia economics. ICLARM, Philippines (in press).
SNIESZKO, S F, 1974. The effects of environmental stress on outbreaks of infectious diseases in fishes. J. Fish. Biol., *6*, 197-208.
SODIKIN, D, 1978. Fish culture in Indonesia: its construction and management. South China Sea Fisheries Development and Co-ordinating Programme. SCS/GEN/77/15: SCSP, Manila, Philippines. 351-357.
SOLOMON, D J and HAWKINS, A D, 1981. Fish capture and transport. *In*: Aquarium Systems. A D Hawkins (ed). Academic Press, London. 197-221.
SOWERBUTTS, B J and FORSTER, J R M 1981. Gaseous exchange and reoxygenation. Schr. Bundesforschungsanst., *16/17* (1), 199-218.
STEVENSON, J P, 1980. Trout farming manual. Fishing News Books, Ltd., Farnham, Surrey. 186 pp.
STEWART, K, 1984. A study on the environmental impact of cage farming in an enclosed sea lough. Unpubl. M.Sc. thesis, University of Stirling. 65 pp.
STICKNEY, R R, 1979. Principles of warm water aquaculture. John Wiley and Sons, New York. 375 pp.
STICKNEY, R R, 1983. Care and handling of live fish. *In*: Fisheries Techniques. L A Nielsen and D L Johnson (eds). American Fisheries Society, Bethesda, Maryland. 85-94.
STIRLING, H P, (ed.) 1985. Chemical and biological methods of water analysis for aquaculturists. Institute of Aquaculture, University of Stirling. 111 pp.
STOKOE, A, 1973. Reed's naval architecture for marine engineers. Thomas Reed and Co. Ltd., Glasgow. 377 pp.
STRASKRABA, M, 1980.The effect of physical variables on freshwater production. *In*: The functioning of freshwater ecosystems. E D Le Cren and R H Love-McConnell (eds). Cambridge University Press. 13-84.
SUFFERN, J S, 1980. The potential of tilapia in United States aquaculture. Aquacult. Mag., *6*, 14-18.
SUGIMOTO, N, ISHIBASHI, S and YONE, Y, 1976. On the effectiveness and safety of sodium nifurstyrarate as a chemotherapeutic agent for pseudotuberculosis of yellowtail. Rep. Fish. Res. Lab. Kyushu Univ., *3*, 33-43.
SUTTERLIN, A M and MERRILL, S P, 1978. Norwegian salmonid farming. Tech. Rep. Fish. Mar. Serv. Can., *779*. 52 pp.
SUTTERLIN, A M, JOKOLA, K J and HOLTE, B, 1979. Swimming behaviour of

salmonid fish in ocean pens. J. Fish. Res. Bd. Can., *36*, 948- 954.
SUWANASART, P, 1972. Effects of feeding, mesh size and stocking size on the growth of *Tilapia aurea* in cages. Annu. Rep. Int. Conf. Aquaculture, Auburn University, Auburn, Alabama. 71-79.
SWINGLE, W E, 1971. A marine cage design. Prog. Fish. Cult., *33*, 102.
SYLVESTER, J R, 1975. Biological considerations on the use of thermal effluents for finfish aquaculture. Aquaculture, *6*, 1-10.
SYMONS, P E K, 1976. Behaviour and growth of juvenile Atlantic salmon (*Salmo salar*) and three competitors at two stream velocities. J. Fish. Res. Bd. Can., *33*, 2766-2773.
SZCZERBOWSKI, J A and MAMCARZ, A, 1984. Rearing of coregonid fishes (Coregonidae) in illuminated lake cages. II. Environmental conditions during fish rearing. Aquaculture, *40*, 147-161.
TABACHEK, J L and YURKOWSKI, M, 1976. Isolation and identification of blue-green algae producing muddy odour metabolites, geosmin and 2-methylisoborneol in saline lakes in Manitoba. J. Fish. Res. Bd. Can., *33*, 25-35.
TABIRA, N, 1980. The fish culture under the optimum density and the fishery right administered by fishermen's cooperative association. Mem. Fac. Fish. Kagoshuna Univ., *29*, 157-167.
TACON, A G J and DE SILVA, S S, 1983. Mineral composition of some commercial fish feeds available in Europe. Aquaculture, *31*, 11-20.
TAKEUCHI, M and NAKAZOE, J, 1981. Effects of dietary phosphorus on lipid content and its composition in carp. Bull. Jap. Soc. Sci. Fish., *47*, 347-352.
TANGEN, K, 1977. Blooms of *Gyrodinium aureolum* in North European waters, accompanied by mortalities of marine organisms. Sarsia, *63*, 123-133.
TANIGUCHI, M, 1983. Feeding experiment of yellowtail using frozen diet. Bull. Jap. Soc. Sci. Fish., *49*, 611-615.
TATTANON, T and MANEEWONGSA, S, 1982. Distribution and transport of seabass fry. *In*: Report of training course on seabass spawning and larval rearing, Songkhla, Thailand. 1-20 June, 1982. SCS/GEN/82/39. South China Sea Development and Co-ordinating Programme Manila, Philippines. 33.
TAYLOR, A L and SOLOMON, D J, 1979. Critical factors in the transport of living freshwater fish. I. General considerations and atmospheric gases. J. Fish. Mgmt., *10*, 27-32.
TAYLOR, D L and SELIGER, H H (eds), 1979. Toxic dinoflagellate blooms. Elsevier/North-Holland, New York. 380 pp.
TAZAKI, S, ISHIDA, Y, URAKAWA, T and TAKAMATSU, T, 1975. Development of a floating breakwater. Oceanol. Int., *75*, 191-194.
THOMAS, W A, SPALDING, H A and PAVLOVICH, Z, 1967. The engineers vest pocket book. National Book Store, Philippines. 192 pp.
THORPE, J E and WANKOWSKI, J W J, 1979. Feed presentation and food particle size for juvenile Atlantic salmon, *Salmo salar* L. Schr. der Burdesforschung fur Fischerei, *14/15*, 501-513.
THORPE, J E, BERN, H A, SAUNDERS, R L and SONIO, A (eds), 1985. Salmonid smoltification II. Proc. EEC Workshop, Stirling University, Stirling, Scotland. 3-6 July, 1984. Aquaculture, *45*, 404 pp.
TILLAPAUGH, D L and EDWARDS, J C, 1980. A permit and licence guide for the prospective aquaculturist. Marine Resources Branch, Ministry of Environment, Province of British Columbia, Canada. 23 pp.
TOMI, W, NAIKI, K and YAMADA, Y, 1979. Investigations into technical development of mariculture on commercial scale applied to offshore region. Proc. Jap-Sovt. Joint Symp. Aquaculture, 7. 111-120.

TOMIYAMA, T (ed), 1973. Fisheries in Japan. Prawns. Japan Marine Producers Photo Materials Association, Tokyo. 182 pp.

TOOR, H S, SEHGAL, H S and SEHDEV, R S, 1983. A case study of acute fish diseases in tanks loaded with high levels of organic manures. Aquaculture, *35*, 277-282.

TREWAVAS, E, 1982. Tilapias: taxonomy and speciation. *In*: Biology and culture of tilapias. R S V Pullin and R H Lowe-McConnell (eds). ICLARM, Philippines. 3-14.

TREWAVAS, E, 1983. Tilapiine fishes of the genera *Sarotherodon*, *Oreochromis* and *Danikilia*. British Museum of Natural History, London. 583 pp.

TSENG, W Y, 1983. Prospects for commercial netcage culture of red grouper (*Epinephelus akaara* T & S) in Hong Kong. J. World Maricult. Soc., *14*, 650-660.

TSENG, W Y and YUEN, K H, 1979. Studies on fouling organisms on mariculture nets and cages in Hong Kong. *In*: Proc. Conf. Aquatic Environment in the Pacific. SCOPE, Academia Sinica, Taipei. 151-159.

TSUTSUMI, H and KIKUCH, T, 1983. Benthic ecology of a small cove with essential oxygen depletion caused by organic pollution. Publ. Amak. Mar. Biol. Lab., Kyushu Univ., 7. 17-40.

TUCHOLSKI, S and WOJNO, T, 1980. Studies on removal of wastes produced during cage-rearing of rainbow trout (*Salmo gairdneri* Richardson) in lakes III. Budgets of material and some nutrient elements. Zesz. Nauk. Akad. Roln. Szcz., *82*, 31-50.

TUCHOLSKI, S, KOK, J and WOJNO, T, 1980a. Studies on removal of wastes produced during cage-rearing of rainbow trout (*Salmo gairdneri* Richardson) in lakes I. Chemical composition of wastes. Zesz. Nauk. Roln, Szcz., *82*, 3-15.

TUCHOLSKI, S, WIECLAWSKI, F and WOJNO, T, 1980b. Studies on removal of wastes produced during cage-rearing of rainbow trout (*Salmo gairdneri* Richardson) in lakes II. Chemical regime of water and bottom sediments. Zesz. Nauk. Akad. Roln. Szcz., *82*, 17-30.

TUNDISI, J G, 1983. A review of basic ecological processes interacting with production and standing stock of phytoplankton in lakes and reservoirs in Brazil. Hydrobiologie, *100*, 223-243.

URYN, B A, 1979. Farming of juvenile whitefish *Coregonus lavaretus* (L.) in submerged, illuminated cages. Spec. Publ. Eur. Maricult. Soc., *4*, 289-297.

US ARMY CORPS OF ENGINEERS, 1984. Shore protection manual, Vol I and II. US Government Printing Office, Washington, D.C. 1222 pp.

VALLENTYNE, J R, 1974. The algal bowl: lakes and man. Misc. Publ. Dep. Environ. Fish. Mar. Serv. Can., *22*. 186 pp.

VASS, K R and SACHLAN, M, 1957. Cultivation of common carp in running water in West Java. Proc. IPFC *6* (1-2), 187-196.

VENKATASAMY, G, 1984. Studies on the culture of *Penaeus indicus* in cages. *In*: Proc. Natl. Seminar on Cage and Pen Culture, Fisheries College, Tamil Nadu Agricultural University, Tuticorin. 18- 19 March, 1983. Ramil Nadu Agricultural University. 99-102.

VOLLENWEIDER, R A, 1968. Scientific fundamentals of the eutrophication of lakes and flowing water with particular reference to nitrogen and phosphorus as factors in eutrophication. Tech. Rep. DA5/SU/68-27. OECD, Paris. 250 pp.

VOLLENWEIDER, R A, 1975. Input-output models with special reference to the phosphorus loading concept in limnology. Schweicz. Z. Hydrol., *37*, 455-472.

WALDOCK, M J and THAIN, J E, 1983. Shell thickening in *Crassostrea gigas*: organotin antifouling or sediment induced? Mar. Pollut. Bull., *14*, 411-415.

WALKER, W W Jr, 1982. An empirical analysis of phosphorus, nitrogen and turbidity effects on reservoir chlorophyll a levels. Can. Water Resour. J., 7, 88-107.

WALMSLEY, R D and THORNTON, J A, 1984. An evaluation of OECD-type phosphorous eutrophication models for predicting the trophic state of southern African man-made lakes. S. Afr. J. Sci.,

WANKOWSKI, J W and THORPE, J E, 1979. The role of food particle size in the growth of juvenile Atlantic salmon (*Salmo salar* L.) J. Fish. Biol., *14*, 351-370.

WARRER-HANSEN, I, 1982. Evaluation of matter discharged from trout farming in Denmark. *In*: Report of the EIFAC Workshop on Fish Farm Effluents, Silkeborg, Denmark, 26-28 May, 1981. J S Alabaster (ed). EIFAC Tech. Pap. *41*, 57-64.

WATANABE, T, MURUKAMI, A, TAKEUCHI, L, NOSE, T and OGINO, C, 1980a. Requirement of chum salmon held in freshwater for dietary phosphorus. Bull. Jap. Soc. Sci. Fish., *46*, 361-367.

WATANABE, T, TAKEUCHI, T, MURUKAMI, A and OGINO, C, 1980b. The availability to *Tilapia nilotica* of phosphorous in white fish meal. Bull. Jap. Soc. Sci. Fish., *46*, 897-900.

WATSON, D J and RAJA, P T, 1979. A review of the present state of cage culture in the Baram district of Sarawak. *In*: Proc. IDRC/SEAFDEC Int. Workshop on pen and cage culture of fish, Tigbauan, Iloilo, Philippines. 11-22 Feb., 1979. SEAFDEC, Philippines. 89-92.

WEBBER, H H and HUGENIN, H E, 1979. Fish feeding technologies. Schr. Bundesforschungsanst., *14/15* (1), 297-316.

WEDEMEYER, G, 1970. The role of stress in the disease resistance of fishes. *In*: A symposium on diseases of fishes and shellfishes. S.F. Snieszko (ed). Special Publication No. 5. American Fisheries Society, Washington, D.C. 30-35.

WEE, K L, 1979. Ventilation of floating cages. Unpubl. M.Sc. thesis, University of Stirling. 42 pp.

WEE, K L, 1982. Snakeheads — their biology and culture. *In*: Recent Advances in Aquaculture, Vol I. J F Muir and R J Roberts (eds). Croom Helm, London. 179-213.

WENT, A E J (ed). 1980. Atlantic salmon: its future. Fishing News Books Ltd., Farnham, Surrey. 253 pp

WHEELER, A, 1969. The fishes of the British Isles and Northwest Europe. MacMillan Ltd., London. 613 pp.

WHEELER, A 1978. Key to the fish of Northern Europe. Frederick Warne Ltd., London. 380 pp.

WHITE, E, PAYNE, G, PICKMORE, S and PICK, F R, 1982. Factors influencing orthophosphate turnover times: A comparison of Canadian and New Zealand lakes. Can. J. Fish. Aquat. Sci., *39*, 469-474.

WHITE, W J, WATT, W D and SCOTT, C D, 1984. An experiment on the feasibility of rehabilitating acidified Atlantic salmon habitat in Nova Scotia by the addition of lime. Fisheries, *9* (1), 25-30.

WIEGEL, R L, 1964. Oceanographical Engineering. Prentice-Hall, Inc., New Jersey. 532 pp.

WONG, C, 1982. Harvesting and marketing of cultured marine fish in Hong Kong. South China Sea Fisheries Development and Coordinating Programme, Manila, Philippines. SCS/GEN/82/34. 137-140.

WONG, S-Y, ONG, B and CHUA, T-E, 1979. Isolation, identification of causitive agent of "red-boil disease" in grouper (*Epinephelus salmoides*) and its possible control by vaccination. *In*: Proc. IDRC/SEAFDEC Workshop on pen and cage culture of fish, Tigbauan, Iloilo, Philippines. 11-22 Feb., 1979. SEAFDEC, Philippines. 81-87.

WOOD, J W, 1974. Diseases of Pacific salmon, their prevention and treatment. Department of Fisheries, Hatchery Division, State of Washington, Olympia, Washington. 297 pp. 2nd Edition.

WOODS HOLE ENGINEERING ASSOCIATES, INC., 1984. Design guide for use of copper alloy expanded metal mesh in marine aquaculture. Contract Report Technology for the Copper Industry. INCRA Proj. 268B. Woods Hole Engineering Associates, Inc., Woods Hole, Massachusetts. 87 pp.

WOOTEN, R, 1979. Tapeworm threat to trout in floating freshwater cages. Fish Farmer, *2* (3), 5.

WORNIALLO, E and MAMCARZ, A, 1985. Rearing of coregonid fishes (Coregonidae) in illuminated Lake cages IV. Gonad development in whitefish (*Coregonus lavaretus* L.)under conditions of cage culture. Aquaculture, *49*, 31-40.

WRIGHT, R F and HENRICKSEN, A, 1979. Regional survey of lakes and streams in southwestern Scotland, April, 1979. Internal report IR72/80. SNSF, Oslo. 63 pp.

WURTSBAUGH, W, VINCENT, W F, TAPIA, R A, VINCENT, C L and RICHERSON, P J, 1985. Nutrient limitation of algal growth and nitrogen fixation in a tropical alpine lake, Lake Titicaca (Peru/Bolivia). Freshwat. Biol., *15*, 185-196.

YAMAHA MOTOR CO., LTD., 1980. Fisheries in Japan. Yamaha Motor Co. Ltd., Japan. 79 pp.

YATER, L and SMITH, I R, 1985. Economics of private tilapia hatcheries in Laguna and Rizal Provinces. *In*: Proc. PCARRD ICLARM Tilapia Economics Workshop, U.P. Los Banos, Laguna, Philippines. August 10-13, 1983. I R Smith, E B Torres and E O Tan (eds). ICLARM, Philippines (in press).

YU, O K, VIZCARRA, A T and SITOY, H S, 1979. Development of circular floating cages for milkfish broodstock at the SEAFDEC Aquaculture Department. *In*: Proc. IDRC/SEAFDEC Int. Workshop on pen and cage culture of fish, Tigbauan, Iloilo, Philippines. 11-22 Feb., 1979. SEAFDEC, Philippines. 107-117.

ZAUGG, B and PEDROLI, J C, 1984. Culture of young fish in natural waters with net cages. Schriftenv. Fisch. Bundesant. Umweltshut, *43*, 53-87.

# Species index

# General index

# Other books published by **Fishing News Books Ltd**

*Free catalogue available on request*